Design of Adhesive Bonded Joints

Design of Adhesive Bonded Joints

Editor

Raul D. S. G. Campilho

Basel • Beijing • Wuhan • Barcelona • Belgrade • Novi Sad • Cluj • Manchester

Editor
Raul D. S. G. Campilho
Polytechnic of Porto
Porto, Portugal

Editorial Office
MDPI
St. Alban-Anlage 66
4052 Basel, Switzerland

This is a reprint of articles from the Special Issue published online in the open access journal *Processes* (ISSN 2227-9717) (available at: https://www.mdpi.com/journal/processes/special_issues/adhesive_bonded_joints).

For citation purposes, cite each article independently as indicated on the article page online and as indicated below:

Lastname, A.A.; Lastname, B.B. Article Title. *Journal Name* **Year**, *Volume Number*, Page Range.

ISBN 978-3-0365-9907-6 (Hbk)
ISBN 978-3-0365-9908-3 (PDF)
doi.org/10.3390/books978-3-0365-9908-3

© 2024 by the authors. Articles in this book are Open Access and distributed under the Creative Commons Attribution (CC BY) license. The book as a whole is distributed by MDPI under the terms and conditions of the Creative Commons Attribution-NonCommercial-NoDerivs (CC BY-NC-ND) license.

Contents

About the Editor vii

Preface ix

Raul D. S. G. Campilho
Design of Adhesive Bonded Joints
Reprinted from: *Processes* 2023, *11*, 3369, doi:10.3390/pr11123369 1

Mateusz Weisbrodt and Agnieszka Kowalczyk
Removable Pressure-Sensitive Adhesives Based on Acrylic Telomer Syrups
Reprinted from: *Processes* 2023, *11*, 885, doi:10.3390/pr11030885 7

Georgino C. G. Serra, José A. M. Ferreira and Paulo N. B. Reis
Static and Fatigue Characterization of Adhesive T-Joints Involving Different Adherends
Reprinted from: *Processes* 2023, *11*, 2640, doi:10.3390/pr11092640 19

Nergizhan Anaç and Zekeriya Doğan
The Effect of Organic Fillers on the Mechanical Strength of the Joint in the Adhesive Bonding
Reprinted from: *Processes* 2023, *11*, 406, doi:10.3390/pr11020406 39

Panagiotis Kormpos, Selen Unaldi, Laurent Berthe and Konstantinos Tserpes
A Laser Shock-Based Disassembly Process for Adhesively Bonded Ti/CFRP Parts
Reprinted from: *Processes* 2023, *11*, 506, doi:10.3390/pr11020506 51

Costanzo Bellini, Vittorio Di Cocco, Francesco Iacoviello, Larisa Patricia Mocanu, Gianluca Parodo, Luca Sorrentino and Sandro Turchetta
Analysis of Hydrothermal Ageing on Mechanical Performances of Fibre Metal Laminates
Reprinted from: *Processes* 2023, *11*, 2413, doi:10.3390/pr11082413 67

Raffaele Ciardiello, Carlo Boursier Niutta and Luca Goglio
Adhesive Thickness and Ageing Effects on the Mechanical Behaviour of Similar and Dissimilar Single Lap Joints Used in the Automotive Industry
Reprinted from: *Processes* 2023, *11*, 433, doi:10.3390/pr11020433 79

Tao Wu, Xin Chen, Shiju Wen, Fangsong Liu and Shengping Li
Coupled Excitation Strategy for Crack Initiation at the Adhesive Interface of Large-Sized Ultra-Thin Chips
Reprinted from: *Processes* 2023, *11*, 1637, doi:10.3390/pr11061637 95

Luís F. R. Neves, Raul D. S. G. Campilho, Isidro J. Sánchez-Arce, Kouder Madani and Chander Prakash
Numerical Modelling and Validation of Mixed-Mode Fracture Tests to Adhesive Joints Using *J*-Integral Concepts
Reprinted from: *Processes* 2022, *10*, 2730, doi:10.3390/pr10122730 119

Peer Schrader, Dennis Domladovac and Stephan Marzi
Influence of Loading Rate on the Cohesive Traction for Soft, Rubber-Like Adhesive Layers Loaded in Modes I and III
Reprinted from: *Processes* 2023, *11*, 356, doi:10.3390/pr11020356 139

Stefania Minosi, Fabrizio Moroni and Alessandro Pirondi
Evaluation of XD 10 Polyamide Electrospun Nanofibers to Improve Mode I Fracture Toughness for Epoxy Adhesive Film Bonded Joints
Reprinted from: *Processes* 2023, *11*, 1395, doi:10.3390/pr11051395 163

Hannes Stolze, Michael Gurnik, Sebastian Kegel, Susanne Bollmus and Holger Militz
Determination of the Bonding Strength of Finger Joints Using a New Test Specimen Geometry
Reprinted from: *Processes* **2023**, *11*, 445, doi:10.3390/pr11020445 . . . . **179**

Jens-David Wacker, Tobias Kloska, Hannah Linne, Julia Decker, Andre Janes, Oliver Huxdorf and Sven Bose
Design and Experimental Analysis of an Adhesive Joint for a Hybrid Automotive Wheel
Reprinted from: *Processes* **2023**, *11*, 819, doi:10.3390/pr11030819 . . . . **193**

José Antonio Butenegro, Mohsen Bahrami, Miguel Ángel Martínez and Juana Abenojar
Reuse of Carbon Fibers and a Mechanically Recycled CFRP as Rod-like Fillers for New Composites: Optimization and Process Development
Reprinted from: *Processes* **2023**, *11*, 366, doi:10.3390/pr11020366 . . . . **211**

About the Editor

Raul D. S. G. Campilho

Prof. Raul D. S. G. Campilho is a researcher and Professor at ISEP-School of Engineering of the Polytechnic Institute of Porto. Prof. Campilho is an expert in testing and modeling composite materials and adhesive joints. Furthermore, Prof. Campilho is a recognized researcher in the application of cohesive zone models and the extended finite element method for the analysis of adhesively bonded joints. Consequently, his ample experience led him to be among the most-cited scientists of the world (top 2% according to Stanford University). His research interests are characterization of materials, material joining, adhesive joints, advanced composite and sandwich structures, finite element method, fracture mechanics, computational mechanics, cohesive zone models, advanced manufacturing systems, automation and robotics, and industrial design.

Preface

Adhesive bonding has emerged as a fundamental technology in various industrial applications, promoting lightweight structures, improved durability, and enhanced performance in joining diverse families of materials. The scientific exploration and understanding of adhesive bonded joints have become imperative for industrialists and academics. This book, *"Design of Adhesive Bonded Joints"*, published by MDPI, is a comprehensive compilation of scientific papers on cutting-edge research, methodologies, and applications on the topic of adhesive bonding.

The book includes a large spectrum of topics within adhesive bonding, namely theoretical frameworks, experimental investigations, fracture mechanics concepts, and innovative applications. The addressed subjects involve different design approaches for adhesive bonded joints, elucidating the underlying principles governing their mechanical behavior, aging effects, and innovative disassembly processes. The interplay between adhesive thickness, loading rates, and the incorporation of novel materials further extends the book scope. The motivation behind assembling this collection of scientific papers lies in the increasing significance of adhesive bonding in contemporary engineering practices. The need for a consolidated resource that synthesizes the latest advancements, experimental findings, and theoretical frameworks in this field prompted the creation of this book. Through collaboration among esteemed researchers and experts, this book aims to contribute to the available literature of adhesive bonding technology. This book is tailored for a diverse audience, including researchers, engineers, students, and professionals engaged in materials science, mechanical engineering, and related disciplines. Its multidisciplinary approach accommodates readers seeking a fundamental understanding of adhesive bonding principles as well as those delving into the intricacies of fracture mechanics and experimental methodologies.

In conclusion, *"Design of Adhesive Bonded Joints"* is a collaborative effort to consolidate and disseminate knowledge in this dynamic field. Each scientific paper of this book is a unique contribution, collectively forming a comprehensive resource on adhesive bonding design. The guest editor deeply thanks all authors for their contribution and for believing in this book project from the start.

Raul D. S. G. Campilho
Editor

Editorial

Design of Adhesive Bonded Joints

Raul D. S. G. Campilho [1,2]

[1] ISEP—School of Engineering, Polytechnic of Porto, R. Dr. António Bernardino de Almeida, 431, 4200-072 Porto, Portugal; raulcampilho@gmail.com; Tel.: +351-939-526-892

[2] INEGI—Institute of Science and Innovation in Mechanical and Industrial Engineering, Pólo FEUP, Rua Dr. Roberto Frias, 400, 4200-465 Porto, Portugal

Adhesive bonded joints have become vital to modern engineering, offering advantages such as weight reduction, enhanced fatigue performance, and improved stress distribution [1]. As a result of these characteristics, the evolution of adhesive technology has significantly influenced engineering practices, leading to widespread adoption in various industries [2]. The design of adhesive joints involves a complex interplay of factors, including adhesive selection, joint configuration, and loading conditions [3]. Despite significant progress, challenges persist, necessitating a critical examination of current design practices [4]. This MDPI Special Issue entitled "Design of Adhesive Bonded Joints" serves as a platform to explore the limitations and opportunities in bonded joint design, emphasizing adhesives, joint characterization, experimental and analytical analyses, and predictive modeling. In this editorial, current design limitations, avenues for improvement, ongoing lines of research, and prospects in bonded joint design are addressed.

Current Design Limitations:

One primary concern is the lack of standardized procedures for adhesive joint characterization. The variability in testing methods and reporting parameters hinders the comparability of results, impeding the establishment of universal design guidelines [5]. Additionally, the absence of a unified approach for predictive modeling and failure analysis poses challenges in ensuring design reliability under diverse loading conditions.

- Experimental limitations: one of the primary challenges in adhesive bonding design is the lack of standardized procedures for adhesive joint characterization. Experimental testing methods vary widely across studies, leading to inconsistencies in results and hindering the establishment of universally applicable design guidelines [5]. Variability in factors such as specimen geometry, loading conditions, and environmental parameters complicates the comparison of results and compromises the reliability of experimental data. Examples of round-robin studies try to mitigate this disadvantage [6]. Moreover, quasi-static testing fails to adequately capture the dynamic behavior of adhesive joints under real-world conditions including dynamic loads and impact, often leading to extrapolations of the material behavior [7]. The impact of factors such as temperature variations, humidity, and loading rates on joint performance remains insufficiently explored [8].
- Numerical limitations: numerical simulations, particularly those based on finite element methods (FEMs), constitute a powerful tool to predict the behavior of bonded joints [9]. However, challenges persist in achieving accurate and reliable simulations. The complexity of adhesive joint behavior, e.g., plasticity, stress concentrations, and initiation and propagation of cracks, requires advanced modeling approaches that surpass the common simplifying assumptions [10]. Cohesive zone models, the most widespread technique to simulate crack propagation in adhesive joints, are limited by the assumptions inherent in their formulations [11]. The estimation of cohesive parameters, such as the cohesive strength and fracture toughness, often relies on trial-and-error procedures, introducing uncertainties in the predictions [12]. Additionally,

Citation: Campilho, R.D.S.G. Design of Adhesive Bonded Joints. *Processes* 2023, 11, 3369. https://doi.org/10.3390/pr11123369

Received: 15 November 2023
Accepted: 1 December 2023
Published: 4 December 2023

Copyright: © 2023 by the author. Licensee MDPI, Basel, Switzerland. This article is an open access article distributed under the terms and conditions of the Creative Commons Attribution (CC BY) license (https://creativecommons.org/licenses/by/4.0/).

the applicability of standardized law shapes, such as triangular, across different adhesive types and joint geometries remains a topic of ongoing investigation due to the known geometry effects on the cohesive properties [13,14]. While the FEM provides valuable insights, the computational cost associated with detailed simulations of large and complex structures poses a challenge [15].

Improving Existing Design Processes:

Addressing current limitations requires a concerted effort to standardize testing protocols and develop comprehensive design guidelines [16]. Robust methodologies for adhesive joint characterization, including experimental testing and numerical analyses, are essential. Advances in non-destructive evaluation techniques can enhance the understanding of joint behavior, contributing to more accurate predictions of performance.

- Standardization of testing protocols: as previously mentioned, a fundamental challenge of bonded joint design is the lack of standardized testing protocols for adhesive joint characterization. Variations in testing methodologies, specimen configurations, and data analysis hinder the comparability of results across studies [17]. To improve design processes, it is necessary to develop and adopt standardized testing procedures [18]. These protocols should encompass quasi-static, dynamic, and environmental factors, ensuring a comprehensive understanding of adhesive joint behavior.
- Advanced numerical techniques: simplistic analytical/numerical models fail to capture the complex behavior of adhesive joints [19]. To address this limitation, designers should embrace advanced techniques, such as FEM simulations, for a more detailed understanding of joint mechanics, especially post-elasticity [20]. Cohesive zone models and damage mechanics can be integrated into numerical frameworks to provide a more accurate representation of crack onset and growth [21].
- Tailoring adhesives for specific applications: advances in materials science offer an opportunity to tailor adhesives at the molecular level, catering to the specific requirements of diverse applications [22]. Designers can collaborate with material scientists to develop adhesives with enhanced properties, such as improved thermal resistance, durability, and flexibility [23].
- Incorporating non-destructive evaluation (NDE) techniques: the integration of NDE techniques into the design process can significantly enhance the monitoring process of adhesive joint performance under service [24]. Techniques such as ultrasonic testing, thermography, and acoustic emission monitoring provide real-time information on the integrity of joints without causing damage [25]. This approach promotes early detection of potential damage, enabling corrective actions to be implemented before loss of structural integrity.

Current Lines of Research:

The field of adhesive bonded joint design is actively evolving. Structural adhesives, with tailored properties to meet specific application requirements, as identified in the previous section, are a focal point of ongoing investigations [26]. Experimental testing of adhesives under extreme conditions, such as high temperatures or corrosive environments, provides insights into the limits of adhesive performance [27,28]. Numerical analyses, including FEM simulations, are becoming more sophisticated.

- Tailoring structural adhesives: structural adhesives can be tailored to meet specific application requirements [29]. The quest for adhesives with customized properties, such as enhanced strength, durability, and environmental resistance, is driving collaborations between material scientists and engineers. Researchers are exploring innovative formulations, including nanocomposite adhesives and bio-inspired adhesives, to achieve superior performance in diverse operating conditions [30,31].
- Experimental testing under extreme conditions: the performance of adhesive joints under high temperatures, in humid and corrosive environments, and under dynamic loading, is an active area of investigation [32,33]. This line of research not only expands the funda-

mental knowledge but also informs the development of adhesive formulations resilient to harsh operating conditions, crucial for industries like aerospace and automotive [34].

- Advancements in numerical analyses: the refinement of numerical analyses, particularly FEM simulations, is a vital point of research [9,35]. Researchers are incorporating more sophisticated modeling techniques to accurately simulate the intricate mechanics of adhesive joints [36,37]. Cohesive zone models are being fine-tuned to enhance their predictive capabilities [38]. This research contributes to a more nuanced understanding of joint behavior, facilitating the design of adhesive joints.
- Dynamic impact and fatigue testing: understanding how adhesive joints respond to dynamic loading, impact forces, and fatigue conditions is another area garnering significant attention [39–41]. Researchers are conducting experiments to elucidate the dynamic behavior of adhesive joints, providing insights into their resilience and failure mechanisms under varying loading rates [42]. This research is pivotal for applications where structures are subjected to cyclic loading or impact events, such as in automotive crash scenarios or structural components in wind turbines [43,44].

Prospects:

The future of adhesive bonding is driven by advancements in materials science, computational modeling, and manufacturing technologies [45]. Tailoring adhesives at the molecular level will ensure high-performance joints for specialized applications. The integration of machine learning algorithms into predictive modeling can enhance the accuracy of strength and failure predictions [46]. Additionally, the exploration of meshless methods and extended finite element methods (XFEMs) can provide a more efficient and accurate representation of complex joint behaviors.

- Tailoring adhesives at the molecular level: one of the most promising prospects lies in the ability to tailor adhesives at the molecular level [47]. This entails designing adhesives with precise properties to meet the specific demands of diverse applications. Adhesives are expected to have enhanced performance characteristics, such as superior strength, durability, and adaptability to challenging environmental conditions [48].
- Integration of machine learning into predictive modeling: the future of adhesive joint design envisions a seamless integration of machine learning algorithms into predictive modeling [39]. By learning from vast datasets of experimental and simulated results, machine learning algorithms can identify patterns and correlations that might elude traditional predictive methods [49]. Thus, the optimization of adhesive joint designs becomes possible for a wide range of applications.
- Exploration of meshless methods and XFEMs: the traditional FEM faces challenges in efficiently representing complex crack initiation and propagation in adhesive joints [50]. Meshless methods and XFEMs offer alternatives that could provide a more accurate and computationally efficient representation of joint behavior [51,52].
- Emerging technologies: as technology advances, so do the tools available for adhesive joint design. Emerging technologies, such as additive manufacturing, present opportunities to create intricate joint geometries and customized adhesive interfaces [53]. These technologies not only enhance the manufacturing process but also open avenues for innovative joint configurations that were previously impractical or impossible to achieve [54].

In conclusion, this MDPI Special Issue entitled "Design of Adhesive Bonded Joints" provides a timely platform to address the challenges and opportunities in this evolving field. Current design limitations necessitate standardized testing procedures and guidelines, while ongoing research explores advanced materials and improved numerical techniques. The future holds exciting prospects, with a focus on tailoring adhesives, integrating advanced modeling approaches, and embracing emerging technologies to drive the design of adhesive bonded joints to new heights. As the field continues to evolve, collaborative efforts among researchers, engineers, and industry professionals will be crucial in advancing the science and practice of adhesive joint design.

Conflicts of Interest: The author declares no conflict of interest.

References

1. Kumar, S.; Mittal, K.L. *Advances in Modeling and Design of Adhesively Bonded Systems*; John Wiley & Sons: Hoboken, NJ, USA, 2013.
2. Fay, P.A. 1—A history of adhesive bonding. In *Adhesive Bonding: Science, Technology and Applications*; Adams, R.D., Ed.; Elsevier: Amsterdam, The Netherlands, 2021; pp. 3–40.
3. Gualberto, H.R.; do Carmo Amorim, F.; Costa, H.R.M. A review of the relationship between design factors and environmental agents regarding adhesive bonded joints. *J. Braz. Soc. Mech. Sci. Eng.* **2021**, *43*, 389. [CrossRef]
4. Wang, S.; Li, J.; Li, S.; Wu, X.; Guo, C.; Yu, L.; Murto, P.; Wang, Z.; Xu, X. Self-contained underwater adhesion and informational labeling enabled by arene-functionalized polymeric ionogels. *Adv. Funct. Mater.* **2023**, *33*, 2306814. [CrossRef]
5. Tserpes, K.; Barroso-Caro, A.; Carraro, P.A.; Beber, V.C.; Floros, I.; Gamon, W.; Kozłowski, M.; Santandrea, F.; Shahverdi, M.; Skejić, D.; et al. A review on failure theories and simulation models for adhesive joints. *J. Adhes.* **2022**, *98*, 1855–1915. [CrossRef]
6. Ceroni, F.; Kwiecie, A.; Mazzotti, C.; Bellini, A.; Garbin, E.; Panizza, M.; Valluzzi, M. The role of adhesive stiffness on the FRP-masonry bond behavior: A round robin initiative. In *Structural Analysis of Historical Constructions: Anamnesis, Diagnosis, Therapy, Controls*; Balen, K.V., Verstrynge, E., Eds.; CRC Press: Boca Raton, FL, USA, 2016; pp. 1061–1069.
7. Valente, J.P.A.; Campilho, R.D.S.G.; Marques, E.A.S.; Machado, J.J.M.; da Silva, L.F.M. Adhesive joint analysis under tensile impact loads by cohesive zone modelling. *Compos. Struct.* **2019**, *222*, 110894. [CrossRef]
8. Viana, G.M.S.O.; Costa, M.; Banea, M.D.; da Silva, L.F.M. A review on the temperature and moisture degradation of adhesive joints. *Proc. Inst. Mech. Eng. L* **2017**, *231*, 488–501. [CrossRef]
9. Campilho, R.D.S.G. *Strength Prediction of Adhesively-Bonded Joints*; CRC Press: Boca Raton, FL, USA, 2017.
10. Lélias, G.; Paroissien, E.; Lachaud, F.; Morlier, J.; Schwartz, S.; Gavoille, C. An extended semi-analytical formulation for fast and reliable mode I/II stress analysis of adhesively bonded joints. *Int. J. Solids Struct.* **2015**, *62*, 18–38. [CrossRef]
11. Rocha, R.J.B.; Campilho, R.D.S.G. Evaluation of different modelling conditions in the cohesive zone analysis of single-lap bonded joints. *J. Adhes.* **2018**, *94*, 562–582. [CrossRef]
12. Karac, A.; Blackman, B.R.K.; Cooper, V.; Kinloch, A.J.; Rodriguez Sanchez, S.; Teo, W.S.; Ivankovic, A. Modelling the fracture behaviour of adhesively-bonded joints as a function of test rate. *Eng. Fract. Mech.* **2011**, *78*, 973–989. [CrossRef]
13. Zhang, J.; Wang, J.; Yuan, Z.; Jia, H. Effect of the cohesive law shape on the modelling of adhesive joints bonded with brittle and ductile adhesives. *Int. J. Adhes. Adhes.* **2018**, *85*, 37–43. [CrossRef]
14. Watson, B.; Worswick, M.J.; Cronin, D.S. Quantification of mixed mode loading and bond line thickness on adhesive joint strength using novel test specimen geometry. *Int. J. Adhes. Adhes.* **2020**, *102*, 102682. [CrossRef]
15. Nicassio, F.; Cinefra, M.; Scarselli, G.; Filippi, M.; Pagani, A.; Carrera, E. Numerical approach to disbonds in bonded composite single lap joints: Comparison between carrera unified formulation and classical finite element modeling. *Thin-Walled Struct.* **2023**, *188*, 110813. [CrossRef]
16. Brunner, A.J. 1—Investigating the performance of adhesively-bonded composite joints: Standards, test protocols, and experimental design. In *Fatigue and Fracture of Adhesively-Bonded Composite Joints*; Vassilopoulos, A.P., Ed.; Woodhead Publishing: Cambridge, UK, 2015; pp. 3–42.
17. Zhang, F.; Yang, X.; Xia, Y.; Zhou, Q.; Wang, H.-P.; Yu, T.-X. Experimental study of strain rate effects on the strength of adhesively bonded joints after hygrothermal exposure. *Int. J. Adhes. Adhes.* **2015**, *56*, 3–12. [CrossRef]
18. Dobrzański, P.; Oleksiak, W. Design and analysis methods for composite bonded joints. *Trans. Aerosp. Res.* **2021**, *2021*, 45–63. [CrossRef]
19. Ramalho, L.D.C.; Campilho, R.D.S.G.; Belinha, J.; da Silva, L.F.M. Static strength prediction of adhesive joints: A review. *Int. J. Adhes. Adhes.* **2020**, *96*, 102451. [CrossRef]
20. Dean, G.; Crocker, L. *The Use of Finite Element Methods for Design with Adhesives*; National Physical Laboratory: Teddington, UK, 2023.
21. Sugiman, S.; Ahmad, H. Comparison of cohesive zone and continuum damage approach in predicting the static failure of adhesively bonded single lap joints. *J. Adhes. Sci. Technol.* **2017**, *31*, 552–570. [CrossRef]
22. Bovone, G.; Dudaryeva, O.Y.; Marco-Dufort, B.; Tibbitt, M.W. Engineering hydrogel adhesion for biomedical applications via chemical design of the junction. *ACS Biomater. Sci. Eng.* **2021**, *7*, 4048–4076. [CrossRef] [PubMed]
23. Chen, S.-W.; Lu, P.; Zhao, Z.-Y.; Deng, C.; Wang, Y.-Z. Recyclable strong and tough polyamide adhesives via noncovalent interactions combined with Energy-Dissipating soft segments. *Chem. Eng. J.* **2022**, *446*, 137304. [CrossRef]
24. Broughton, W.R. *Durability Performance of Adhesive Joints*; National Physical Laboratory: Teddington, UK, 2023.
25. Li, W.; Palardy, G. Damage monitoring methods for fiber-reinforced polymer joints: A review. *Compos. Struct.* **2022**, *299*, 116043. [CrossRef]
26. Gonçalves, F.A.M.M.; Santos, M.; Cernadas, T.; Alves, P.; Ferreira, P. Influence of fillers on epoxy resins properties: A review. *J. Mater. Sci.* **2022**, *57*, 15183–15212. [CrossRef]
27. dos Reis, M.O.; Nascimento, H., Jr.; Monteiro, E.C.; Leão, S.G.; Ávila, A.F. Investigation of effects of extreme environment conditions on multiwall carbon nanotube-epoxy adhesive and adhesive joints. *Polym. Compos.* **2022**, *43*, 7500–7513. [CrossRef]
28. Fan, Y.; Zhao, G.; Liu, Z.; Yu, J.; Ma, S.; Zhao, X.; Han, L.; Zhang, M. An experimental study on the mechanical performance of BFRP-Al adhesive joints subjected to salt solutions at elevated temperature. *J. Adhes. Sci. Technol.* **2023**, *37*, 1937–1957. [CrossRef]
29. Zheng, Y.; Cui, T.; Wang, J.; Ge, H.; Gui, Z. Unveiling innovative design of customizable adhesive flexible devices from self-healing ionogels with robust adhesion and sustainability. *Chem. Eng. J.* **2023**, *471*, 144617. [CrossRef]

30. Panta, J.; Rider, A.N.; Wang, J.; Yang, C.H.; Stone, R.H.; Taylor, A.C.; Brack, N.; Cheevers, S.; Zhang, Y.X. High-performance carbon nanofiber reinforced epoxy-based nanocomposite adhesive materials modified with novel functionalization method and triblock copolymer. *Compos. B Eng.* 2023, *249*, 110401. [CrossRef]
31. Lutz, T.M.; Kimna, C.; Casini, A.; Lieleg, O. Bio-based and bio-inspired adhesives from animals and plants for biomedical applications. *Mater. Today Bio* 2022, *13*, 100203. [CrossRef] [PubMed]
32. Ulus, H.; Kaybal, H.B.; Berber, N.E.; Tatar, A.C.; Ekrem, M.; Ataberk, N.; Avci, A. An experimental evaluation on the dynamic response of water aged composite/aluminium adhesive joints: Influence of electrospun nanofibers interleaving. *Compos. Struct.* 2022, *280*, 114852. [CrossRef]
33. Korminejad, S.A.; Golmakani, M.E.; Kadkhodayan, M. Experimental and numerical analyses of damaged-steel plate reinforced by CFRP patch in moisture and the acidic environment under tensile test. *Structures* 2022, *39*, 543–558. [CrossRef]
34. Madrid, M.; Turunen, J.; Seitz, W. General industrial adhesive applications: Qualification, specification, quality control, and risk mitigation. In *Advances in Structural Adhesive Bonding*; Dillard, D.A., Ed.; Elsevier: Amsterdam, The Netherlands, 2023; pp. 849–876.
35. Qureshi, A.; Guan, T.; Alfano, M. Finite element analysis of crack propagation in adhesive joints with notched adherends. *Materials* 2022, *16*, 391. [CrossRef]
36. Sun, L.; Tie, Y.; Hou, Y.; Lu, X.; Li, C. Prediction of failure behavior of adhesively bonded CFRP scarf joints using a cohesive zone model. *Eng. Fract. Mech.* 2020, *228*, 106897. [CrossRef]
37. Orsatelli, J.-B.; Paroissien, E.; Lachaud, F.; Schwartz, S. Bonded flush repairs for aerospace composite structures: A review on modelling strategies and application to repairs optimization, reliability and durability. *Compos. Struct.* 2023, *304*, 116338. [CrossRef]
38. Li, W.; Liang, Y.; Liu, Y. Failure load prediction and optimisation for adhesively bonded joints enabled by deep learning and fruit fly optimisation. *Adv. Eng. Inform.* 2022, *54*, 101817. [CrossRef]
39. Fernandes, P.H.E.; Silva, G.C.; Pitz, D.B.; Schnelle, M.; Koschek, K.; Nagel, C.; Beber, V.C. Data-driven, physics-based, or both: Fatigue prediction of structural adhesive joints by artificial intelligence. *Appl. Mech.* 2023, *4*, 334–355. [CrossRef]
40. Gollins, K.; Elvin, N.; Delale, F. Characterization of adhesive joints under high-speed normal impact: Part II—Numerical studies. *Int. J. Adhes. Adhes.* 2020, *98*, 102530. [CrossRef]
41. Gollins, K.; Elvin, N.; Delale, F. Characterization of adhesive joints under high-speed normal impact: Part I—Experimental studies. *Int. J. Adhes. Adhes.* 2020, *98*, 102529. [CrossRef]
42. Li, L.; Jiang, H.; Zhang, R.; Luo, W.; Wu, X. Mechanical properties and failure behavior of flow-drilling screw-bonding joining of dissimilar aluminum alloys under dynamic tensile and fatigue loading. *Eng. Fail. Anal.* 2022, *139*, 106479. [CrossRef]
43. Mishnaevsky, L., Jr. Root causes and mechanisms of failure of wind turbine blades: Overview. *Materials* 2022, *15*, 2959. [CrossRef] [PubMed]
44. Silva, A.F.M.V.; Peres, L.M.C.; Campilho, R.D.S.G.; Rocha, R.J.B.; Silva, F.J.G. Cohesive zone parametric analysis in the tensile impact strength of tubular adhesive joints. *P. I. Mech. Eng. E-J. Pro.* 2022, *237*, 26–37. [CrossRef]
45. Xiao, Z.; Zhao, Q.; Niu, Y.; Zhao, D. Adhesion advances: From nanomaterials to biomimetic adhesion and applications. *Soft Matter* 2022, *18*, 3447–3464. [CrossRef]
46. Freed, Y.; Salviato, M.; Zobeiry, N. Implementation of a probabilistic machine learning strategy for failure predictions of adhesively bonded joints using cohesive zone modeling. *Int. J. Adhes. Adhes.* 2022, *118*, 103226. [CrossRef]
47. Das, A.; Dhar, M.; Manna, U. Small molecules derived tailored-superhydrophobicity on fibrous and porous substrates—With superior tolerance. *Chem. Eng. J.* 2022, *430*, 132597. [CrossRef]
48. Guan, W.; Jiang, W.; Deng, X.; Tao, W.; Tang, J.; Li, Y.; Peng, J.; Chen, C.L.; Liu, K.; Fang, Y. Supramolecular adhesives with extended tolerance to extreme conditions via water-modulated noncovalent interactions. *Angew. Chem. Int. Ed.* 2023, *62*, e202303506. [CrossRef]
49. Samaitis, V.; Yilmaz, B.; Jasiuniene, E. Adhesive bond quality classification using machine learning algorithms based on ultrasonic pulse-echo immersion data. *J. Sound. Vib.* 2023, *546*, 117457. [CrossRef]
50. Gonçalves, D.C.; Sánchez-Arce, I.J.; Ramalho, L.D.C.; Campilho, R.D.S.G.; Belinha, J. Fracture propagation based on meshless method and energy release rate criterion extended to the Double Cantilever Beam adhesive joint test. *Theor. Appl. Fract. Mech.* 2022, *122*, 103577. [CrossRef]
51. Resende, R.F.P.; Resende, B.F.P.; Sanchez Arce, I.J.; Ramalho, L.D.C.; Campilho, R.D.S.G.; Belinha, J. Elasto-plastic adhesive joint design approach by a radial point interpolation meshless method. *J. Adhes.* 2022, *98*, 2396–2422. [CrossRef]
52. Djebbar, S.C.; Madani, K.; El Ajrami, M.; Houari, A.; Kaddouri, N.; Mokhtari, M.; Feaugas, X.; Campilho, R.D.S.G. Substrate geometry effect on the strength of repaired plates: Combined XFEM and CZM approach. *Int. J. Adhes. Adhes.* 2022, *119*, 103252. [CrossRef]
53. Pizzorni, M.; Lertora, E.; Parmiggiani, A. Adhesive bonding of 3D-printed short-and continuous-carbon-fiber composites: An experimental analysis of design methods to improve joint strength. *Compos. B Eng.* 2022, *230*, 109539. [CrossRef]
54. Kowalczyk, J.; Ulbrich, D.; Sędłak, K.; Nowak, M. Adhesive joints of additively manufactured adherends: Ultrasonic evaluation of adhesion strength. *Materials* 2022, *15*, 3290. [CrossRef]

Disclaimer/Publisher's Note: The statements, opinions and data contained in all publications are solely those of the individual author(s) and contributor(s) and not of MDPI and/or the editor(s). MDPI and/or the editor(s) disclaim responsibility for any injury to people or property resulting from any ideas, methods, instructions or products referred to in the content.

Article
Removable Pressure-Sensitive Adhesives Based on Acrylic Telomer Syrups

Mateusz Weisbrodt and Agnieszka Kowalczyk *

Faculty of Chemical Technology and Engineering, West Pomeranian University of Technology, Piastów Ave. 42, 71-065 Szczecin, Poland
* Correspondence: agnieszka.kowalczyk@zut.edu.pl

Abstract: Removable pressure-sensitive adhesives (PSAs) are used in the production of self-adhesive materials such as protective films, masking tapes or biomedical electrodes. This work presents a new and environmentally friendly method of obtaining this type of adhesive materials, i.e., photochemically induced free radical telomerization. Adhesive binders to removable PSAs, i.e., the photoreactive acrylic telomer syrups (ATS) were prepared from n-butyl acrylate, acrylic acid, and 4-acrylooxybenzophenone. Tetrabromomethane (CBr_4) or bromotrichloromethane ($CBrCl_3$) were used as the telogens. ATS was modified with unsaturated polybutadiene resin and a radical photoinitiator. Adhesive compositions were coated onto a carrier and UV cross-linked. The effects of the chemical nature of telomers (i.e., terminal Br or Cl atoms) and their molecular weight (K-value), as well as the cross-linking degree on adhesive properties of PSAs, were studied. It was found that with the increase in telogen content in the system, the dynamic viscosity of ATS and K-value of acrylic telomers decrease, and the conversion of monomers increases. CBr_4 turned out to be a more effective chain transfer agent than $CBrCl_3$. Moreover, telomers with terminal Br-atoms (7.5 mmol of CBr_4), due to slightly lower molecular weights and viscosity, showed a higher photocrosslinking ability (which was confirmed by high cohesion results at 20 and 70 °C, i.e., >72 h). Generally, higher values of the temperature at which adhesive failure occurred were noted for PSAs based on ATS with lower telogen content (7.5 mmol), both CBr_4 and $CBrCl_3$. The excellent result for removable PSA was obtained in the case of telomer syrup Br-7.5 crosslinked with a 5 J/cm^2 dose of UV-radiation (adhesion ca.1.3 N/25 mm, and cohesion > 72 h).

Keywords: pressure-sensitive adhesives; removable PSA; telomerization; photopolymerization; adhesion

Citation: Weisbrodt, M.; Kowalczyk, A. Removable Pressure-Sensitive Adhesives Based on Acrylic Telomer Syrups. *Processes* **2023**, *11*, 885. https://doi.org/10.3390/pr11030885

Academic Editor: Raul D. S. G. Campilho

Received: 15 February 2023
Revised: 10 March 2023
Accepted: 14 March 2023
Published: 15 March 2023

Copyright: © 2023 by the authors. Licensee MDPI, Basel, Switzerland. This article is an open access article distributed under the terms and conditions of the Creative Commons Attribution (CC BY) license (https://creativecommons.org/licenses/by/4.0/).

1. Introduction

Pressure-sensitive adhesives (PSAs) are viscoelastic materials that remain permanently adhesive and can adhere even under light pressure [1]. Of the many base polymer materials, polyacrylates have enjoyed the fastest growth in commercial applications [2]. This is due to high resistance to oxidation and water, and lack of yellowing and transparency, which allows the use of acrylic PSAs in many industries, i.e., packaging tape, medical pads, protective films, optically clear adhesives, masking tapes, and hydrogels [3]. The most important properties of PSAs include tack (the adhesive's ability to adhere quickly), peel adhesion (a force required to remove a coated flexible sheet material from a test panel at a specified angle and rate of removal), and cohesion (resistance to static shear load). These values result from the nature of adhesives (chemical composition and state of the adhesives) molecular weight of the base polymer, crosslinking yields, coating weight, temperature, time, as well as a test method and conditions and face stock materials [4]. Pressure-sensitive adhesives are usually classified as permanent or removable (e.g., masking tapes where a low release force is required so as not to damage the surface) [5]. Permanent PSAs should be characterized by high peel adhesion, high tack, and high cohesion. While a limited peel value and a high shear strength are required for removable adhesives. According to peel adhesion values, PSAs can be divided into excellent permanent (>14 N/25 mm), permanent (from 10 to 14 N/25 mm), semi-removable (from 6 up to 8 N/25 mm), removable (from 2 to 4 N/25 mm) and excellent removable (<1 N/25 mm) [6]. While I. Benedek points out that removable PSAs should exhibit a peel adhesion value of 1.5 N/25 mm and cohesion values of at least 100 min at room temperature [7].

To obtain removable PSAs with such properties, the following methods are used: physical modification of the polymer matrix using inorganic additives (e.g., glass spheres or calcium carbonate) [8], use of mixtures of sticky and non-sticky ingredients (resins) [9] and polymer mixtures, in which one has a glass transition temperature below room temperature and the other slightly above room temperature [10]. Other methods consist of obtaining PSAs with an appropriate cross-linking degree (high cohesion and low tack) [11,12], using a mixture of "hard" and "soft" monomers [13] or segments, mainly based on mixtures of polyurethanes and polyethylene glycols [14–16]. The method of the greatest industrial importance in the production of the adhesive binder for removable PSA is the emulsion polymerization of acrylic monomers. An interesting way is also the chemical modification of the polyacrylate emulsion with isobornyl methacrylate, thanks to which it is possible to obtain PSA with adhesion below 0.3 N/25 mm and cohesion above 100 h [17]. In contrast, the use of the UV technique in the removable PSAs preparation mainly concerns the step of cross-linking the polymer matrix (photoreactive or not) modified with unsaturated cross-linking monomers (multifunctional), i.e., ethylene glycol dimethylacrylate, bisphenol-A ethoxylate diacrylate, or trimethylolpropane triacrylate. Achieving dense polymer networks results in obtaining PSA with outstanding properties (peel adhesion <1 N/25 mm) [18]. Removable PSAs from UV technology is used as attachments of ultra-thin electrodes (1 to 2 μm) to human skin (peel adhesion below 0.01 N/25 mm and ultra-high optical transparency) [19]. This article describes, for the first time, a new method of preparing the polymer matrix to removable PSAs, i.e., photochemically induced free radical telomerization.

Telomerization is a method of obtaining macromolecules/oligomers characterized by low polydispersity. In telomerization, taxogen (also called monomer) reacts with telogen (chain transfer agent) according to the radical mechanism, but also ionic (anionic or cationic) [20,21]. The telomerization process can be initiated by thermal initiators (such as organic peroxides, hydroperoxides, azo compounds, etc.), UV radiation, γ radiation accompanying beta decay of ^{60}Co to nonradioactive ^{60}Ni, and redox processes involving metal ions with variable valence [22]. Particularly noteworthy is the process of photoinduced telomerization, in which it is possible to obtain permanent PSAs without the need to use organic solvents that have particular chemical structures, for example, terminal Si atoms [23,24].

The work aimed to present a new method of preparation of removable PSAs from acrylic telomer syrups (as a product of photoinduced telomerization) and polybutadiene resin. In addition, the influence of telogen content (CBr_4 or $CBrCl_3$) as well as the K-value of acrylic telomers on the thermal and mechanical properties of PSAs were determined. Potentially, the presented method could be used in the production of masking tapes or protective films because it is ecologically safe (processes without the use of organic solvents), fast (up to several dozen minutes), and enables obtaining adhesive binders as solutions of oligomers with a low content of volatile organic compounds.

2. Materials and Methods

2.1. Materials

The acrylic telomer syrups (ATS) were prepared using the following components:

- monomers: n-butyl acrylate (BA), acrylic acid (AA BASF; Ludwigshafen, Germany), 4-acryloylooxybenzophenone (ABP, Chemitec, Scandicci, Italy)
- telogens: tetrabromomethane (CBr_4), bromotrichlomethane ($CBrCl_3$), Merck, Warsaw, Poland)

- radical photoinitiator: ethyl (2,4,6-trimethyl benzoyl)-phenyl-phosphinate (Omnirad TPOL, IGM Resins, Waalwijk, The Netherlands), i.e., acylphosphine-type (APO).

The adhesive compositions were prepared using the acrylic telomers syrups and:
- hydroxyl-terminated polybutadiene resin Hypro 1200X90 HTB (HTB, CVC Thermoset Specialties, Emerald Kalama Chemical, Kalama, WA, USA)
- radical photoinitiator: ethyl (2,4,6-trimethyl benzoyl)-phenyl-phosphinate (Omnirad TPOL, IGM Resins, Waalwijk, The Netherlands).

The components were used without additional purification. The chemical structures of the compounds are shown in Figure 1.

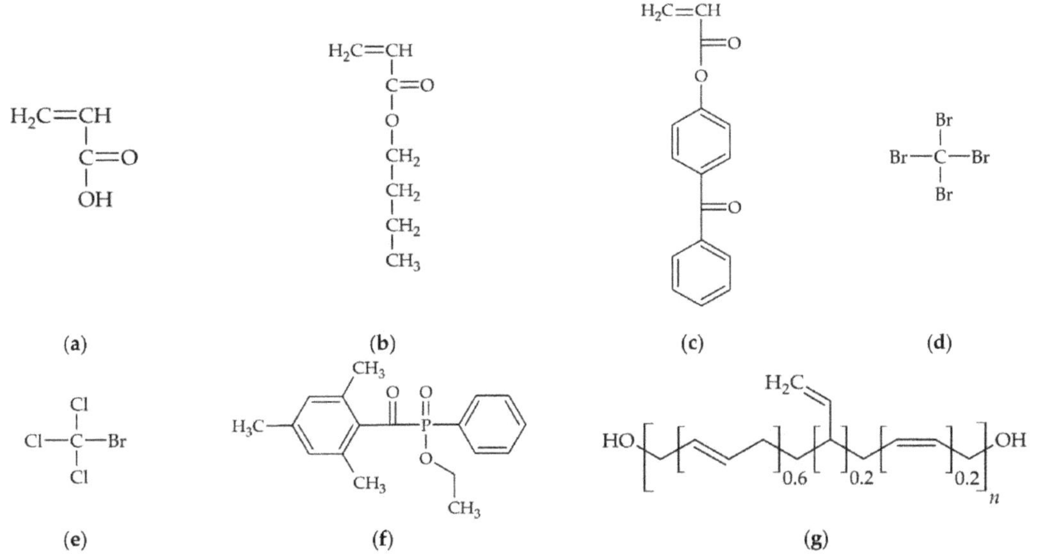

Figure 1. Chemical structures of the compounds: (**a**) acrylic acid, (**b**) n-butyl acrylate, (**c**) 4-acrylooxybenzophenone, (**d**) tetrabromomethane, (**e**) bromotrichloromethane, (**f**) TPOL radical photoinitiator (APO-type), (**g**) polybutadiene resin HTB.

2.2. Synthesis and Characterization of Acrylic Telomer Syrups

Acrylic telomer syrups (ATSs) were obtained by photo-induced telomerization of BA, AA, and ABP initiated by a radical photoinitiator TPOL and CBr_4 or $CBrCl_3$ as a telogen (7.5 or 15 mmol/100 wt. parts of monomers mixture). The compositions of the reaction mixtures are shown in Table 1.

Table 1. Compositions of monomers, photoinitiator, and telogen for acrylic telomer syrups.

Syrups Acronym	Monomers (wt.%)			TPOL (wt. part) [2]	CBr_4		$CBrCl_3$	
	BA	AA	ABP		(wt. parts) [2]	mmol [2]	(wt. parts) [2]	mmol [2]
RS [1]					—	—	—	—
Br-7.5					2.5	7.5	—	—
Br-15	91.5	7.5	1	0.2	5.0	15	—	—
Cl-7.5					—	—	1.5	7.5
Cl-15					—	—	3	15

[1] reference sample; [2] per 100 wt. parts of the monomer mixture.

The telomerization process was carried out in a glass reactor equipped with a mechanical stirrer (mixing speed 300 rpm) and a thermocouple under argon as inert gas at a temperature of 20 °C for 15 min. A mixture of monomers (50 g) was introduced into the reactor and purged with argon. The high-intensity UV lamp emitting UV-A radiation

(UVAHAND 250, Dr. Hönle AG UV Technology, Gräfelting, Germany) as a UV light source was used. The UV irradiation inside the reactor (15 mW/cm^2) was controlled with UV-radiometer SL2W (UV-Design, Brachttal, Germany). A schematic diagram of the photochemically induced telomerization process is shown in Figure 2.

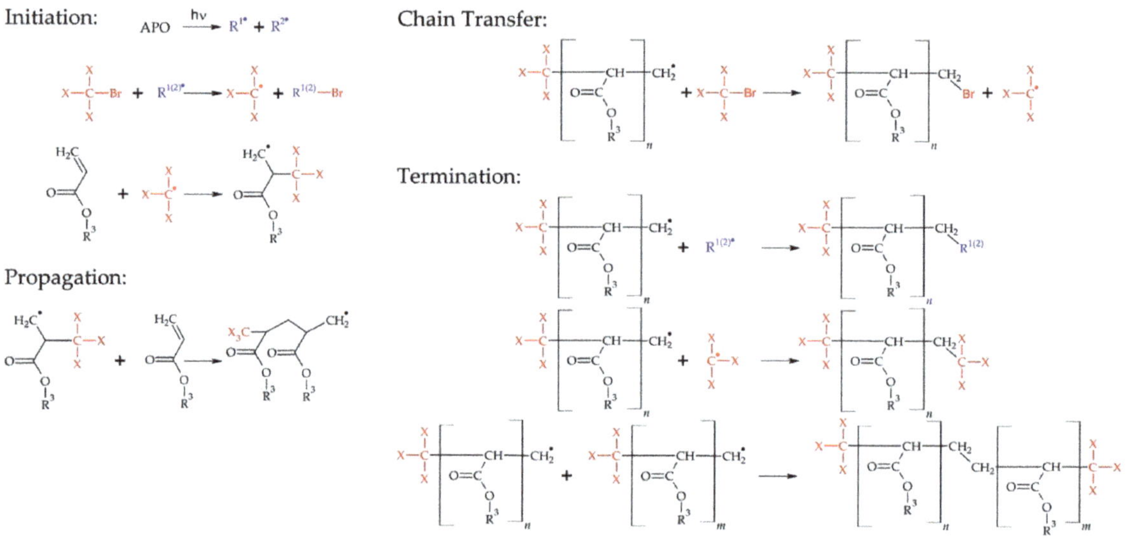

Figure 2. Scheme of the photochemically induced telomerization process where: $R^{1(2)}$ is a radical formed from the decomposition of APO photoinitiator, X is a Br or Cl atom, R^3 is H, C_3H_9 or $C_6H_6COC_6H_6$.

The solid content (SC) in ATS was determined using a thermobalance (MA 50/1.X2.IC.A; Radwag, Radom, Poland). In an aluminum pan, syrup samples (ca. 2 g) were heated at 105 °C for 4 h. Equation (1) was used to calculate the SC value:

$$SC = \frac{m_2}{m_1} \cdot 100 (\text{wt\%}) \qquad (1)$$

where: m_1 is the initial weight of a sample and m_2 is the residual weight after an evaporation process.

K-values were determined for the dry acrylic telomers based on the EN ISO 1628-1:1998 standard and the Fikentscher Equation (2).

$$K = 1000 \cdot k = 1000 \cdot \frac{1.5 \log \eta_r - 1 + \sqrt{1 + \left(\frac{2}{c} + 2 + 1.5 \log \eta_r\right) 1.5 \log \eta_r}}{150 + 300c} \qquad (2)$$

where: $\eta_r = \eta/\eta_0$; η is the viscosity of a telomer/copolymer solution; η_0 is the viscosity of a pure auxiliary diluent (i.e., tetrahydrofuran); c is the telomer concentration (g/cm^3).

2.3. Preparation and Characterization of Pressure-Sensitive Adhesives

Pressure-sensitive adhesives were obtained from acrylic telomer syrups (100 wt. parts), polybutadiene resin HTB (7.5 wt. parts/100 wt. parts of ATS), and APO photoinitiator (2.5 wt. parts/100 wt. parts of ATS). Adhesive compositions were homogenized with a high-speed mechanical mixer (T10 Basic Ultra-Turrax, IKA, Königswinter, Germany), applied onto polyester foils, and UV-irradiated (UV-doses were 2, 3, 4, 5, or 6 J/cm^2; UV-irradiation time was 32 s, 48 s, 64 s, and 96 s, respectively) using the medium-pressure mercury lamp (UV-ABC; Hönle UV-Technology, Gräfelfing, Germany). The base weight of the PSA layers was 15 g/m². The UV exposure was controlled with the radiometer (Dynachem 500; Dynachem Corp., Westville, IL, USA). During UV irradiation of adhesive films, two independent processes take place. The first is the formation of a polymer network from telomeric chains with hanging benzophenone moieties (from the APB structure) that are capable of abstracting hydrogen (for example from the hanging acrylate chains of other telomeric chains) (Figure 3a). The second process consists of photocrosslinking with the participation of the radical photoinitiator APO, unsaturated HTB resin and unreacted monomers from ATS (Figure 3b).

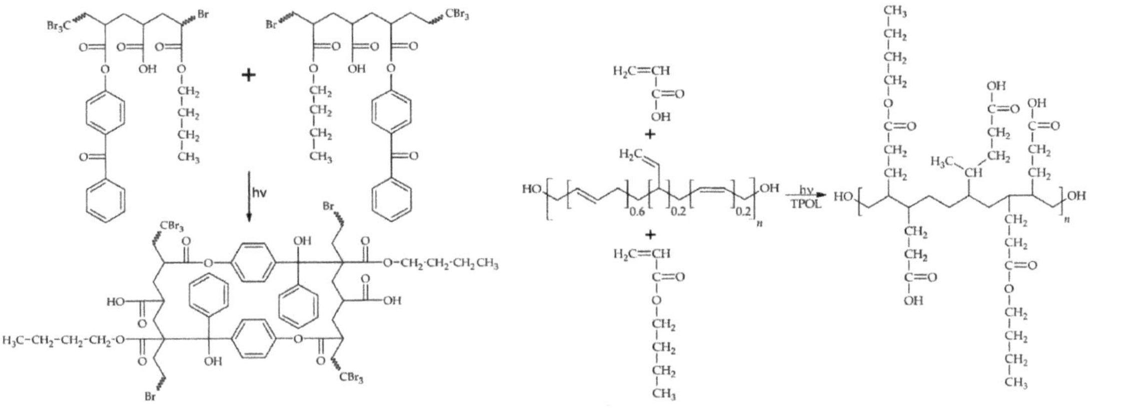

Figure 3. Photocrosslinking of pressure-sensitive adhesives: (**a**) via copolymerizable photoinitiator ABP (hydrogen abstractor) and (**b**) via HTB and unreacted monomers [23].

Self-adhesive tests (i.e., adhesion to steel at 180°, tack, cohesion at 20 °C and 70 °C, and shear adhesion failure test) of UV-crosslinked PSAs were performed. Adhesion to steel at a 180° (also called peel adhesion) was determined at room temperature using the Zwick/Roell Z010 testing machine (Zwick/Roell, Ulm, Germany) according to the AFERA 5001 (standard developed by the European Association des Fabricants Europeens de Rubans Auto-Adhesifs—AFERA). The degreased steel plate was applied with a one-sided PSA film measuring 175 × 25 mm and pressed with a 2 kg rubber roller. The test was performed 20 min after the application of the film to the plate with a peeling speed of 300 mm/min. Tack values were determined using the Zwick/Roell Z010 testing machine (Zwick/Roell, Ulm, Germany) according to AFERA 5015 standard. PSA film with dimensions of 175 × 25 mm was mounted in the upper jaws to obtain loops with the adhesive layer on the outside. In the lower jaws, a degreased steel plate was placed perpendicularly to the sample, which was lowered perpendicularly at a speed of 100 mm/min. The contact area was about 6.25 cm². The machine recorded the force needed to detach the adhesive film after a short contact with the steel surface, without external forces. The cohesion of PSAs (shear resistance) was measured according to the AFERA 5012 standard, with a device developed by the West Pomeranian University of Technology's International Laboratory of Adhesives and Self-Adhesive Materials in Szczecin that measures the time when joint cracks occur automatically. A one-sided adhesive film was applied to the degreased steel plate to form a 25 × 25 mm (6.25 cm²) joint and pressed with a 2 kg rubber roller to improve wettability. A 1 kg weight was attached to the free end of the film. The setup was then placed in a tripod so that the force of gravity was exerted on the weld at an angle of 180°. The cohesion value was defined as the time needed for the weld to crack. The test was carried out at a temperature of 20 °C and 70 °C. The shear adhesion failure test (SAFT) was carried out to determine the resistance to increased temperature. For this purpose, the PSA samples were prepared analogously to the cohesion test, however, they were heated in the range of 20 ÷ 250 °C with a temperature increase of 0.5 °C/min. Three samples of each adhesive film were evaluated for each test. Material damage may occur in any of the previously mentioned tests, i.e., adhesive failure (af), when the adhesive layer remains on the carrier (cohesion forces > adhesion forces); cohesive failure (cf) when the adhesive remains on both the carrier and substrate (cohesion forces < adhesion forces) and mixed failure (mf) when both adherent and cohesive failures occur at the same time.

To determine the glass transition temperatures (T_g) of UV-crosslinked PSAs the differential scanning calorimeter method was used (DSC250, TA Instruments, New Castle, DE, USA). Samples (ca. 10 mg) were analyzed using hermetic aluminum pans at temperatures from −80 to 200 °C (heating rate of 10 °C/min).

3. Results and Discussion

3.1. The Physicochemical Properties of Acrylic Telomer Syrups and Dry Telomers

The physicochemical properties of acrylic telomer syrups (dynamic viscosity and solid content), as well as the K-value of dry acrylic telomers, are presented in Table 2.

Table 2. The physicochemical properties of obtained acrylic telomer syrups and dry telomers.

ATS Acronym	SC (%)	Dynamic Viscosity (Pa·s)	K-Value (a.u.)
RS	n.d.	n.d.	n.d.
Br-7.5	79.2	13.8	26.3
Br-15	81.5	7.3	18.1
Cl-7.5	75.0	48.1	42.2
Cl-15	79.4	23.5	30.8

n.d.—no data.

In the beginning, it should be noted that carrying out the mass photopolymerization of BA, AA, and APB monomers (reference sample, Table 1) under the same reaction conditions as in phototelomerization, resulted in the gel formation in the entire volume of the reaction mixture. The use of chain transfer agents (i.e., CBr_4 and $CBrCl_3$) in the mass photopolymerization process allows for obtaining liquid reaction products (acrylic telomer syrups) in a relatively short time. As can be seen in Table 2, ATS with CBr_4 is characterized by higher solid content (higher monomers conversion) than ATS with $CBrCl_3$ telogen (2–4%). An increase in the telogen content in the reactive system caused a decrease in the solid content and the dynamic viscosity values of the ATS, as expected. A higher concentration of chain transfer agents (CBr_4 or $CBrCl_3$) causes the formation of more radicals, initiating the propagation step, as well as more frequent chain transfer and termination. This was similar to previously published results [25,26]. It is worth noting that telomer syrups with CBr_4 were characterized by a much lower dynamic viscosity (<14 Pa·s) and are perfect for coating the carrier. In the case of ATS with $CBrCl_3$, dynamic viscosity values were almost three times higher than their CBr_4 counterparts. This is related to the molecular weights of the resulting telomeres. Based on the K-value results, which express the molecular weight of polymeric materials, it was proved that Br-telomeres (based on CBr_4) was characterized by almost two times smaller molecular weights than Cl-telomeres (based on $CBrCl_3$). The above results are due to the greater reactivity of CBr_4 as a chain transfer agent. It is known that CBr_4 has a high kinetic chain transfer constant [27].

3.2. Properties of UV-Crosslinked PSA

Adhesion to steal at 180° and tack of prepared PSAs based on Br- or Cl-telomer syrups and polybutadiene resin HTB, depending on the UV dose used in the UV-crosslinking step (in fact on the cross-linking degree) are shown in Figure 4.

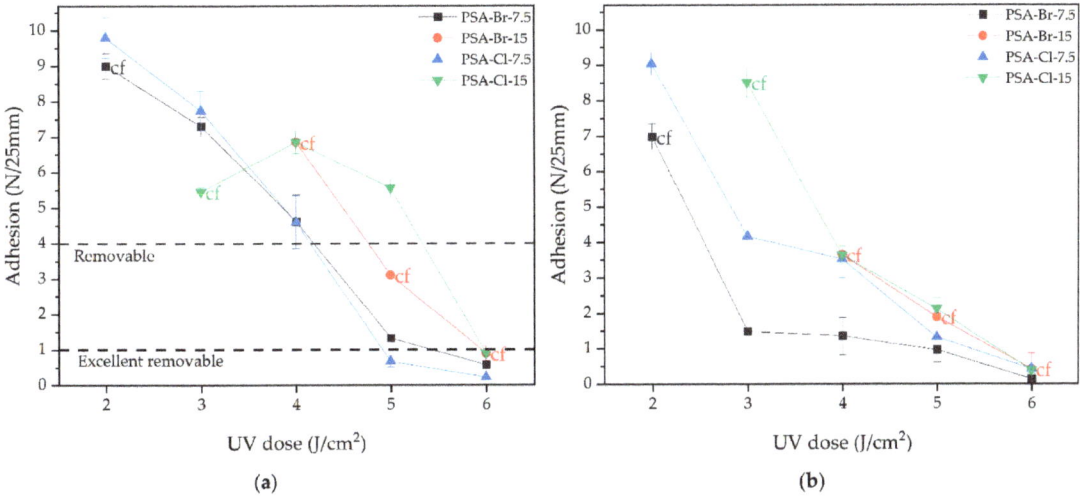

Figure 4. Adhesion to steal at 180° (**a**) and tack (**b**) tack of PSAs based on acrylic telomer syrups.

As mentioned earlier, the 3D polymer network is formed in presented PSAs because of the formation of cross-links between the telomers'/oligomers' chains and the polymer network made of polybutadiene resin and unreacted acrylate monomers (20–25 wt.% of unreacted BA or AA). The tests results showed that adhesion to steel and tack significantly depended on the UV dose (i.e., from the density of the polymer network), as well as the type and number of terminal groups in the polyacrylate chains, which interacted with the polar surface of the stainless steel. The adhesion values significantly decreased as the UV dose increased (from about 10 N/25 mm to even 0.2 N/25 mm for PSA-Cl-7.5). A similar dependence was shown for tack values (from 9.5 N to 0.1 N). Generally, cross-linking degree reduces chain mobility, thus decreasing adhesion and tack. Another factor affecting the adhesion (and tack) values is the chemical nature of the polymer forming the PSAs. This article presents new adhesive binders, i.e., acrylic telomers with terminal groups, namely: Br- (PSAs based on CBr_4 telomers) or Br- and -Cl_3 (samples from $CBrCl_3$). In addition, the weight fraction of telomeres in the adhesive binder is significant. Based on SC measurements it was 75 to 81.5 wt.% (Table 2). The other additives are polybutadiene resin (but only in the amount of 7.5 wt. parts/100 wt. parts of telomer syrup). It can therefore be concluded that the PSAs were mainly based on telomers. Thus, their chemical nature/physicochemical properties (i.e., molecular weight characterize as K-value and special built-in terminal groups) determine the adhesive properties of PSAs. The molar content of telogens in the system was the same (7.5 or 15 mmol/100 g of monomers mixture). Therefore, the properties of PSAs are influenced by the chemical nature of telogens. The adhesion values can be partly explained by the electrostatic theory of adhesion. According to this theory, the presence of more electronegative groups in the PSAs (electronegativity of bromine is 2.8 and chlorine is 3.0 on the Pauling scale) [28] should cause higher adhesion and tack [29]. This theory explains the positive influence of halogen concentration on adhesion and tack (PSAs based on Br-15 and Cl-15 telomer syrups exhibited higher adhesion and tack, than those with lower content of telogens). It is worth noting that only at low doses of UV radiation (2 or 3 J/cm^2) the adhesion values for PSA-Cl-7.5 (Cl atoms) were slightly higher than for PSA-Br-7.5 (only Br atoms). In the case of 4 J/cm^2 of UV dose, the adhesion values for the discussed PSAs samples were equal (4.8 N/25mm). On the other hand, a further increase in the UV dose (increase in cross-linking density) caused the PSA-Br-7.5 films to be characterized by higher adhesion (1.5 and 0.8 N/25 mm) than the PSA-Cl-7.5 (<1 N/25 mm). This was related to the K-values of the telomeres (i.e., their molecular weights). Br-telomeres exhibited a lower K-value (26.3 a.u.) than Cl-telomeres

(42.2 a.u.). Therefore, Br-telomers exhibited higher chain mobility and thus higher adhesion values. The same phenomenon explains the dependence of the tack values of the chemical nature of telomers. It is worth noting that only in the case of the PSA-Cl-15 films, no material damage was noted during the adhesion and tack tests. This was because the Cl-telomere had an appropriate K-value (ca. 42 a.u.). In the case of Cl-telomers, a higher dose of UV radiation was required to obtain properly cross-linked adhesive films (without cf) (>3 J/cm^2). Based on ATS with a higher content of telogens (15 mmol), both CBr_4 and $CBrCl_3$, it was impossible to obtain PSAs at a low UV dose (2 or 3 J/cm^2) (no results). In turn, all samples with Br-15 (the lowest K-value, 18 a.u.), regardless of the UV dose used, were insufficiently cross-linked (telomeric chains were too short). According to the previously indicated criteria and based on the tests carried out, it can be concluded that the PSA-Br-7.5 film, cross-linked by the UV dose of 5 J/cm^2, showed the adhesion and tack appropriate for removable PSA, i.e., 1.3 N/25 mm and 1 N, respectively. In turn, cohesion tests at 20 and 70 °C (Figure 5) revealed that this sample also exhibited excellent shear resistance (>72 h). Cohesion tests confirmed that the values of this parameter increased with cross-linking density (with UV dose) for all tested PSAs. The weakest results were confirmed for the PSA-Br-15 (the lowest K-value for telomer), i.e., <5 h.

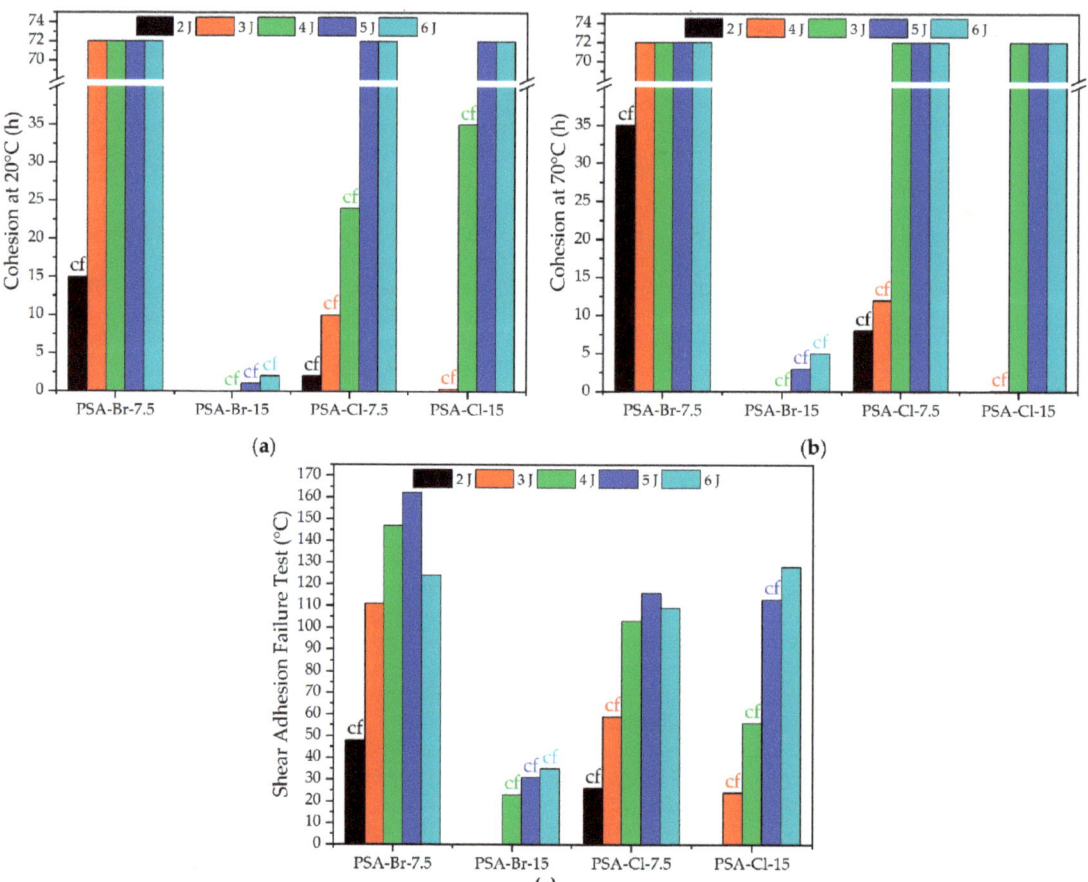

Figure 5. Cohesion at 20 °C (**a**) and 70 °C (**b**) and the SAFT (**c**) of PSAs based on acrylic telomer syrup.

Interestingly, higher values of cohesion at 70 °C were noticed. This may be due to the better wettability of the steel substrate by PSAs. In particular, it concerned PSA-Cl samples and cross-linked with a UV dose of 4 J/cm^2 (increase in cohesion at 70 °C to 72 h). It can also be stated that the best cohesion results (both at 20 and 70 °C) were recorded for the adhesive film based on Br-7.5 telomer syrup, as already at the UV dose of 3 J/cm^2 the cohesion values were 72 h. However, the adhesion was too high for removable PSA products (7.3 N/25 mm; Figure 4a). This may be due to the lower viscosity of this telomer (13.8 Pa·s) and relatively high K-value (26 a.u.), thus facilitating the migration of radicals and formation of cross-linked structure. Considering the shear adhesion failure test (SAFT) results (Figure 3c) it can be concluded that the high thermal resistance was demonstrated for the PSA-Br-7.5 sample cross-linked by 5 J/cm^2 (162 °C). Generally, higher values of the temperature at which adhesive failure occurred were noted for PSAs based on ATS with lower telogen content (7.5 mmol of CBr_4 or $CBrCl_3$). However, PSA-Br adhesive films (based on CBr_4) turned out to be slightly better, even though Br-telomeres were characterized by slightly lower molecular weights than Cl-telomers (lower K-value, 26 a.u, and 42 a.u., respectively). It's known that the shear measured as SAFT is directly dependent on the softening point (i.e., the molecular weight of the resin/polymer). In the case of PSAs based on ATS, it turns out that their photocrosslinking ability is also important. Systems with CBr_4, due to slightly lower molecular weights and viscosity, show a higher photocrosslinking ability (already at 3 J/cm^2, which was confirmed by cohesion tests at 20 and 70 °C).

The self-adhesive properties of PSAs are also affected by the glass transition temperature of the cross-linked system (resin/polymer). Figure 6 shows the DSC thermograms of the PSAs cross-linked using a UV dose of 6 J/cm^2.

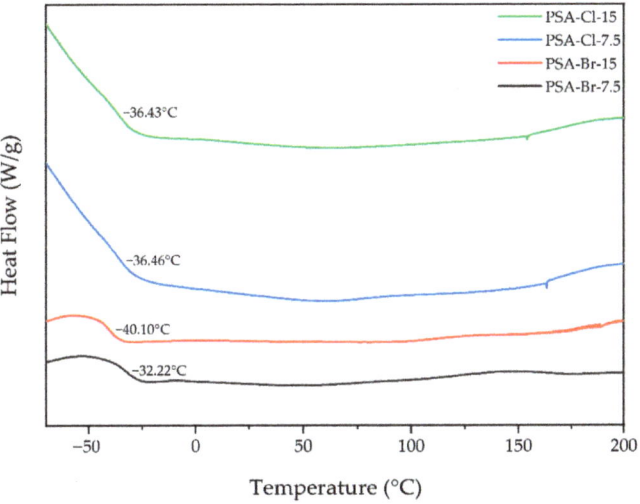

Figure 6. DSC thermograms of the cross-linked PSAs.

The T_g values for all acrylic PSAs were below −20 °C. Additionally, the T_g values for the PSA-Br-7.5 and PSA-Cl-7.5 and PSA-Cl-15 samples were very similar (−32 °C and −36 °C, respectively). Only PSA-Br-15 exhibited a lower T_g (−40 °C). These results correspond very well with the results of cohesion at 20 and 70 °C for these adhesive films (cross-linked with a UV dose of 6 J/cm^2). Namely, samples with higher T_g (of the order of −30 °C) were characterized by high cohesion, resulting from a high cross-linking degree with the UV dose, hence lower mobility of polymer chains and higher T_g values. The opposite is for the PSA-Br-15 sample with the lowest cross-linking degree and short, more mobile telomeric chains.

4. Conclusions

Acrylic telomer syrups (ATS) with terminal Br or Cl atoms (Br-telomers/Cl-telomers) were prepared via a UV-phototelomerization process using n-butyl acrylate, acrylic acid, 4-acrylooxybenzophenone. Two kinds of telogens, i.e., tetrabromomethane (CBr_4) and bromotrichloromethane ($CBrCl_3$) and radical photoinitiator (acylphosphine-type, APO) were tested as photoinitiating systems. Adhesive compositions (ATS compounded with the polybutadiene resin and APO photoinitiator) were used for the creation of removable pressure-sensitive adhesives (PSA) by UV cross-linking. The main conclusions are as follows:

- CBr_4 telogen allowed preparing of acrylic telomer syrup with significantly lower dynamic viscosity and slightly higher solids content than $CBrCl_3$;
- tBr-telomers are characterized by lower K-value compared to Cl-telomers.
- A greeter concentration of telogens (both CBr_4 and $CBrCl_3$) results in the formation of much shorter telomer chains, which determines the use of a higher dose of UV radiation at the stage of cross-linking the adhesive films, but despite this, cohesive failure still occurs, and the cohesion values of such systems are low, especially in case of PSA based on Br-telomers.
- The influence of terminal Br and Cl atoms and their amounts are consistent with the electrochemical theory of adhesion, but only in the range of low cross-linking densities of systems (small UV doses).
- The excellent removable PSA was obtained using 7.5 mmol of CBr_4/100 g of monomer mixtures, and after cross-linking of the adhesive film by 5 J/cm^2 of UV dose, i.e., adhesion to steal was 1.3 N/25 mm, tack 1 N, cohesion at 20 and 70 °C >72 h and shear adhesion failure test ca. 160 °C.

Author Contributions: Conceptualization, M.W. and A.K.; methodology, validation, M.W. and A.K.; formal analysis, A.K.; investigation, M.W.; writing—original draft preparation, A.K. and M.W.; writing—review and editing, A.K.; visualization, M.W.; supervision, A.K. All authors have read and agreed to the published version of the manuscript.

Funding: This research received no external funding.

Data Availability Statement: Not applicable.

Conflicts of Interest: The authors declare no conflict of interest.

References

1. Benedek, I.; Feldstein, M.M. *Fundamentals of Pressure Sensitivity*, 1st ed.; CRC Press: Boca Raton, FL, USA, 2008; ISBN 9781420059380.
2. Taghizadeh, S.M.; Ghasemi, D. Rheological and Adhesion Properties of Acrylic Pressure-Sensitive Adhesives. *J. Appl. Polym. Sci.* **2011**, *120*, 411–418. [CrossRef]
3. Park, H.W.; Seo, H.S.; Lee, J.H.; Shin, S. Adhesion Improvement of the Acrylic Pressure-Sensitive Adhesive to Low-Surface-Energy Substrates Using Silicone Urethane Dimethacrylates. *Eur. Polym. J.* **2020**, *137*, 109949. [CrossRef]
4. Benedek, I.; Feldstein, M.M. *Technology Pressure-Sensitive Adhesives and Products*; CRC Press: Boca Raton, FL, USA, 2019; ISBN 9781420059397.
5. Ortega-Iguña, M.; Chludzinski, M.; Sánchez-Amaya, J.M. Comparative Mechanical Study of Pressure Sensitive Adhesives over Aluminium Substrates for Industrial Applications. *Polymers* **2022**, *14*, 4783. [CrossRef] [PubMed]
6. Czech, Z. Solvent-Based Pressure-Sensitive Adhesives for Removable Products. *Int. J. Adhes. Adhes.* **2006**, *26*, 414–418. [CrossRef]
7. Benedek, I. *Pressure-Sensitive Adhesives and Applications*, 2nd ed.; Marcel Dekker, Inc.: New York, NY, USA, 2004; ISBN 0-8247-5059-4.
8. Graham, P.D.; Lu, Y.-Y.; Romsos, J.D. Adhesive Composition Having Non-Tacky Microspheres and Sheets Made Therefrom. US Patent 2009/024647, 1 October 2009.
9. Müller-Buschbaum, P.; Bauer, E.; Wunnicke, O.; Stamm, M. The Control of Thin Film Morphology by the Interplay of Dewetting, Phase Separation and microphase Separation. *J. Phys. Condens. Matter* **2005**, *17*, S363. [CrossRef]
10. Müller-Buschbaum, P.; Ittner, T.; Maurer, E.; Körstgens, V.; Petry, W. Pressure-Sensitive Adhesive Blend Films for Low-Tack Applications. *Macromol. Mater. Eng.* **2007**, *292*, 825–834. [CrossRef]
11. Chivers, R.A. Easy Removal of Pressure Sensitive Adhesives for Skin Applications. *Int. J. Adhes. Adhes.* **2001**, *21*, 381–388. [CrossRef]

12. Czech, Z. Crosslinking of Pressure Sensitive Adhesive Based on Water-Borne Acrylate. *Polym. Int.* **2003**, *52*, 347–357. [CrossRef]
13. Xue, J.; Wang, J.; Huang, H.; Wang, M.; Zhang, Y.; Zhang, L. Feasibility of Processing Hot-Melt Pressure-Sensitive Adhesive (HMPSA) with Solvent in the Lab. *Processes* **2021**, *9*, 1608. [CrossRef]
14. Bae, J.H.; Won, J.C.; Lim, W.B.; Kim, B.J.; Lee, J.H.; Min, J.G.; Seo, M.J.; Mo, Y.H.; Huh, P. Tacky-Free Polyurethanes Pressure-Sensitive Adhesives by Molecular-Weight and HDI Trimer Design. *Materials* **2021**, *14*, 2164. [CrossRef]
15. Lobo, S.; Sachdeva, S.; Goswami, T. Role of Pressure-Sensitive Adhesives in Transdermal Drug Delivery Systems. *Ther. Deliv.* **2015**, *7*, 33–48. [CrossRef] [PubMed]
16. Chen, X.; Liu, W.; Zhao, Y.; Jiang, L.; Xu, H.; Yang, X. Preparation and Characterization of PEG-Modified Polyurethane Pressure-Sensitive Adhesives for Transdermal Drug Delivery. *Drug Dev. Ind. Pharm.* **2009**, *35*, 704–711. [CrossRef]
17. Zhang, L.; Cao, Y.; Wang, L.; Shao, L.; Bai, Y. Polyacrylate Emulsion Containing IBOMA for Removable Pressure Sensitive Adhesives. *J. Appl. Polym. Sci.* **2016**, *133*, 42886. [CrossRef]
18. Chu, H.H.; Wang, C.K.; Sein, C.K.; Chang, C.Y. Removable Acrylic Pressure-Sensitive Adhesives Activated by UV-Radiation. *J. Polym. Res.* **2014**, *21*, 472. [CrossRef]
19. Kim, S.W.; Ju, Y.H.; Han, S.; Kim, J.S.; Lee, H.J.; Han, C.J.; Lee, C.R.; Jung, S.B.; Kim, Y.; Kim, J.W. A UV-Responsive Pressure Sensitive Adhesive for Damage-Free Fabrication of an Ultrathin Imperceptible Mechanical Sensor with Ultrahigh Optical Transparency. *J. Mater. Chem. A Mater.* **2019**, *7*, 22588–22595. [CrossRef]
20. Chen, J.; Chalamet, Y.; Taha, M. Telomerization of Butyl Methacrylate and 1-Octadecanethiol by Reactive Extrusion. *Macromol. Mater. Eng.* **2003**, *288*, 357–364. [CrossRef]
21. Bolshakov, A.I.; Kuzina, S.I.; Kiryukhin, D.P. Tetrafluoroethylene Telomerization Initiated by Benzoyl Peroxide. *Russ. J. Phys. Chem. A* **2017**, *91*, 482–489. [CrossRef]
22. Boutevin, B. From Telomerization to Living Radical Polymerization. *J. Polym. Sci. A Polym. Chem.* **2000**, *38*, 3235–3243. [CrossRef]
23. Kowalczyk, A.; Weisbrodt, M.; Schmidt, B.; Gziut, K. Influence of Acrylic Acid on Kinetics of UV-Induced Cotelomerization Process and Properties of Obtained Pressure-Sensitive Adhesives. *Materials* **2020**, *13*, 5661. [CrossRef]
24. Weisbrodt, M.; Kowalczyk, A. Self-Crosslinkable Pressure-Sensitive Adhesives from Silicone-(Meth)Acrylate Telomer Syrups. *Materials* **2022**, *15*, 8924. [CrossRef]
25. Weisbrodt, M.; Kowalczyk, A.; Kowalczyk, K. Structural Adhesives Tapes Based on a Solid Epoxy Resin and Multifunctional Acrylic Telomers. *Polymers* **2021**, *13*, 3561. [CrossRef] [PubMed]
26. Kraśkiewicz, A.; Kowalczyk, A.; Kowalczyk, K.; Schmidt, B. Novel Solvent-Free UV-Photocurable Varnish Coatings Based on Acrylic Telomers—Synthesis and Properties. *Prog. Org. Coat.* **2023**, *175*, 107365. [CrossRef]
27. Starks, C.M. *Free Radical Telomerization*; Academic Press: Cambridge, MA, USA, 1974; ISBN 9780126636505.
28. Ouellette, R.J.; Rawn, J.D. 1-Structure of Organic Compounds. In *Principles in Organic Chemistry*; Elsevier: Amsterdam, The Netherlands, 2015; pp. 1–32.
29. Chen, R.; Huang, Y.; Tang, Q.; Bai, L. Modelling and Analysis of the Electrostatic Adhesion Performance Considering a Rotary Disturbance between the Electrode Panel and the Attachment Substrate. *J. Adhes. Sci. Technol.* **2016**, *30*, 2301–2315. [CrossRef]

Disclaimer/Publisher's Note: The statements, opinions and data contained in all publications are solely those of the individual author(s) and contributor(s) and not of MDPI and/or the editor(s). MDPI and/or the editor(s) disclaim responsibility for any injury to people or property resulting from any ideas, methods, instructions or products referred to in the content.

Article

Static and Fatigue Characterization of Adhesive T-Joints Involving Different Adherends

Georgino C. G. Serra [1], José A. M. Ferreira [2] and Paulo N. B. Reis [2,*]

[1] Department of Electromechanical Engineering, University of Beira Interior, 6201-001 Covilhã, Portugal; georgino.serra@sapo.pt
[2] University of Coimbra, CEMMPRE, ARISE, Department of Mechanical Engineering, 3030-788 Coimbra, Portugal; martins.ferreira@dem.uc.pt
* Correspondence: paulo.reis@dem.uc.pt

Abstract: It is very important to understand the damage mechanisms as well as the mechanical response of T-joints involving different materials on the base plate. For this purpose, two configurations were studied. In one, the joint is composed of a base plate and a T-element, both in Al 6063-T5, while in the other one, the aluminum base plate was replaced by a glass fiber composite. Finally, each configuration was divided into two batches, where in one, the elements were bonded with a stiff adhesive (Araldite® AV 4076-1/HY 4076) while in the other, a more ductile adhesive (Araldite® AW 106/HV 953 U) was used. The static and fatigue strength of all configurations was evaluated in bending. In all cases, the damage occurred at the end of the T-element, where a crack appeared and propagated toward the interior of the T-joint. The bending strength is highest for joints involving aluminum and the ductile adhesive, which is 2.8 times higher than the same configuration involving composite base plates and 1.7 times higher than that using the stiff adhesive. Finally, the highest fatigue lives were obtained for T-joints involving Al 6063-T5 base plates, and regardless of the base plate material, the ductile adhesive promoted the highest fatigue strength.

Keywords: structural adhesives; T-joints; static characterization; fatigue strength; damage mechanisms; mechanical testing

1. Introduction

Compared to traditional joining methods (bolted, riveted or welded joints), adhesive joints have significant advantages due to the absence of fretting between materials, better fatigue response, and easier adaptation to complex shapes, among others. In addition to these advantages, adhesives are also increasingly reliable and durable. Therefore, it is not surprising that adhesive joints are increasingly being adopted by different industrial sectors [1].

In this context, and depending on the specific application or loading mode, designers have a wide variety of joint architectures at their disposal, where the single lap joints are the most used due to their simplicity and low cost [2,3]. However, they are responsible for promoting high shear and peeling stresses despite the various strategies that can be adopted to minimize them. Some examples include changing the strength and modulus of the adhesives and adherends [4–7], the thickness of adhesives and adherends [4–6], the overlap length [2,6], and adding fillets to the overlapped edges [8]. While these strategies increase the efficiency of single-lap joints, they cannot change the preferred loading mode for which they were designed. Therefore, to overcome this problem, T-joints are used to transfer bending, compressive, shear, and tensile loads between the leg panel and the base panel [9].

In terms of industrial applications, they can be used in the aircraft, automotive and marine sectors. In the first one, Johnson and Kardomateas [10] studied an adhesively bonded insert type T-joint for use in a composite space frame due to its high specific stiffness and large bond area. In this study, a finite element analysis was performed to obtain the stress distribution along the joint. Regarding naval applications, the University of Southampton developed extensive work on T-joint design and performance in which, for example, Shenoi and Violette [11] studied the influence of T-joint geometry on the ability to transfer out-of-plane loads in small boats. They concluded that geometry and material have a significant influence on the T-joint strength, but the radius of the fillet and the thickness of the overlaminate are the most determining parameters. Finally, in terms of aeronautic applications, T-joints are very favorable due to their structural efficiency, simplicity, and lightness. In this context, Moreno et al. [12] proposed a system to take advantage of the energy that could be dissipated by the structural bonded joints.

However, complete knowledge of joint strength and damage mechanisms is required to allow widespread use. From the different studies available in the literature [13–16], it is possible to conclude that the failure initiates in regions where high stresses occur, so it is important to find these critical points for specific design solutions and geometric parameters. For example, Shenoi and Hawkins [17] observed that although the geometry and material influence the strength and failure modes of T-joints, the fillet radius and the thickness of the overlaminate are determining parameters. On the other hand, the gap between the panels and the edge preparation of the T-piece showed less expressiveness. According to Dodkins et al. [18], there are two critical variables for T-joints: the thickness of the overlamination and the fillet radius. While in the first case, its increase affects the performance of the joint, in the case of the fillet, it allows the joint to support higher loads. However, Shenoi et al. [19] found that the influence of key variables is very dependent on the loading mode. Moreover, the effect of geometric parameters on stiffness, strength, and failure region is also significantly dependent on loading mode [15,16,20]. Finally, Chaves et al. [21] compared the strength of adhesive T-joints with screwed ones and observed that the adhesive joints have a similar or better mechanical performance than conventional screwed T-joints. However, they also noted that the strength of adhesive T-joints could be further improved by using spew fillets at the ends of the overlap.

According to Zhan et al. [22], the joint's dimensions, the size of the adherend and adhesive bondline, as well as the mechanical properties of the constituents (adherend and adhesive) significantly affect the strength of the T-joints. Therefore, based on these considerations, they studied experimentally and numerically the effect of different geometries subjected to a tensile load, concluding that the geometry of the bondline has a strong effect on the stress distributions, stress concentrations, and load-bearing capacity. Furthermore, it was also observed that the displacement of the Y axis tends to decrease with the increase in the average bondline area, while the average failure load increases with the increase in the average bonding area. Finally, the asymmetry of the stringer can cause unequal Von Mises equivalent stress distributions, higher Von Mises equivalent stresses, and deflection deformations. Ferreira et al. [23], using the finite element method (FEM) and cohesive zone models (CZM), studied the mechanical performance of different T-stiffener configurations under peel loads. For this purpose, authors considered the following geometrical parameters: flat adherend thickness (t_P), stiffener thickness (t_0), overlap length (L_O), and curved deltoid radius (R). They found a significant effect of all the parameters on both the stress distribution and maximum load. For example, the maximum load increased by around 94.1% when t_P was increased from 1 to 4 mm, decreased by about 27.3% when t_0 changed from 0.5 mm to 2.5 mm, increased by around 94.1% when L_0 was increased from 10 to 20 mm, and increased by around 135.4% when R was increased from 1 to 3 mm.

In T-joints, the stiffener can be a single part or composed of two L-shaped elements joined by co-curing or a structural adhesive. Using the last configuration (two L-shaped elements), Ma et al. [24] numerically (using the extended finite element method combined with cohesive zone model) and experimentally studied the damage mechanisms of carbon fiber reinforced polymer (CFRP) T-stiffeners subjected to pull-off loads. Contrary to what is reported in the literature, in which the final failure begins with the debonding in the radius region and then spreads to the stringer-skin interface, these authors observed a crack that begins in the filler region and near the fillet apex due to the stress concentration in this region. Subsequently, the crack propagates vertically, generating a debonding between the two L-shaped elements (stringer/stringer) but also towards the skin. Finally, when the crack reaches the skin, it promotes debonding at the stringer-skin interface, which moves towards both ends until the final rupture. The authors also found that the crack in the filler region started at the location of the maximum principal strain and that the large strain concentration region was limited to the filler region and the composite laminates. Therefore, improving the mechanical performance of T-joints can be achieved by reducing the stress concentration at the critical points of the joint and/or distributing the stresses more evenly along the bondline, among other solutions reported by Ravindran et al. al. [25]. In the first case, for example, Carvalho et al. [26] proposed the dual-adhesive joining technique, in which flexible adhesives are used in regions of high stress and stiff adhesives in regions of low stress. Different adhesive ratios were considered, namely 12.5/75/12.5 mm and 33.3/33.3/33.3 mm, and a numerical study was developed using the CZM technique (cohesive zone modeling) in the ABAQUS® software (ABAQUS 2017, Dassault Systèmes. RI, USA). These authors observed that failure occurred at the bondline and in the transition zone between adhesives, with the maximum load not only changing position but also decreasing in magnitude. In this context, improvements in the strength of the T-joints were obtained compared to those using only one adhesive, although more significant for the 33.3/33.3/33.3 mm ratio. With regard to the more uniform distribution of stresses along the bondline, Morano et al. [27] very recently suggested an alternative approach that does not compromise the integrity of the skin and is based on the use of corrugated stiffeners. Compared to the conventional configuration (flat stiffeners), the corrugated ones promoted improvements of around 65% in terms of pull-out strength and about 416% for the absorbed energy. The authors observed that the modified stiffeners promoted a redistribution of the stress along the bondline, with a consequent reduction in peak stresses at the free edge.

In terms of fatigue life, literature does not present many studies on this subject. Shenoi et al. [9], for example, noted that for higher load values, the fatigue strength significantly depends on the fillet radius (larger radii promote longer fatigue lives), but when the load decreases, this effect is lost, and the fatigue strength decreases in both cases to a fatigue threshold. The damage accumulation was assessed in terms of global stiffness loss, and three different regimes were observed. Initially, the stiffness decreased very rapidly during the first 20% of the fatigue life, followed by an almost linear regime up to 80% of the fatigue life, after which a very rapid degradation occurred again until the final collapse. In another study, Read and Shenoi [28] observed that the fatigue life increases both with the increase of the fillet radius if the overlaminate thickness (number of layers) is kept constant and with the increase of the overlaminate thickness if the fillet radius remains constant. Studies developed by Marcadon et al. [29] showed that fatigue life is strongly influenced by frequency due to the viscous behavior of the different materials, especially for higher load levels. Loureiro et al. [30] studied T-joints and compared the fatigue strength of two different adhesives (an epoxy adhesive and a polyurethane adhesive). They observed that the slope of both fatigue curves and the dispersion of the data are very similar. Although elastomeric adhesives have better fatigue behavior, this phenomenon was explained by heating the adhesive during the fatigue tests (greater for the elastomeric adhesive than for the epoxy). Finally, more recently, Cullinan et al. [31] studied T-joints repaired by cyanoacrylate adhesive systems and found that the fatigue life was lower than that obtained in control (unrepaired) samples.

Therefore, from the available literature, it is very important to understand the damage mechanisms and the mechanical response of adhesive T-joints to expand their application in the most diverse industrial sectors. If this subject is already reasonably understood for static loads, in terms of fatigue life, it is still limited to a very small number of studies, which does not allow for well-consolidated knowledge. For this purpose, the present study intends to analyze the fatigue performance of adhesive T-joints involving aluminum adherends and adherends that combine aluminum and glass fiber-reinforced composites.

2. Materials and Methods

Figure 1 shows the T-joint geometry and respective dimensions used in this study. The first configuration analyzed used only aluminum elements and, for this purpose, $150 \times 30 \times 3$ mm^3 Al 6063-T5 bars (Supplied by Alu-Stock, Madrid, Spain) were used as the base plate and T-elements of Al 6063-T5 with $30 \times 30 \times 2$ mm^3 as the stiffener. The nominal chemical composition and average tensile properties of this alloy are summarized in Tables 1 and 2, respectively.

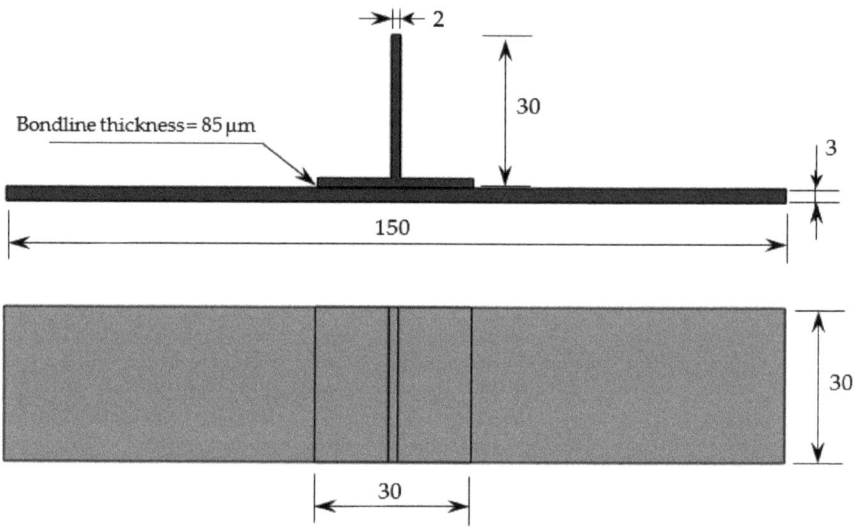

Figure 1. T-joint geometry and respective dimensions (in mm).

Table 1. Chemical composition (wt.%) of aluminum alloy 6063-T5 [32].

Si	Fe	Cu	Mn	Mg	Cr	Zn	Ti	Al
0.3–0.6	0.35	0.1	0.1	0.4–0.85	0.1	0.1	0.1	>96.9

Table 2. Principal mechanical properties of aluminum alloy 6063-T5 [32].

Elastic Limit (MPa)	UTS (MPa)	Elongation (%)	Young Modulus (GPa)	Hardness Brinell
145	187	≈12	68.9	60

The second configuration used composites in the base plates and stiffeners similar to the first configuration. Composite laminates were previously produced by hand lay-up using a Sicomin SR 8100 epoxy resin with SD 8824 hardener (both supplied by Sicomin, Chateauneuf les Martigues, France) and eighteen layers of bidirectional glass fiber fabric (taffeta with 195 g/m^2). Details regarding the mechanical/physical properties of the resin and manufacturing process can be found in [33]. Finally, the $150 \times 30 \times 3$ mm^3 base plates were cut from $330 \times 330 \times 3$ mm^3 composite plates. The maximum error observed in the thickness of the base plates was 0.35 mm for the composite ones and 0.04 mm for the aluminum ones.

Subsequently, two batches of samples were produced. In one of them, the base plates and T-stiffeners were bonded with the adhesive "Araldite® (Lausanne, Suisse) AV 4076-1 resin/HY 4076 hardener", while in the other one, the adhesive "Araldite® AW 106 resin/HV 953 U hardener" was used. The main mechanical properties of these adhesives are reported in [34,35]. These two-part paste adhesives were selected because the first one (Araldite® AV 4076-1/HY 4076) is a very stiff and brittle epoxy, and the last one (Araldite® AW 106/HV 953 U) is flexible/ductile. Before bonding, all surfaces to be joined (both base plates and T-stiffeners) were abrasively prepared with P220 silicon carbide paper and then cleaned with dry air and alcohol. This methodology was successfully applied in several previous studies [2,36–39], where the passive mechanical method used does not actively alter the chemical nature of the surface but only cleans the substrate and removes weak boundary layers in the form of contamination. Finally, to ensure a constant bondline thickness, all specimens were subjected to constant pressure during the adhesive curing process, and for this purpose, black metal dovetail clips were used. As reported in Figure 2 (a representative roughness profile), a pressure of 0.11 MPa applied to the specimens leads to an average bondline thickness of 85 μm without significant dispersion. Based on studies available in the literature [39–41], this value was obtained/measured after testing using a Mitutoyo (Kawasaki, Japan) SURFTEST SJ-500 surface measuring system.

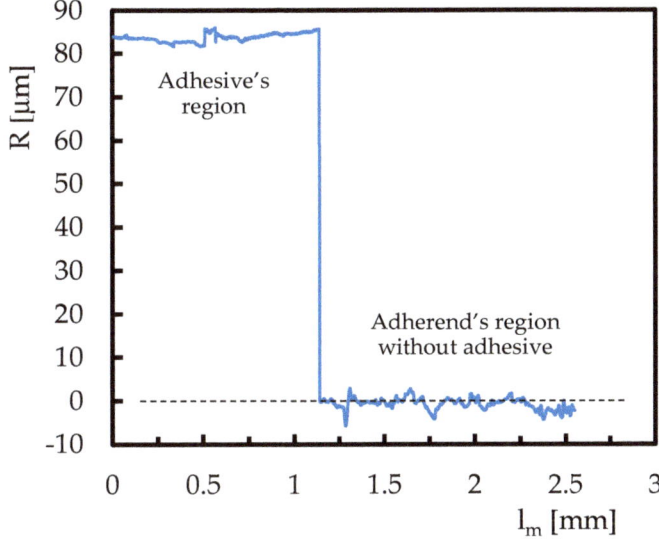

Figure 2. Typical roughness profile (R = roughness size; l_m = evaluated length).

The curing process was carried out at 40 °C for 16 h. According to Serra et al. [42], these parameters do not maximize the mechanical properties of the adhesives, but they were the most appropriate considering the T_g (glass transition temperature) of the resin used in the composite laminates. Therefore, for comparability of results between specimens with different base plates, a temperature of 40 °C was used for all batches analyzed.

These specimens are used to study the effect of materials (at the level of base plates) and adhesive type on the mechanical performance of T-joints. For this purpose, static three-point bending (3PB) tests are carried out at room temperature on a Shimadzu universal testing machine, model Autograph AGS-X (Shimadzu, Kyoto, Japan), equipped with a 100 kN load cell. As can be seen in Figure 3, the span used in the 3PB tests was 100 mm, with a displacement rate of 5 mm/min, and for each condition, three specimens were tested. Regarding the fatigue tests, they were carried out in an E 10000 Instron Electropulse (Norwood, MA, USA) uniaxial fatigue testing machine equipped with a 10 kN load cell and controlled by a computer with data acquisition. These tests were performed at room temperature, under a constant amplitude sinusoidal waveform loading, a stress ratio (R) of 0.05, and a frequency of 10 Hz. The load levels used in this study were selected to obtain fatigue lives between 10^3 and 10^6 cycles, and similar to the static tests, the load was also applied according to the schematic representation shown in Figure 3. Finally, the failure surface morphologies were also analyzed in detail using different techniques and equipment, such as a Mitutoyo SURFTEST SJ-500 surface measuring system, a Hirox (Hackensack, NJ, USA) RH 2000 microscope, and a Hitachi (Tokyo, Japan) Scanning Electron Microscopes SU3800.

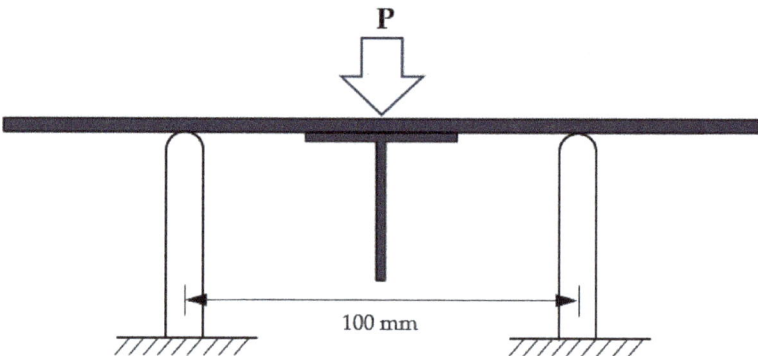

Figure 3. Apparatus and load mode used in static and fatigue testing.

3. Results and Discussion

The static response of the different T-joints was analyzed by 3PB tests according to Figure 3, and the results are shown in Figure 4. As reported by Loureiro et al. [30], due to non-uniform shear stresses and significant peel stresses that occur in this geometry, it is preferable to indicate the load rather than the stress. Therefore, this analysis avoids extremely misleading average shear stresses.

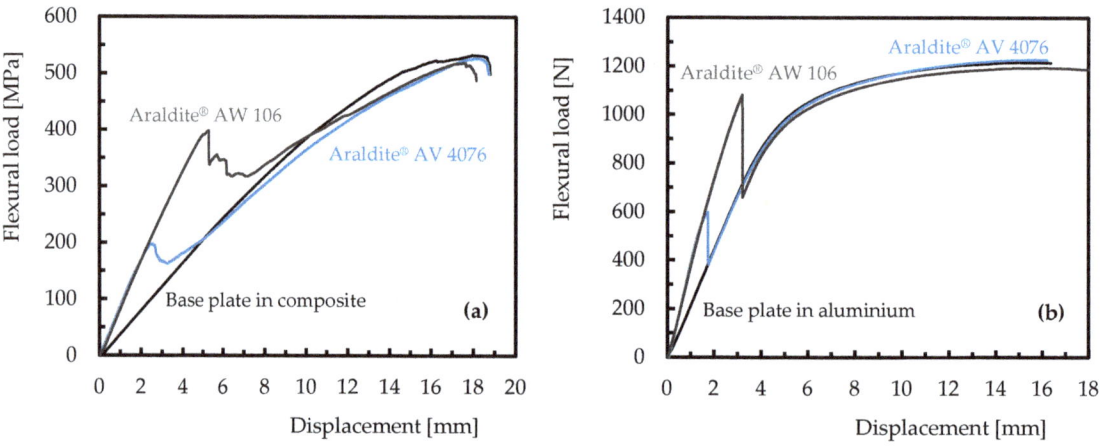

Figure 4. Flexural load-displacement curves for (**a**) T-joints involving base plates in the composite and (**b**) T-joints involving base plates in aluminum.

It is possible to observe that, for all the configurations studied (involving different base plate materials and adhesive types), the load increases linearly up to a maximum value, from which it starts to decrease until it reaches values close to those of the load displacement curve of the base plate material. Subsequently, the load-displacement curves practically overlap those of the base plates. It is also noticeable that the first peak load strongly depends on the adhesive type used in the T-joints as well as the material of the base plates. In fact, ductile adhesives can deform plastically and provide higher elongations than brittle adhesives before their collapse. Consequently, the highest peak loads observed for the ductile adhesive can be explained by the lower stress concentrations at the ends of the adhesive edges and better redistribution of stresses as the load increases [43,44]. On the other hand, regardless of the adhesive, peak loads are higher for T-joints involving Al 6063-T5 base plates because a higher stiffness of the adherends promotes a more uniform distribution of stresses in the adhesive [7]. In this context, the material that is less stiff determines the strength of the joint [7], and different failure mechanisms can be expected. Finally, for both adhesives, the second peak load is strictly related to the maximum load value observed for the base material. After the second peak load, the load always decreases more or less rapidly depending on the type of base material.

From Figure 4, it is also possible to obtain the main static properties, which are summarized in Table 3 in terms of average values and respective standard deviation. Stiffness was defined as the slope in the linear region of the load-displacement response, and the displacements are the values obtained for the different peak loads.

Table 3. Main static properties obtained for the different T-joints studied.

Base Plate Material	Adhesive	First Peak			Second Peak	
		Load [N]	Disp. [mm]	Stiffness [N/mm]	Load [kN]	Disp. [mm]
Composite plate	-	498.7 (±49.8)	18.1 (±2.31)	43.8 (±5.3)	-	-
Al 6063-T5 plate	-	1213 (±11.0)	15.6 (±0.20)	225.7 (±2.0)	-	-
Composite	AW 106	389.6 (±7.3)	5.8 (±0.49)	79.5 (±4.5)	519.1 (±50.5)	17.5 (±0.85)
	AV 4076-1	228.6 (±7.4)	3.5 (±0.14)	79.4 (±10.4)	491.8 (±49.8)	18.8 (±0.82)
Al 6063-T5	AW 106	1080.1 (±65.5)	3.6 (±0.51)	386.3 (±13.2)	1204.5 (±32.8)	15.8 (±0.20)
	AV 4076-1	626.7 (±25.2)	1.9 (±0.13)	399.6 (±6.3)	1200.9 (±22.3)	15.5 (±0.55)

(±SD)—Standard deviation values.

It is possible to observe that Al 6063-T5 base plates have the highest maximum load (about 2.4 times higher) and stiffness (about 5.1 times higher) and the lowest displacement at maximum load (about 13.8% lower) compared to the values obtained for the composite base plates. Regarding the adhesive joints, and regardless of the adhesive used, it is clearly noticeable that the second peak of load practically coincides with the maximum load value observed for each type of material used in the base plate. The dispersion is very small and, as can be seen in the results and in Figure 3, after a certain value, the curves of the base materials almost overlap with those of the T-joints. Finally, the effect of the T-element on the mechanical performance of the adhesive joints is significant only up to the first peak, proving to be dependent on the base plate material and type of adhesive used, after which the joint strength is similar to that of the base plate. The highest load peaks are obtained with the ductile adhesive (AW 106), where the values obtained for T-joints involving Al 6063-T5 base plates are 2.8 times greater than those involving composite base plates. On the other hand, the opposite is observed for displacement, where the maximum values are 1.6 times greater for T-joints involving composite base plates. However, comparing the adhesive type for joints with composite base plates, the ductile one accounts for 70.4% higher peak loads, while in terms of displacement, it is around 65.7%. For joints with Al 6063-T5 base plates, these values are 72.3% and 89.5% respectively. Lastly, regardless of the adhesive type and base plate material, it is quite evident that the bending stiffness values of the T-joints are much higher than those observed for the base plates (81.5% for composites and 74.1% for Al 6063-T5) due to the reinforcing effect introduced by the T-element, but very similar to each other despite the adhesives being different. For example, for T-joints involving composite base plates, the average bending stiffness is around 79.5 N/mm, while for those involving Al 6063-T5, it is about 393 N/mm. In the last case (T-joints involving Al 6063-T5 base plates), the stiff adhesive promotes a small difference of 3.3% in relation to the ductile one, but this value is statistically insignificant in relation to the observed dispersion. Therefore, the stiffness introduced by the T-element is clearly more important than the effect of the adhesive type, which would be expected given the bondline thickness. To complement the results described above, the failure surface morphologies were also analyzed and are shown in Figure 5.

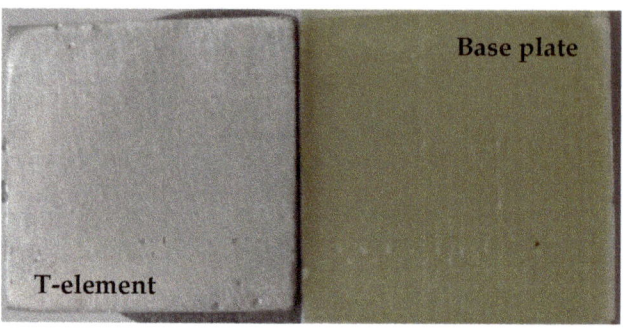

(a)

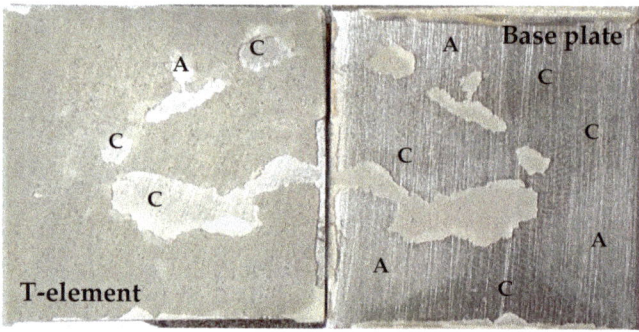

(b)

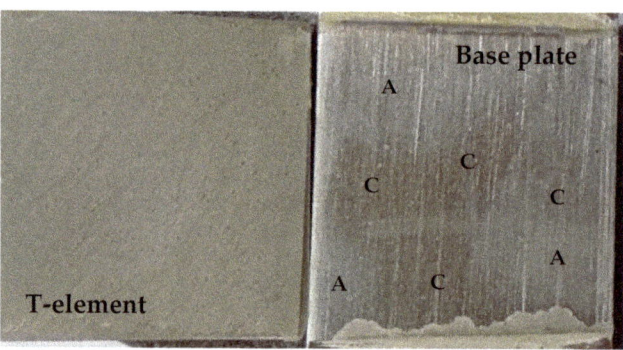

(c)

Figure 5. Typical failure morphologies for T-joints involving: (**a**) Composite base plates and ductile adhesive; (**b**) Al 6063-T5 base plates and ductile adhesive; (**c**) Al 6063-T5 base plates and stiff adhesive. (A = adhesive failure, C = cohesive failure).

A first analysis of Figure 5 shows that there are two failure modes, one for T-joints with composite base plates and another for those involving Al 6063-T5 base plates. In the first case, although Figure 5a shows only the damage morphology for T-joints with ductile adhesive, it is also representative of those involving the stiff adhesive. Therefore, for these joints, the failure is typically adhesive, with all the adhesive (whether ductile or stiff) remaining in the T-element (see Figure 5a). On the other hand, all T-joints involving Al 6063-T5 base plates revealed a mixed adhesive/cohesive failure mode (see Figure 5b,c). Adhesive failure (represented in Figure 5 by the letter "A") occurs when the forces exerted on the joint are greater than those between the adherend and adhesive, while cohesive failures (represented in Figure 5 by the letter "C") occur when the bond between molecules within the adhesive fails due to the greater external force. In this context, cohesive failure occurs when the maximum adhesive strain exceeds its limit [45].

To confirm the occurrence of mixed adhesive/cohesive failure mode, the authors used different techniques to assess the damage in different failure regions. Figure 6, for example, shows the roughness profile along a line (L) covering different failure modes and for a T-joint involving the Al 6063-T5 base plate and the ductile adhesive (Figure 5b). From the roughness profile shown in Figure 6, it is possible to identify three distinct regions: Region 1, which corresponds to an adhesive failure but with the adhesive layer completely on the T-element; Region 2, where the adhesive failure is also identified but with the adhesive layer completely over the base plate; and Region 3 where a cohesive failure occurs. In this case, a part of the adhesive remains on the T-element (about one-third of the thickness) and the rest on the base plate (the remaining two-thirds). As reported above, this defect is due to the strain having exceeded the maximum strain of the adhesive or, in other words, the external force is greater than the internal forces between molecules of the adhesive [45].

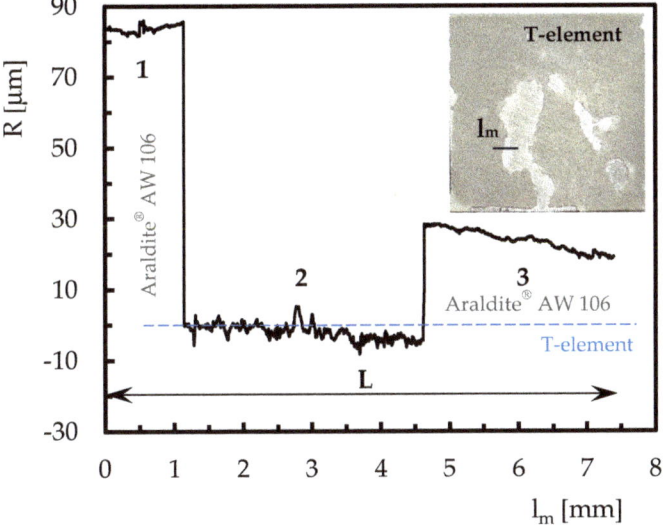

Figure 6. Roughness profile along the indicated line (l_m) for a T-joint involving aluminum base plate and ductile adhesive (R = roughness size; l_m = evaluated length).

Subsequently, because Figure 5b,c shows darker regions, especially at the level of base plates, they were analyzed to determine what type of damage would be underlying them. In this context, Figure 7 shows the analysis developed by digital microscopy (Hirox RH 2000 microscope) of a T-joint involving the Al 6063-T5 base plate and the stiff adhesive (Figure 5b), where it is evident that the darker regions are very thin films of adhesive.

Figure 7. Analysis developed by digital microscopy of a T-joint involving the Al 6063-T5 base plate and the stiff adhesive.

Although this technique is not as informative as the previous one because it does not allow thicknesses to be assessed, it clearly shows the existence of a thin film of adhesive on the base plate, with the remainder on the other adherend (i.e., adhered to the T-element). In this context, despite its simplicity, it proved the presence of adhesive traces in the adherend and, consequently, the existence of a cohesive failure. However, to deepen this analysis even further, scanning electron microscopy (SEM) and energy dispersive X-ray spectroscopy (EDS) were used to obtain more precise and magnified images of the fracture surface, as well as to identify its elemental chemical composition. Figure 8 shows, in this case, the results obtained from the SEM/EDS analysis for a specific region of Figure 5b.

This region A was selected because it apparently represents an adhesive failure, with one part of the adhesive on the base plate and the other part on the T-element. However, in detail, Figure 8a shows the existence of two regions, where the darker one represents the adhesive while the lighter one represents the surface of the T-element with eventual traces of adhesive. To prove this evidence, Figure 8b shows the analysis of the elemental chemical composition carried out along area *A* represented in Figure 8a. From this figure (Figure 8b), it can be observed that the darker region of Figure 8a is essentially constituted by hydrogen (H) and oxygen (O), chemical elements typical of the adhesive, while the lighter one is dominated by Al (aluminum), an element that underlies the chemical composition of the 6063-T5 aluminum alloy (>96.9%, according to Table 1) of the T-element. In addition to this chemical element, (Al), H (hydrogen), and O (oxygen) are also present, which confirms the presence of adhesive traces on the surface of the T-element. This evidence proves that an optical microscopy analysis similar to that carried out in Figure 7, based on light and a combination of lenses to magnify an image, is not entirely effective for assessing failure modes in adhesive joints. In fact, it does not allow for detection with complete assertiveness of the existence of adhesive traces on fracture surfaces because many of them can be confused with roughness or other defects in the adherends (see Figures 7 and 8a,b). Therefore, what initially appeared to be an adhesive failure region, this technique showed the existence of adhesive traces, changing the failure mode to mixed but with a larger predominance of the adhesive mode. On the other hand, when compared with the roughness analysis that supports Figure 6, the latter also does not clarify that in regions 1 and 2, there are adhesive traces on both adherends. Consequently, this technique should be used together with SEM/EDS to have a complete analysis of the damage mechanism. Finally, Figure 8c shows the elemental chemical composition between two points and along a trajectory (points 1 to 2 shown in Figure 8b). It can be seen that point 1 is on the adhesive due to the strong predominance of carbon (C), and point 2 is on the adherend where Al (aluminum) predominates, confirming the analysis reported above.

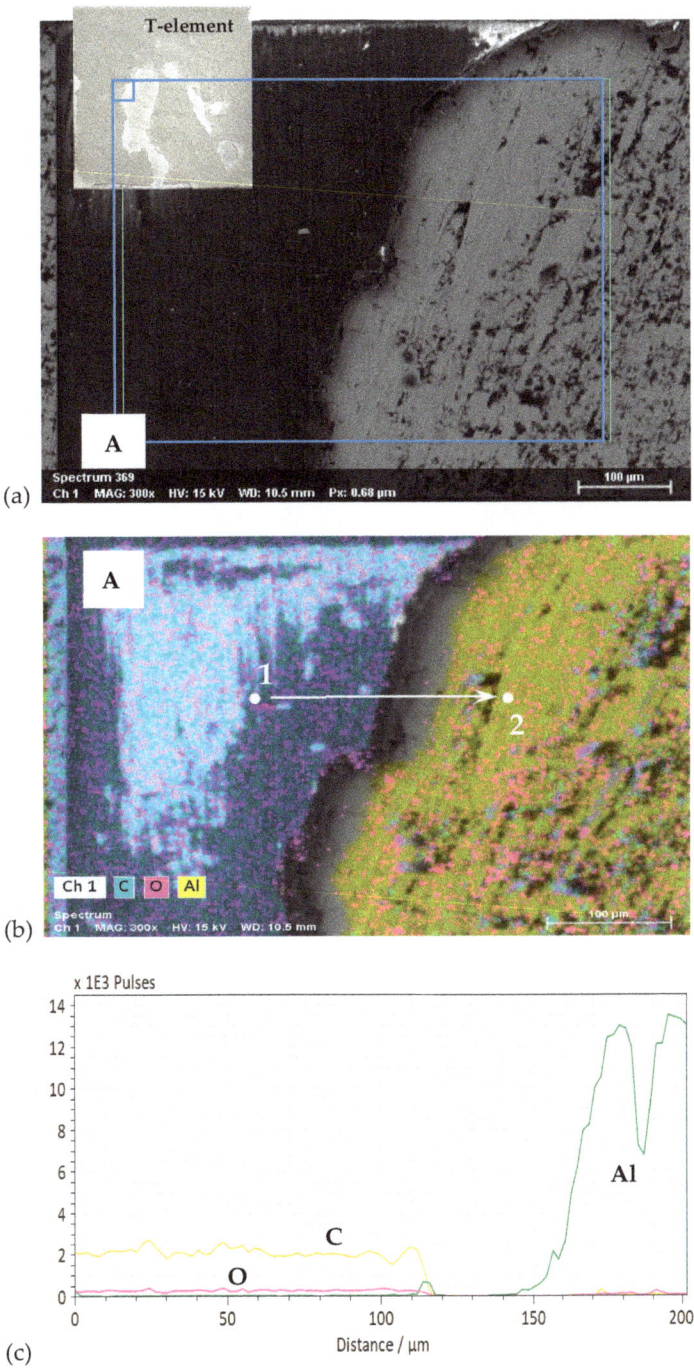

Figure 8. SEM-EDS analysis showing: (**a**) Fracture surface; (**b**) elemental chemical composition map; (**c**) distribution of elemental chemical composition between two points.

Based on the representativeness of the failure modes presented in Figure 5 and the various techniques that promoted the discussion supported by Figures 6–8, the areas related to the adhesive and cohesive failures were obtained using the ImageJ software (a graphic design tool dedicated to analyzing images). From this assessment, it was possible to observe that while in T-joints with the ductile adhesive, the cohesive area represents about 48.3% of the total bonded area, this value is only 3.8% lower for those involving the stiff adhesive. Another piece of evidence that can be taken from this study is the fact that adhesion between adhesives and aluminum is higher than between adhesives and composite. In fact, it would be possible to increase the adhesion using the various surface treatments suggested in the literature, but the authors chose not to adopt them in order to bring the study as close as possible to the reality of some industries where assemblies are made along the production line and from a mass production perspective [46]. However, according to Loureiro et al. [30], these conditions promote a higher scatter of failure loads.

To conclude this study and understand the typical profile of the curves shown in Figure 4, the bending tests were monitored by a high-speed video camera to assess the damage as a function of load. Figure 9 shows some damages for specific load values, and although the image sequence was obtained for a T-joint involving only aluminum and the stiff adhesive, it is representative of all other configurations. Therefore, based on the repeatability observed for all configurations, it can be noted that the load increases almost linearly up to a certain value that depends on the base plate material and the adhesive used (representative point—A). Subsequently, when the load reaches its first maximum (B), a crack appears at the end of the T-element between the adhesive and adherend (adhesive failure) or within the adhesive (cohesive failure). As reported above, adhesive failure occurs when the forces exerted on the joint are greater than those between adherend/adhesive, and cohesive failures occur when the maximum adhesive strain exceeds its limit [37]. Subsequently, after the first load peak (B), the load decreases more or less abruptly due to the propagation of the crack towards the interior of the reinforcement element until it reaches a value that coincides with the load-displacement curve of the base plate (D). For this load value (D), the damage has already reached half the length of the T-element. Thereafter, the damage propagation is negligible, and the load-displacement curves of the T-joints practically overlap with those of the base plate material (E). In this case, the load increases until it reaches a second peak load, which coincides with the maximum load of the base plate material. Therefore, for this geometry and loading, the structural integrity of the T-joints as a whole is only guaranteed up to the first peak load, after which it is ensured only by the base plate.

These results are in line with those observed by Hirulkar et al. [47], where the same behavior (significant load drop) was observed and explained by the appearance of cracks at the end of the overlap. Compared to the in-plane loading, the bending that is imposed here is responsible for a significant deflection in the joint, and consequently, higher peel stress concentrations occur [48,49]. In this case, the simultaneous action of tensile peel and shear stresses is responsible for the first cracks that appear at the end of the T-element (B) [42,50].

Regarding the fatigue response of the different T-joints, the results are shown in Figure 10 in terms of load versus the number of cycles to failure on a logarithmic scale for both adhesives and materials. The typical representation for a fatigue analysis (SN curves) was adopted, but instead of stress, loading was implemented due to the non-uniform nature of shear stresses and the existence of significant debonding stresses in these joints [30]. As already mentioned, these results were obtained for different constant amplitude loads, whose values were selected to obtain fatigue lives between 10^3 and 10^6 cycles. Moreover, it was ensured that all of them were lower than the first peak load observed in the static curves (Table 3) to guarantee the structural integrity of the adhesive joints at the beginning of each test and the existence of a load value common to all configurations for comparability of the results (in this case 150 N). Based on the static analysis performed around Figure 9 (damage initiation and its propagation), the fatigue failure criterion adopted considered the instant when the crack reaches half the length of the T-element (see Figure 9). However, because this methodology incorporates some subjectivity or the crack front may not be detectable by optical methods, it was necessary to adopt a more accurate methodology. Therefore, from the results collected by the data acquisition system, it is possible to plot, for example, the maximum load versus number of cycles and the maximum displacement versus number of cycles curves, as shown in Figure 10a. Although these curves were obtained for the T-joint involving Al 6063-T5 base plates, the ductile adhesive (AW 106), and a maximum load of 350 N, they are representative of all the others.

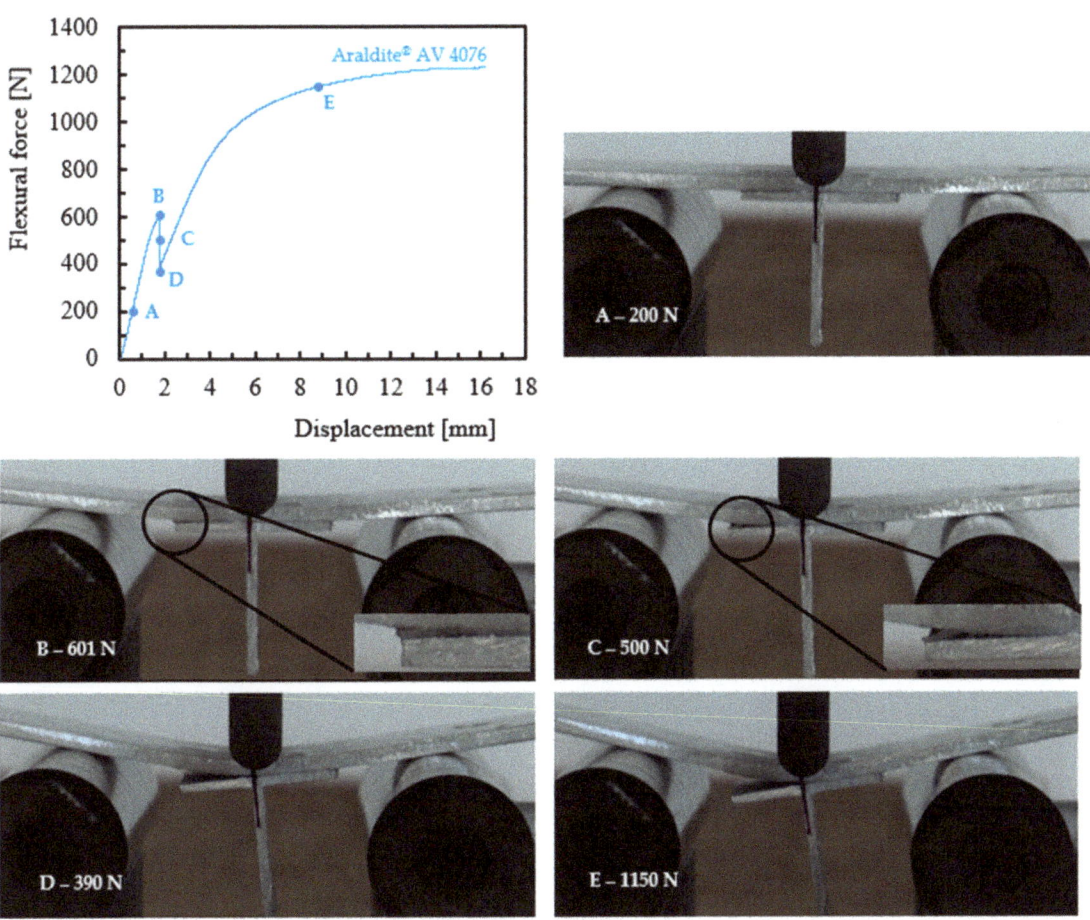

Figure 9. Damage evolution observed for T-joints involving only aluminum and a stiff adhesive.

In this context, because the load is constant, the fatigue failure criterion considers the number of cycles obtained for the first plateau point (N_f), from which the displacement is controlled by the stiffness of the base plate. Consequently, based on this methodology, the load versus number of cycles curves for the different configurations are shown in Figure 10b,c, where the mean curve fitted to the experimental results is also superimposed.

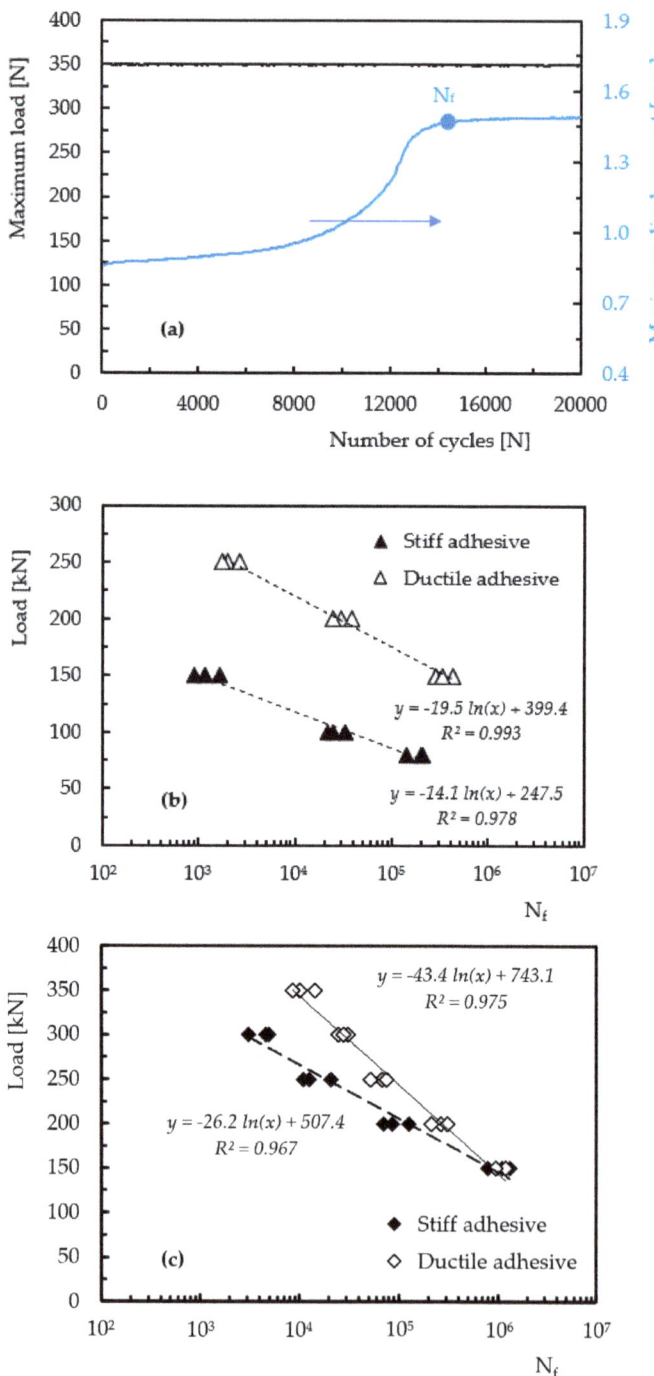

Figure 10. (a) Maximum load versus the number of cycles and maximum displacement versus the number of cycles; (b) fatigue life curves for T-joints involving composite base plates; (c) fatigue life curves for T-joints involving aluminum base plates.

From these figures, it is possible to observe that, when comparing T-joints involving composite base plates and the same load value (150 N), the ductile adhesive (AW 106) promotes fatigue lives about 280 times longer than those observed for the stiff adhesive (AV 4076–1). The slope of the fatigue curves shows some convergence for very long lives (Figure 10b), although T-joints involving the ductile adhesive always denote higher fatigue strength. On the other hand, regarding the T-joints involving Al 6063-T5 base plates (Figure 10c), it is noted that the curves converge for the 150 N load and with very close fatigue lives, i.e., the adhesive type has no effect on the fatigue strength for this load level. This behavior is very similar to that noticed for T-joints involving composite base plates, but with much greater convergence and fatigue lives about 871 and 3 times longer than those observed with brittle and ductile adhesive, respectively. Furthermore, increasing the load leads to greater differences in fatigue lives between the T-joints with the different adhesives used, reaching around 6.6 times for the 300 N load (see Figure 10c). Therefore, it is evident from Figure 10b,c that the ductile adhesive is responsible for longer fatigue lives due to higher percentages of elongation before failure and lower stress concentrations [43,51].

According to the bibliography, stiff adhesives experience higher stress concentrations and early failure compared to ductile ones [43]. For example, Temiz [52] observed that the use of ductile adhesives decreases the stress concentrations at the overlap ends and, consequently, increases the strength as well as delays the beginning of the failure due to their high strain to failure [44]. Moreover, the slow rate of stiffness degradation in a ductile adhesive allows the redistribution of stresses within it, whereas in a brittle one, after the damage initiation, uncontrolled crack growth leads to a more catastrophic failure. In addition to this, the lower strength of a ductile adhesive is compensated for its higher fracture toughness.

Therefore, if the stress and strain fields are more advantageous for T-joints using ductile adhesive and, consequently, responsible for longer fatigue lives, the adhesion between adhesives and adherends, as well as the stiffness of the adherends, cannot be neglected in the fatigue resistance of the joints. In the first case, it is well documented in the literature that an adhesive joint with low adhesion strength between constituents can fail unpredictably and cause adhesive failure [53,54], which is clearly visible in this study for adhesive joints involving composite base plates (Figure 5a). The adhesive always adheres to the T-element, and the fatigue lives are much shorter than those observed in joints with aluminum base plates, as mentioned above. On the other hand, regarding the adhesive joints involving Al 6063-T5 base plates, Reis et al. [7] report that increasing the stiffness of the adherends promotes a more uniform distribution of stresses in the adhesive and the less stiff material determines the strength of the joint. In this context, and because the composite used in the base plate is less stiff (43.8 N/mm) than Al 6063-T5 (225.7 N/mm), it justifies the longer fatigue lives observed for these T-joints.

Finally, in terms of damage mechanisms observed for the cyclic loads, they were similar to those observed in the static response. Therefore, to avoid the repetition of images, the authors chose not to display them. Nevertheless, as can be seen in Figure 9 and for all the configurations studied, the crack started at the edge of the T-element and propagated towards its center with the application of the cyclic load due to the mixed-mode stress on the adhesive layer. The peeling stresses (σ_y) predominate over the smaller stresses τ_{xy}, whose distribution has peaks at both edges of the adhesive layer and which will be higher for the stiff adhesive. More detailed analysis also revealed that, in the T-joints involving composite base plates, the crack propagation occurred between the adhesive and the base plate due to the poor adhesive/composite adhesion strength, while in those involving Al 6063-T5 base plates, the failure mode was mixed (mixed adhesive/cohesive failure). In the latter case, and in places where cohesive failure occurs, the adhesion between adhesive/aluminum is very high and even exceeds the values of the applied load or the internal forces between the adhesive molecules [45].

In order to complement this damage analysis, the literature is consensual that residual stiffness is an adequate methodology to assess fatigue damage [41]. The damage initiation and propagation cause changes in stiffness that can be monitored non-destructively, and for this purpose, the corresponding load and displacement values were also collected during the fatigue tests. Therefore, Figure 11 plots E/E_0 versus N/N_f, where E is the stiffness modulus (N/mm) at any given moment of the test, E_0 is the initial value of E (N/mm), N is the current number of cycles, and N_f is the number of cycles to failure. It is possible to observe that the damage accumulation previously discussed for all configurations can be corroborated by the global stiffness loss observed in Figure 11 and simultaneously confirms the findings of Shenoi et al. [22]. However, the profile of the stiffness loss curves contradicts what is described in the literature [22] and shows to be very dependent on the applied load and materials involved in the T-joints (adhesives and adherends).

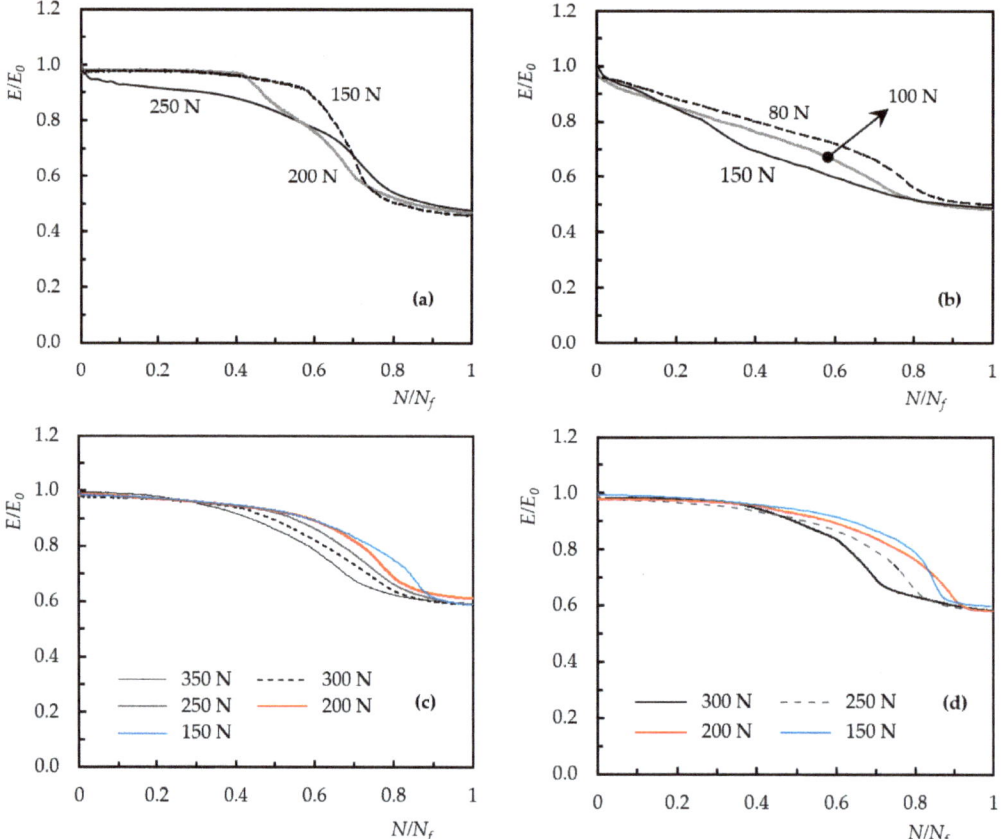

Figure 11. E/E_0 against the normalized number of cycles N/N_f for T-joints involving: (**a**) Composite base plate and ductile adhesive; (**b**) composite base plate and stiff adhesive; (**c**) Al 6063-T5 base plate and ductile adhesive; (**d**) Al 6063-T5 base plate and stiff adhesive.

In terms of T-joints involving composite base plates, for example, the stiff adhesive is responsible for curves in which the stiffness decreases almost linearly until about 60–70% of fatigue life, followed by a rapid degradation that culminates in a slower decrease until the adopted failure criterion is reached. These three regimes are repeated for the different load values, but the slope of the curve in the quasi-linear regime increases with increasing applied load. Therefore, this type of profile shows that the damage starts and propagates rapidly, leading to shorter lives. On the other hand, when these T-joints involve the ductile adhesive, the first regime is characterized by an almost unchanged stiffness (more or less constant plateau) until a sharp loss of stiffness occurs, followed by a slower decrease until the adopted failure criterion is reached (third regime). It is noticeable that, for the lowest loads (200 N and 150 N), the second regime started between 40% and 60% of the fatigue life, while for the highest load (250 N), the curve is similar to those observed for the stiff adhesive. Therefore, longer primary regimes promote slower damage propagation and longer fatigue lives.

Finally, for T-joints involving the aluminum base plates, the curves show a very similar profile for both adhesives used and are characterized by three regimes similar to those observed for the lowest loads shown in Figure 4a. In this context, analogous to what was observed before, the extension of the primary regime is also a determining factor in fatigue life. For the stiff adhesive, it is around 40% of the fatigue life, while for the ductile one, it represents between 30% and 60% of the fatigue life. Subsequently, the second regime is smoother than that observed for the brittle adhesive, revealing slower damage propagation and, consequently, longer fatigue lives.

4. Conclusions

The main goal of this study was to analyze the damage mechanisms as well as the mechanical behavior of T-joints involving different adhesives and base plate materials. For this purpose, four configurations involving base plates of Al 6063-T5 and fiberglass composite with a T-element of Al 6063-T5 were studied, which were bonded with a stiff adhesive (Araldite® AV 4076-1/HY 4076) and a more ductile one (Araldite® AW 106/HV 953 U).

In terms of static response, load-displacement curve profiles common to all configurations were found, in which the load increases up to a certain value and, subsequently, decreases until it reaches the load-displacement curve of the respective base plate. This is explained by the initiation of a crack that begins at the edge of the T-element and propagates more or less rapidly into its interior, depending on the failure surface morphology. T-joints with aluminum base plates experienced mixed adhesive/cohesive failure, while those with composite base plates experienced an adhesive failure. Finally, after decreasing, the curves practically overlap those of the base plate material, reaching a new peak load that coincides with the maximum load obtained for the base plate material. Consequently, the first highest peak loads are obtained with the ductile adhesive (AW 106), where the values obtained for T-joints involving Al 6063-T5 base plates are 2.8 times higher than those involving composite base plates.

Regarding fatigue strength, for both base plate materials, it was observed that the ductile adhesive promotes higher fatigue lives due to higher elongation percentages before failure and lower stress concentrations. For example, the fatigue life of aluminum-based T-joints is around 871 and 3 times longer than that of joints using composite base plates and brittle and ductile adhesive, respectively. Finally, damage accumulation can be corroborated by the stiffness loss curves, where three regimes were found to exist. Longer first regimens and less abrupt second regimens promoted longer fatigue lives.

Author Contributions: Conceptualization, P.N.B.R. and J.A.M.F.; methodology, G.C.G.S. and P.N.B.R.; validation, G.C.G.S. and P.N.B.R.; formal analysis, G.C.G.S., J.A.M.F. and P.N.B.R.; investigation, G.C.G.S. and P.N.B.R.; data curation, G.C.G.S., J.A.M.F. and P.N.B.R.; writing—original draft preparation, G.C.G.S. and P.N.B.R.; writing—review and editing, P.N.B.R. and J.A.M.F. All authors have read and agreed to the published version of the manuscript.

Funding: This research received no external funding.

Data Availability Statement: Not applicable.

Acknowledgments: This research was sponsored by national funds through FCT—Fundação para a Ciência e a Tecnologia, under the project UIDB/00285/2020 and LA/P/0112/2020.

Conflicts of Interest: The authors declare no conflict of interest.

References

1. Ciardiello, R. The mechanical performance of re-bonded and healed adhesive joints activable through induction heating systems. *Materials* **2021**, *14*, 6351. [CrossRef]
2. Reis, P.N.B.; Antunes, F.J.V.; Ferreira, J.A.M. Influence of superposition length on mechanical resistance of single-lap adhesive joints. *Compos. Struct.* **2005**, *67*, 125–133. [CrossRef]
3. Da Silva, L.F.M.; Critchlow, G.W.; Figueiredo, M.A.V. Parametric study of adhesively bonded single lap joints by the Taguchi method. *J. Adhes. Sci. Technol.* **2008**, *22*, 1477–1494. [CrossRef]
4. Sawa, T.; Liu, J.; Nakano, K.; Tanaka, J.J. A two-dimensional stress analysis of single-lap adhesive joints of dissimilar adherends subjected to tensile loads. *J. Adhes. Sci. Technol.* **2000**, *14*, 43–66. [CrossRef]
5. Liu, J.; Sawa, T.; Toratani, H. A two-dimensional stress analysis and strength of single-lap adhesive joints of dissimilar adherends subjected to external bending moments. *J. Adhes.* **1999**, *69*, 263–291. [CrossRef]
6. Da Silva, L.F.M.; Carbas, R.J.C.; Critchlow, G.W.; Figueiredo, M.A.V.; Brownc, K. Effect of material, geometry, surface treatment and environment on the shear strength of single lap joints. *Int. J. Adhes. Adhes.* **2009**, *29*, 621–632. [CrossRef]
7. Reis, P.N.B.; Ferreira, J.A.M.; Antunes, F. Effect of adherend's rigidity on the shear strength of single lap adhesive joints. *Int. J. Adhes. Adhes.* **2011**, *31*, 193–201. [CrossRef]
8. Amaro, A.M.; Neto, M.A.; Loureiro, A.; Reis, P.N.B. Taper's angle influence on the structural integrity of single-lap bonded joints. *Theor. Appl. Fract. Mech.* **2018**, *96*, 231–246. [CrossRef]
9. Shenoi, R.A.; Read, P.J.C.L.; Hawkins, G.L. Fatigue failure mechanisms in fibre-reinforced plastic laminated tee joints. *Int. J. Fatigue* **1995**, *17*, 415–426. [CrossRef]
10. Johnson, N.L.; Kardomateas, G.A. Structural Adhesive Joints for Application to a Composite Space Frame—Analysis and Testing. *SAE Tech. Pap.* **1988**, 880892. [CrossRef]
11. Shenoi, R.A.; Violette, F.L.M. A Study of Structural Composite Tee Joints in Small Boats. *J. Compos. Mater.* **1990**, *24*, 644–666. [CrossRef]
12. Moreno, M.C.S.; Cela, J.J.L.; Vicente, J.L.M.; Vecino, J.A.G. Adhesively bonded joints as a dissipative energy mechanism under impact loading. *Appl. Math. Model.* **2015**, *39*, 3496–3505. [CrossRef]
13. Li, W.; Blunt, L.; Stout, K.J. Analysis and design of adhesive-bonded tee joints. *Int. J. Adhes. Adhes.* **1997**, *17*, 303–311. [CrossRef]
14. Li, W.; Blunt, L.; Stout, K.J. Stiffness analysis of adhesive bonded Tee joints. *Int. J. Adhes. Adhes.* **1999**, *19*, 315–320. [CrossRef]
15. Apalak, Z.G.; Apalak, M.K.; Davies, R. Analysis and design of adhesively bonded tee joints with a single support plus angled reinforcement. *J. Adhes. Sci. Technol.* **1996**, *10*, 681–724. [CrossRef]
16. Apalak, Z.G.; Apalak, M.K.; Davies, R. Analysis and design of tee joints with double support. *Int. J. Adhes. Adhes.* **1996**, *16*, 187–214. [CrossRef]
17. Shenoi, R.A.; Hawkins, G.L. Influence of material and geometry variations on the behaviour of bonded tee connections in FRP ships. *Composites* **1992**, *23*, 335–345. [CrossRef]
18. Dodkins, A.R.; Shenoi, R.A.; Hawkins, G.L. Design of Joints and Attachments in FRP Ships' Structures. *Mar. Struct.* **1994**, *7*, 365–398. [CrossRef]
19. Shenoi, R.A.; Read, P.J.C.L.; Jackson, C.L. Influence of joint geometry and load regimes on sandwich tee joint behaviour. *J. Reinf. Plast. Compos.* **1998**, *17*, 725–740. [CrossRef]
20. Crocker, L.E.; Gower, M.R.L.; Broughton, W.R. Finite Element Assessment of Geometric and Material Property Effects on the Strength and Stiffness of Bonded and Bolted T-Joints. NPL Report MATC(A)124. 2003. Available online: https://eprintspublications.npl.co.uk/2563/ (accessed on 9 July 2023).
21. Chaves, F.J.P.; da Silva, L.F.M.; de Castro, P.M.S.T. Adhesively bonded T-joints in polyvinyl chloride windows. *Proc. Inst. Mech. Eng. Part L J. Mater. Des. Appl.* **2008**, *222*, 159–174. [CrossRef]
22. Zhan, X.; Gu, C.; Wu, H.; Liu, H.; Chen, J.; Chen, J.; Wei, Y. Experimental and numerical analysis on the strength of 2060 Al–Li alloy adhesively bonded T joints. *Int. J. Adhes. Adhes.* **2016**, *65*, 79–87. [CrossRef]
23. Ferreira, J.A.M.; Campilho, R.D.S.G.; Cardoso, M.G.; Silva, F.J.G. Numerical simulation of adhesively-bonded T-stiffeners by cohesive zone models. *Procedia Manuf.* **2020**, *51*, 870–877. [CrossRef]
24. Ma, X.; Bian, K.; Liu, H.; Wang, Y.; Xiong, K. Numerical and experimental investigation of the interface properties and failure strength of CFRP T-Stiffeners subjected to pull-off load. *Mater. Des.* **2020**, *185*, 108231. [CrossRef]
25. Ravindran, A.R.; Ladani, R.B.; Wang, C.H.; Mouritz, A.P. Strengthening of composite T-joints using 1D and 2D carbon nanoparticles. *Compos. Struct.* **2021**, *255*, 112982. [CrossRef]
26. Carvalho, P.M.D.; Campilho, R.D.S.G.; Sánchez-Arce, I.J.; Rocha, R.J.B.; Soares, A.R.F. Adhesively-bonded T-joint cohesive zone analysis using dual-adhesives. *Procedia Struct. Integr.* **2022**, *41*, 24–35. [CrossRef]
27. Morano, C.; Wagih, A.; Alfano, M.; Lubineau, G. Improving performance of composite/metal T-joints by using corrugated aluminum stiffeners. *Compos. Struct.* **2023**, *307*, 116652. [CrossRef]
28. Read, P.J.C.L.; Shenoi, R.A. Fatigue behaviour of single skin FRP tee joints. *Int. J. Fatigue* **1999**, *21*, 281–296. [CrossRef]
29. Marcadon, V.; Nadot, Y.; Roy, A.; Gacougnolle, J.L. Fatigue behaviour of T-joints for marine applications. *Int. J. Adhes. Adhes.* **2006**, *26*, 481–489. [CrossRef]
30. Loureiro, L.; da Silva, L.F.M.; Sato, C.; Figueiredo, M.A.V. Comparison of the mechanical behaviour between stiff and flexible adhesive joints for the automotive industry. *J. Adhes.* **2010**, *86*, 765–787. [CrossRef]

31. Cullinan, J.F.; Velut, P.; Michaud, V.; Wisnom, M.R.; Bond, I.P. In-situ repair of composite sandwich structures using cyanoacrylates. *Compos. Part A Appl. Sci. Manuf.* **2016**, *87*, 203–211. [CrossRef]
32. Almaraz, G.M.D.; Ambriz, J.L.A.; Calderón, E.C. Fatigue endurance and crack propagation under rotating bending fatigue tests on aluminum alloy AISI 6063-T5 with controlled corrosion attack. *Eng. Fract. Mech.* **2012**, *93*, 119–131. [CrossRef]
33. Santos, P.S.; Maceiras, A.; Valvez, S.; Reis, P.N.B. Mechanical characterization of different epoxy resins enhanced with carbon nanofibers. *Frat. Ed. Integrità Strutt.* **2020**, *15*, 198–212. [CrossRef]
34. Negru, R.; Marsavina, L.; Hluscu, M. Experimental and Numerical Investigations on Adhesively Bonded Joints. *IOP Conf. Ser. Mater. Sci. Eng.* **2016**, *123*, 012012. [CrossRef]
35. Mitchell, A.J. The Optimization of Stress Transfer Characteristics in Adhesively Bonded Vehicular Armour by Modification of the Adhesive Phase and by Engineering the Adhesive-to-Metal and Adhesive-to-Composite Interphases. Ph.D. Thesis, Loughborough University, Loughborough, UK, 2016.
36. Pereira, A.M.; Reis, P.N.B.; Ferreira, J.A.M.; Antunes, F.V. Effect of saline environment on mechanical properties of adhesive joints. *Int. J. Adhes. Adhes.* **2013**, *47*, 99–104. [CrossRef]
37. Reis, P.N.B.; Monteiro, J.F.R.; Pereira, A.M.; Ferreira, J.A.M.; Costa, J.D.M. Fatigue behaviour of epoxy-steel single lap joints under variable frequency. *Int. J. Adhes. Adhes.* **2015**, *63*, 66–73. [CrossRef]
38. Reis, P.N.B.; Pereira, A.M.; Ferreira, J.A.M.; Costa, J.D.M. Cyclic creep response of adhesively bonded steel lap joints. *J. Adhes.* **2017**, *93*, 704–715. [CrossRef]
39. Pereira, A.M.; Reis, P.N.B.; Ferreira, J.A.M. Effect of the mean stress on the fatigue behaviour of single lap joints. *J. Adhes.* **2017**, *93*, 504–513. [CrossRef]
40. Ferreira, J.A.M.; Reis, P.N.B.; Costa, J.D.M.; Richardson, M.O.W. Fatigue behaviour of composite adhesive lap joints. *Compos. Sci. Technol.* **2002**, *62*, 1373–1379. [CrossRef]
41. Reis, P.N.B.; Ferreira, J.A.M.; Richardson, M.O.W. Effect of the surface preparation on PP reinforced glass fiber adhesive lap joints strength. *J. Thermoplast. Compos. Mater.* **2012**, *25*, 3–11. [CrossRef]
42. Serra, G.C.G.; Reis, P.N.B.; Ferreira, J.A.M. Effect of cure temperature on the mechanical properties of epoxy-aluminium single lap joints. *J. Adhes.* **2023**, *99*, 1441–1455. [CrossRef]
43. Akhavan-Safar, A.; Ramezani, F.; Delzendehrooy, F.; Ayatollahi, M.R.; da Silva, L.F.M. A review on bi-adhesive joints: Benefits and challenges. *Int. J. Adhes. Adhes.* **2022**, *114*, 103098. [CrossRef]
44. Kadioglu, F.; Adams, R. Non-linear analysis of a ductile adhesive in the single lap joint under tensile loading. *J. Reinf. Plast. Compos.* **2009**, *28*, 2831–2838. [CrossRef]
45. Da Silva, L.F.M.; Neves, P.J.C.; Adams, R.D.; Wang, A.; Spelt, J.K. Analytical Models of Adhesively Bonded joints—Part II: Comparative Study. *Int. J. Adhes. Adhes.* **2009**, *29*, 331–341. [CrossRef]
46. Ciardiello, R.; Niutta, C.B.; Goglio, L. Adhesive thickness and ageing effects on the mechanical behaviour of similar and dissimilar single lap joints used in the automotive industry. *Processes* **2023**, *11*, 433. [CrossRef]
47. Hirulkar, N.S.; Jaiswal, P.R.; Reis, P.N.B.; Ferreira, J.A.M. Bending strength of single-lap adhesive joints underhygrothermal aging combined with cyclic thermal shocks. *J. Adhes.* **2021**, *97*, 493–507. [CrossRef]
48. Da Silva, L.F.M.; Pirondi, A.; Öchsner, A. *Hybrid Adhesive Joints*; Springer: Berlin/Heidelberg, Germany; London, UK, 2011; p. 304.
49. Barbosa, N.G.C.; Campilho, R.D.S.G.; Silva, F.J.G.; Moreira, R.D.F. Comparison of different adhesively-bonded joint types for mechanical structures. *Appl. Adhes. Sci.* **2018**, *6*, 15. [CrossRef]
50. Karachalios, E.F.; Adams, R.D.; da Silva, L.F.M. The behaviour of single lap joints under bending loading. *J. Adhes. Sci. Technol.* **2013**, *27*, 1811–1827. [CrossRef]
51. Meredith, H.J.; Wilker, J.J. The interplay of modulus, strength, and ductility in adhesive design using biomimetic polymer chemistry. *Adv. Funct. Mater.* **2015**, *25*, 5057–5065. [CrossRef]
52. Temiz, Ş. Application of bi-adhesive in double-strap joints subjected to bending moment. *J. Adhes. Sci. Technol.* **2006**, *20*, 1547–1560. [CrossRef]
53. De Barros, S.; Kenedi, P.P.; Ferreira, S.M.; Budhe, S.; Bernardino, A.J.; Souza, L.F.G. Influence of mechanical surface treatment on fatigue life of bonded joints. *J. Adhes.* **2017**, *93*, 599–612. [CrossRef]
54. Pereira, A.M.; Ferreira, J.M.; Antunes, F.V.; Bártolo, P.J. Study on the fatigue strength of AA 6082-T6 adhesive lap joints. *Int. J. Adhes. Adhes.* **2009**, *29*, 633–638. [CrossRef]

Disclaimer/Publisher's Note: The statements, opinions and data contained in all publications are solely those of the individual author(s) and contributor(s) and not of MDPI and/or the editor(s). MDPI and/or the editor(s) disclaim responsibility for any injury to people or property resulting from any ideas, methods, instructions or products referred to in the content.

Article

The Effect of Organic Fillers on the Mechanical Strength of the Joint in the Adhesive Bonding

Nergizhan Anaç [1,*] and Zekeriya Doğan [2]

1 Department of Mechanical Engineering, Engineering Faculty, Zonguldak Bulent Ecevit University, Zonguldak 67000, Turkey
2 Department of Civil Engineering, Engineering Faculty, Zonguldak Bulent Ecevit University, Zonguldak 67000, Turkey
* Correspondence: nergizhan.kavak@beun.edu.tr; Tel.: +90-372-291-1214

Abstract: The most important advantages of adding additives to adhesives are increasing the bonding strength and reducing the adhesive cost. The desire to reduce costs as well as the need for environmentally friendly and health-friendly products have paved the way for the recycling of waste materials and the use of cheaper natural materials as additives. In this study, mussel, olive pomace, and walnut powders in different ratios (5%, 15%, and 30% by weight) and in different sizes (38 and 45 µm) were added to an epoxy adhesive. The steel materials were joined in the form of single-lap joints by using the obtained adhesives with additives. These joints were subjected to the tensile test and the strengths of these joints were examined. SEM images of the bonding interface were taken, and the distribution of the powders was examined. When the powder size was 45 µm, bond strengths increased in all additive ratios compared to the pure adhesive, while for 38 µm powders, the strength value increased only at the 5% additive ratio. In joints with 45 µm powder additives, the strength increased by up to 38% compared to the pure adhesive, while this rate was determined as 31% for 38 µm.

Keywords: adhesive bonding; organic filler; waste material; the single lap joint; mechanical strength

Citation: Anaç, N.; Doğan, Z. The Effect of Organic Fillers on the Mechanical Strength of the Joint in the Adhesive Bonding. *Processes* 2023, 11, 406. https://doi.org/10.3390/pr11020406

Academic Editor: Raul D.S.G. Campilho

Received: 30 December 2022
Revised: 16 January 2023
Accepted: 22 January 2023
Published: 30 January 2023

Copyright: © 2023 by the authors. Licensee MDPI, Basel, Switzerland. This article is an open access article distributed under the terms and conditions of the Creative Commons Attribution (CC BY) license (https://creativecommons.org/licenses/by/4.0/).

1. Introduction

Although both adhesive materials and the application methods have changed over the years, adhesive bonding is a traditional method that has been used since ancient times to join various materials [1]. There are many factors that must be taken into account in order for a bonding application to be considered successful. Since the sectors in which adhesives are used are quite different from each other, it is important to use a strong, economical adhesive appropriate for the application purpose [2]. Additionally, adhesive materials have the potential to be developed according to their intended use. Adhesive materials are very diverse since they are made of plastic and rubber material groups [3]. Moreover, due to the combinations of the properties and amount of organic and inorganic additives in nano or micron size that can be added into additives, the types of new composite adhesives and their application areas are constantly increasing [4–11].

When the literature is examined, it is seen that there are various studies conducted by adding ceramic/glass [12–14], metal [15–18], and plastic-based [19–21] additives into adhesive materials. These studies have focused on reducing the processing cost; increasing the mechanical strength; improving the viscosity, electrical, and/or thermal conductivity; and improving the water/moisture absorption properties [22]. By adding materials found in nature or produced in the laboratory into adhesives, the researchers aim to increase the life expectancy of the joint and its resistance to the forces that the joint is exposed to.

Difficulties in supplying raw materials and increased product costs are serious problems worldwide. The increase in raw material costs and the need for environmentally friendly and biodegradable products have also affected the adhesive industry. In the countries that do not have adhesive manufacturers or whose adhesive suppliers are dependent on foreign countries, it becomes a much bigger problem to reach these products.

The fact that natural (bio-based) additives have advantages such as ease of availability, low cost, and being environmentally friendly support the interest in studies on the additives and their effects on the bonding process. Materials such as rye, wheat, walnut shell and wood flours, flour, soybean powder, wood powder, and bark powder can be organic additives. Similarly, agricultural industrial waste materials such as palm kernel and starch material are specimens for such additives [23]. Some organic materials are used as composite additives [24], and some are used in the form of fibers in the adhesive [25,26].

Kumar et al. [27] investigated the mechanical properties of particle-filled composites produced using biowaste horn powder (HP) and epoxy resin. The HP particles and matrix were mixed and molded in an appropriate ratio and cured at room temperature to produce the specimens. The properties of the samples such as tensile strength, tensile modulus, elongation percentage at break, flexural strength, flexural modulus, impact strength, and microstructure were investigated.

Alireza Akhavan-Safar et al. [28] examined the effects of date palm fibers on the mode I fracture energy of adhesives. For this purpose, they added fibers collected from four different parts of a date palm tree (bunch, rachis, petiole, and mesh) to the adhesive in three various weight ratios (2%, 5%, and 10%). The results showed that date palm fibers had the ability to increase the tensile fracture energy of adhesives. It was also found that the mode I fracture energy of the adhesive reinforced by 10% weight of rachis fiber was 7.6 times higher than that of the pure adhesive. The same authors, in another study [29], improved the static strength of the bonded joints by factors such as the type and size of natural fibers/particles, alkali treatment, and weight ratio. Fibers collected from a date palm tree were added to the adhesive in different weight ratios (2%, 5%, and 10%), in short fiber (0.5–2 mm) and long fiber (30 mm) sizes. They found that the strength of single-lap joints reinforced by 2% weight of rachis fiber treated with 6% by weight NaOH solution increased by 140%.

Barbosa et al. [30] used natural micro cork particles ranging in size from 125 to 250 mm to increase the ductility of a brittle epoxy adhesive. The amount of cork varying between 0.5% and 5% in weight was added to Araldite 2020 epoxy adhesive and the effect of the amount of cork particles on the joint was investigated. As a result of the evaluation conducted using tensile tests, it was seen that higher adhesive ductility and joints containing 1% cork had higher bond strength.

The reuse of waste materials both prevents the pollution of natural resources (soil, water, etc.) and can provide new high value-added products at affordable costs. When the literature was reviewed, no study was found on the reintroduction of waste materials into production and adding them to adhesives as reinforcement.

In this study, waste mussel, olive pomace, and walnut shells were added to an adhesive material in different ratios (5%, 15%, and 30% by weight) and in different sizes (38 and 45 μm) after being recycled. Then, they were used as an adhesive in single-lap joints for the experimental investigation of joint strength. Finally, the tensile test was applied to examine the joint strength and the effects of the additives on the joints were interpreted.

2. Materials and Methods

2.1. Materials and Properties

DX51D+Z galvanized steel material (EN 10346:2015) with dimensions of $100 \times 25 \times 1.5$ mm was used for the experimental study. The chemical compositions of the test samples are given in Table 1, and their mechanical properties are given in Table 2.

Araldite 2015 Huntsman was used as the adhesive material (an intermediate-stiffness epoxy adhesive). The properties of the adhesive are given in Table 3.

Table 1. Chemical composition of steel (% by weight).

C	Mn	P	S	Si	Al	Cu	Ti
0.06	0.3	0.019	0.022	0.02	0.032	0.04	0.002

Table 2. Mechanical properties of the steel.

Hardness	Yield Strength (MPa)	Tensile Strength (MPa)	Elongation at Break %
56 HRB	319	409	25

Table 3. Mechanical properties of Araldite 2015 [31].

Young's modulus (MPa)	1850 ± 0.21
Poisson ratio	0.33
Tensile yield strength (MPa)	12.63 ± 0.61
Tensile strength (MPa)	21.63 ± 1.61
Shear modulus (MPa)	560 ± 0.21
Shear yield strength (MPa)	14.6 ± 1.3
Shear strength (MPa)	17.9 ± 1.8

Three different types of powders, namely mussels, olive pomace, and walnut powders, in two different sizes of 38 μm and 45 μm were used as additives. The grinded powders of 38 μm and 45 μm are given in Figure 1 as olive pomace, walnut, and mussel, respectively. For mussel powders, waste mytilus galloprovincialis shells were used. A hardness of 3.5 Mohs to 4.0 Mohs is acceptable for mussel shells [32].

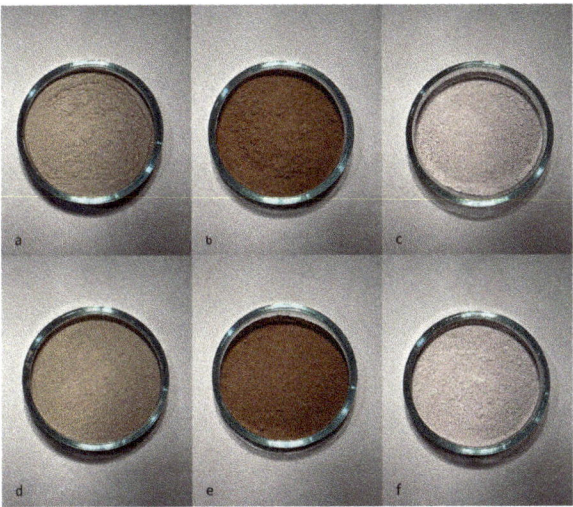

Figure 1. Ground 38 μm (**a**) olive pomace, (**b**) walnut, and (**c**) mussel powders and ground 45 μm (**d**) olive pomace, (**e**) walnut, and (**f**) mussel powders.

Calcium carbonate is a commonly used filler in polymer material. While the chemical composition of mussel shells contains 95.7% CaO, this ratio is 99.1% in commercial $CaCO_3$. Since mussel shells contain a similar amount of CaO as commercial $CaCO_3$, it is appropriate to be used as an additive material [33,34].

Olive pomace was obtained from a company operating in the Aegean region. Walnut shells were also collected from people who consumed walnuts. Olive seeds and walnut shell are lignocellulosic in chemical structure. Olive pomace contains 40% cellulose and 19% lignin in its structure [35]. Lignin content in walnut shells is around 30% [36]. Walnut Sheel has a specific gravity of 1.2–1.4 and a hardness of 3–3.5 MOH [37,38].

After the waste materials were dried in the oven, they were ground in a ring mill and sieved in a sieve shaker. The amounts of organic additives added to the adhesive were determined as 5%, 15%, and 30% by weight.

Infrared spectra were recorded on a Perkin–Elmer Spectrum 100 FTIR spectrophotometer with an attenuated total reflection (ATR) accessory featuring a zinc selenide (ZnSe) crystal at room temperature.

Since walnuts and olive pomace are organic structures, -OH and -CH groups are observed (3300 cm^{-1} OH and 2900 cm^{-1} CH). For organic powders, the OH groups can be hydrogen bond-promoting groups. Likewise, alkene groups (C=C) in the 1610s, amine groups in the 1230s, and aliphatic CO groups in the 1028s are observed in organic structures (Figure 2).

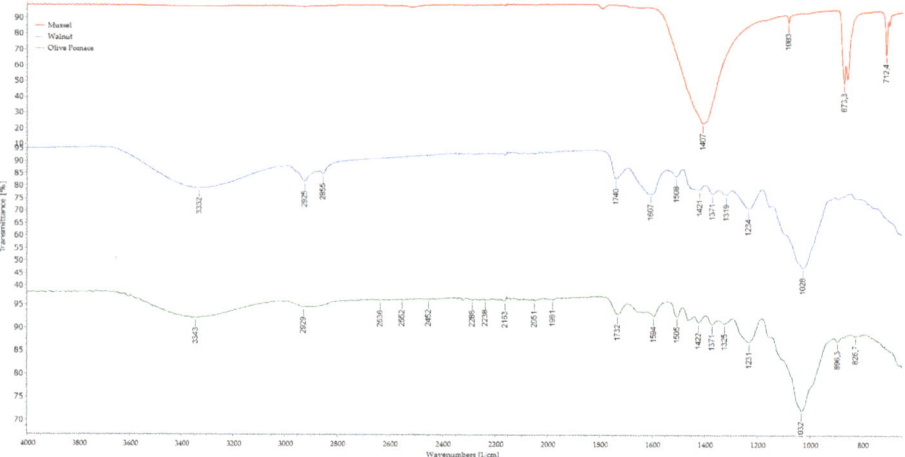

Figure 2. FTIR spectroscopy of organic powders (mussel, walnut, and olive pomace).

For mussels, the carbonate CO_3^{-2} groups are observed around 1407 cm^{-1} and 873 cm^{-1} band. These findings support the calcium carbonate structure for mussel powder.

2.2. Joint Geometry

The type of joint used in the experiments was the single-lap joint model given in Figure 3. The usual test for this type of joint is the ASTM D 1002 [39].

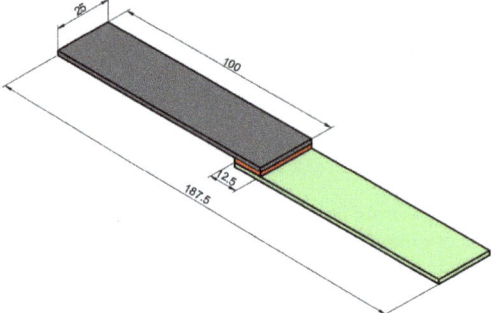

Figure 3. Single-lap joint type (mm).

2.3. Surface Preparation

Mechanical cleaning was conducted to prepare the surface of the adhesive joints. The surfaces of the samples were sanded with 120 SiC sandpaper. To roughen the entire bonding area, the sanding process was carried out in the bonding area in the horizontal and vertical directions, respectively. The sanded surfaces were wiped with acetone, washed with distilled water, and dried. The adhesive material and the filler powder were mixed manually in a plastic plate and then applied on the joint surfaces with the help of a spatula [40–42]. In the bonding area, metal paper clips were placed opposite each other, and the required pressure was provided. The bonding thickness of the joints was measured as 0.1 mm using mechanical caliper.

2.4. Surface Roughness and Tensile Testing

A Mitutoyo brand SJ-301 type desktop profilometer device with a digital display was used for surface roughness measurements. Average surface roughness values Ra were obtained by taking the arithmetic average of the five measurement values taken from the surfaces, according to the EN ISO 21920-2 standard [43].

Tensile tests were carried out at room temperature at a constant crosshead rate of 1 mm/min using the ALSA tensile test machine. All experiments were performed in triplicate.

2.5. Surface Morphology Analysis and Characterization (SEM, Joint Interfaces)

In order to better understand the strength results of the joint formed as a result of the bonding, images were taken of the powder materials and joint regions, and they were examined. All the samples were coated with gold palladium. Images were taken with a scanning electron microscope (SEM). SEM images of the powders are given in Figure 4, and joint interface images are given in Figure 5.

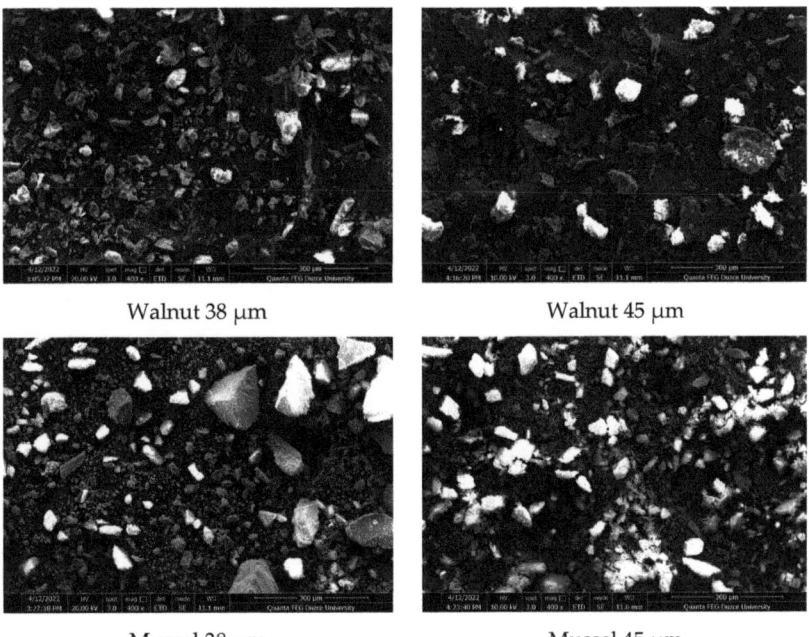

Walnut 38 μm   Walnut 45 μm

Mussel 38 μm   Mussel 45 μm

Figure 4. *Cont.*

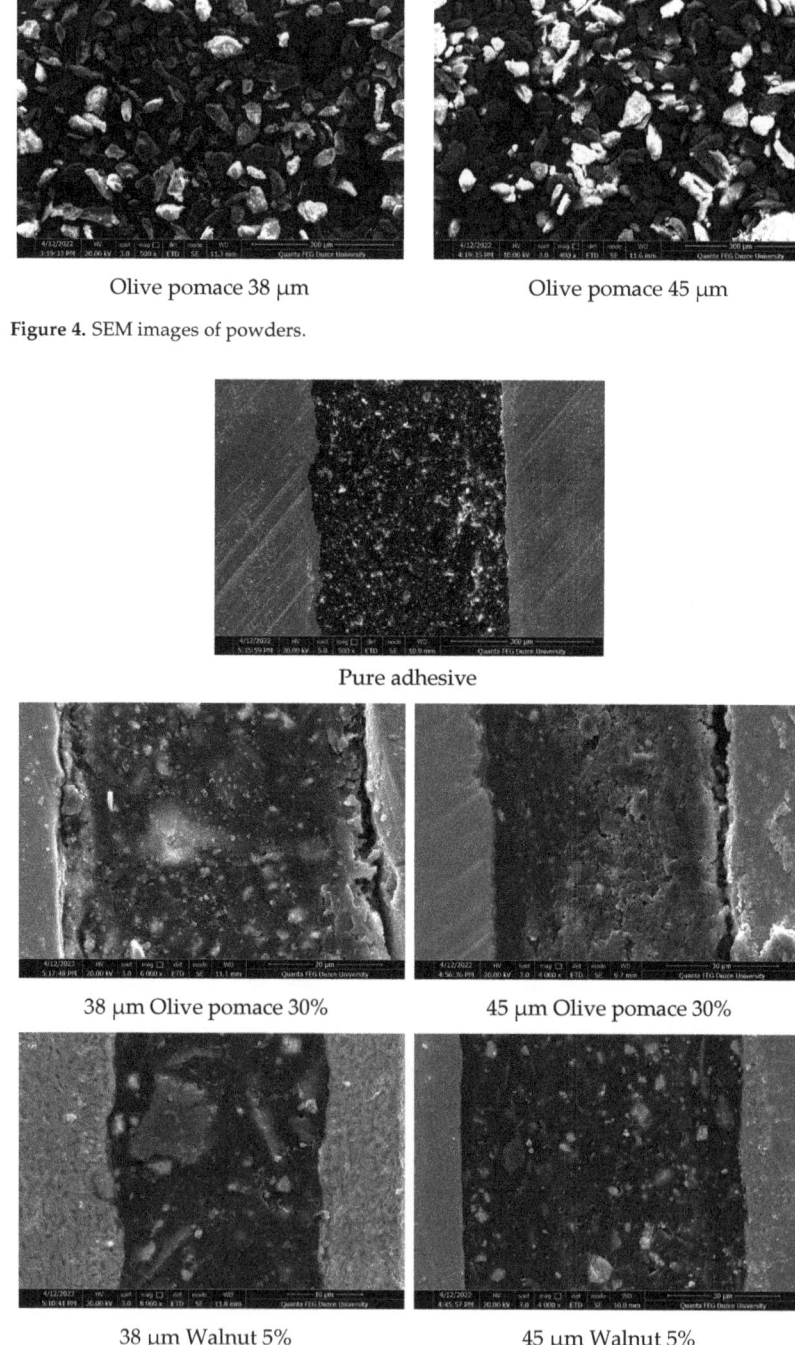

Olive pomace 38 µm Olive pomace 45 µm

Figure 4. SEM images of powders.

Pure adhesive

38 µm Olive pomace 30% 45 µm Olive pomace 30%

38 µm Walnut 5% 45 µm Walnut 5%

Figure 5. *Cont.*

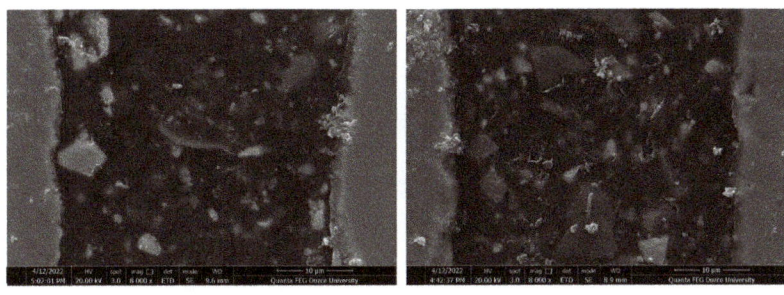

38 µm Mussel 5%   45 µm Mussel 30%

Figure 5. Interface images of bonded joints.

The distributions of the powders at the interfaces of the bonded joints are shown in Figure 5. The SEM images given in Figure 5 were selected from the experiments to explain the joint strength values.

3. Results and Discussion

3.1. Single-Lap Shear Tests

The tensile tests were carried out for the bonded test samples. The tensile test graphs are given in Figures 6 and 7. The average shear strength of the bonded joint obtained using pure adhesive was found as 12.24 N/mm². It is known that the bond strength decreases when the amount of powder added to the pure adhesive is higher than a certain amount (threshold value). In the studies verifying this statement, metal powders are generally used as the additives [44]. The results of the experiments performed using olive pomace, walnuts, and mussels with a size of 38 µm are consistent with the literature. On the other hand, when the powder size increases to 45 µm, there is an effect of increasing-decreasing-increasing strength as the amount of additive increases. Accordingly, it is understood that there is a threshold value for the amount of powder added to the adhesive.

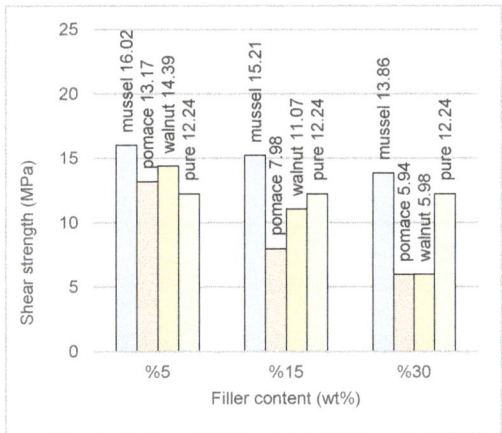

Figure 6. Average shear strengths of the 38 µm adhesive joints.

When the size of the powder added to the adhesive was 45 µm, bond strengths increased for all additive ratios compared to the pure adhesive. The powder-added bonded joints (in 45 µm size) provided better bond strength overall.

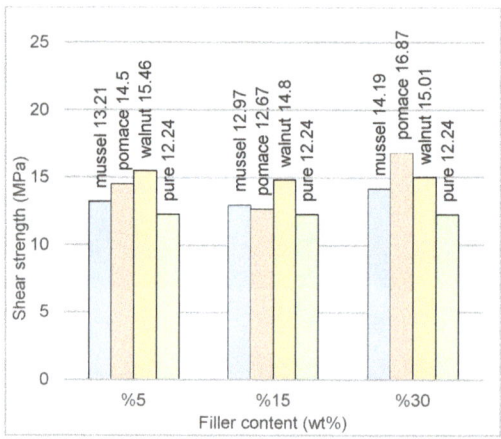

Figure 7. Average shear strengths of the 45 μm adhesive joints.

In the experiments, the lowest strength value was measured as 5.94 MPa in joints where 38 μm olive pomace was added in the ratio of 30% by weight, and the highest strength value was measured as 16.87 MPa in joints where 45 μm olive pomace was added in the ratio of 30% by weight. In addition, when the effect of olive pomace and walnut additives added by 30% on the joint strengths was examined, it was seen that the increases in the joint strength were close to each other. It can be assumed that the organic structural similarity of olive pomace and walnut materials resulted in similar increase amounts.

The mussel additive produced a higher strength value compared to the pure adhesive in all ratios without any size difference. This indicates that a strong bond was formed between the mussel shell and the adhesive. It can be said that the layered natural structure of the mussel shell strengthens this bond [26].

As can be seen from the graphics in Figures 6 and 7, the bond strength varies considerably depending on the type of additive (especially for 38 μm powder). Although the same grinding processes were carried out, there were changes in shape due to the type (structural properties) of the powders. The shape changes can be seen from the SEM images in Figure 4. This may be a parameter that changes the ability of the additive powders to adhere to the adhesive. This situation explains or is affected by the change in the strength values of the joints.

3.2. Surface Roughness Measurement

A surface roughness measurement was taken from the surfaces of the galvanized steel material that was mechanically abrasive using 120 SiC sandpaper, and the average Ra value was found to be 1 μm.

3.3. Surface Morphology Analysis and Characterization

In the visual examination of the powders, there was no difference other than color separation. However, it was seen that there were significant differences when SEM images were taken. Mussel powders are brittle, olive pomace powders are round, and walnut powders have a fringed fractured surface. When the SEM images in Figure 5, taken from the joints where 38 μm olive pomace was added in the ratio of 30% by weight, are examined, it is seen that there is a large void where the olive pomace added adhesive contacts the lower and upper surfaces of the base metal. This void is also seen along the joint interface. Therefore, it is understood that adhesion is not fully realized. Moreover, it is understood that the void spreads over the base metal surface into the bonding zone. Mixing the adhesive and the 38 μm olive pomace with each other at this rate reduced the adhesion effect of the adhesive.

When the SEM images taken from the joints where 45 μm olive pomace was added in the ratio of 30% by weight are examined, it is seen that there is a void where the adhesive contacts the metal surface. However, the width of this void is small.

4. Conclusions

By modifying the adhesives with a natural filler, a strong joint can be obtained, which can improve the parameters of the adhesive joint in terms of mechanical properties. The use of waste materials as additives after recycling makes it possible to reduce the production cost. The research follows the future trend in the field of ecological composites with fillers (or powder) based on waste material. Olive pomace, mussels, and walnut shells are recyclable organic waste materials. In this study, recycled organic materials in 38 μm and 45 μm sizes were added to the adhesive in the ratios of 5%, 15%, and 30% by weight. Single-overlap joints were formed by combining the steel materials with the modified adhesives which the ground powders were added to.

The results of tests to determine the mechanical properties of the adhesive filled with organic powders are presented. All results were compared with those obtained in the adhesive without powder additions.

The findings obtained as a result of the experimental studies are summarized as follows:

- The bond strength decreased when the amount of the powder (in 38 μm size) added to adhesive was more than 5%.
- The strength of all joints obtained by 5% powder additives (for 38 μm) increased compared to that with pure adhesive.
- The bond strength at all additive ratios (for 45 μm powder) increased compared to that with pure adhesive.
- In the experiments, the lowest strength value was obtained in joints where 38 μm olive pomace was added in the ratio of 30% by weight, and the highest strength value was obtained in joints where 45 μm olive pomace was added in the ratio of 30% by weight.
- When the additives in 38 μm powder size were used, it was seen that the change in bond strength varied depending on the powder type, while the effect of powder type on the bond strength did not make a difference in 45 μm powder sizes.

Author Contributions: The study was carried out as a two-person team. Conceptualization and methodology was introduced by N.A.; data curation, N.A. and Z.D.; writing—original draft preparation, N.A. and Z.D.; writing—review and editing, N.A. and Z.D.; visualization, Z.D.; supervision, N.A.; project administration, N.A. and Z.D.; funding acquisition, N.A. and Z.D. All authors have read and agreed to the published version of the manuscript.

Funding: This research received no external funding.

Data Availability Statement: Not applicable.

Conflicts of Interest: The authors declare no conflict of interest.

References

1. Petrie, E.M. *Handbook of Adhesives and Sealants*; McGraw-Hill Education: New York, NY, USA, 2007.
2. Adams, R.D. *Adhesive Bonding: Science, Technology and Applications*; Woodhead Publishing: Sawston, UK, 2021.
3. Pocius, A.V. *Adhesion and Adhesives Technology: An Introduction*; Carl Hanser Verlag GmbH Co., KG: Munich, Germany, 2021.
4. Taylor, A. Advances in nanoparticle reinforcement in structural adhesives. In *Advances in Structural Adhesive Bonding*; Elsevier: Amsterdam, The Netherlands, 2010; pp. 151–182.
5. Hoque, M.E.; Kumar, R.; Sharif, A. *Advanced Polymer Nanocomposites: Science, Technology and Applications*; Woodhead Publishing: Sawston, UK, 2022.
6. Thomas, S.; Gopi, S.; Amalraj, A. *Biopolymers and Their Industrial Applications: From Plant, Animal, and Marine Sources, to Functional Products*; Elsevier: Amsterdam, The Netherlands, 2020.
7. Alam, M.O.; Bailey, C. *Advanced Adhesives in Electronics: Materials, Properties and Applications*; Elsevier: Amsterdam, The Netherlands, 2011.
8. Ghosh, P.; Nukala, S. Properties of adhesive joint of inorganic nano-filler composite adhesive. *Indian J. Eng. Mater. Sci.* **2008**, *15*, 68–74.

9. Müller, M.; Tichý, M.; Šleger, V.; Hromasová, M.; Kolář, V. Research of hybrid adhesive bonds with filler based on coffee bean powder exposed to cyclic loading. *Manuf. Technol.* **2020**, *20*, 1–9. [CrossRef]
10. Gonçalves, F.A.; Santos, M.; Cernadas, T.; Alves, P.; Ferreira, P. Influence of fillers on epoxy resins properties: A review. *J. Mater. Sci.* **2022**, *57*, 15183–15212. [CrossRef]
11. Hrabě, P.; Kolář, V.; Müller, M.; Hromasová, M. Service Life of Adhesive Bonds under Cyclic Loading with a Filler Based on Natural Waste from Coconut Oil Production. *Polymers* **2022**, *14*, 1033. [CrossRef]
12. Yetgin, H.; Veziroglu, S.; Aktas, O.C.; Yalçinkaya, T. Enhancing thermal conductivity of epoxy with a binary filler system of h-BN platelets and Al2O3 nanoparticles. *Int. J. Adhes. Adhes.* **2020**, *98*, 102540. [CrossRef]
13. Kwon, D.-J.; Kwon, I.-J.; Kong, J.; Nam, S.Y. Investigation of impediment factors in commercialization of reinforced adhesives. *Polym. Test.* **2021**, *93*, 106995. [CrossRef]
14. Hunter, R.; Möller, J.; Vizán, A.; Pérez, J.; Molina, J.; Leyrer, J. Experimental study of the effect of microspheres and milled glass in the adhesive on the mechanical adhesion of single lap joints. *J. Adhes.* **2017**, *93*, 879–895. [CrossRef]
15. Hamrah, Z.S.; Lashgari, V.; Mohammadi, M.D.; Uner, D.; Pourabdoli, M. Microstructure, resistivity, and shear strength of electrically conductive adhesives made of silver-coated copper powder. *Microelectron. Reliab.* **2021**, *127*, 114400. [CrossRef]
16. Kahraman, R.; Al-Harthi, M. Moisture diffusion into aluminum powder-filled epoxy adhesive in sodium chloride solutions. *Int. J. Adhes. Adhes.* **2005**, *25*, 337–341. [CrossRef]
17. Kahraman, R.; Sunar, M.; Yilbas, B. Influence of adhesive thickness and filler content on the mechanical performance of aluminum single-lap joints bonded with aluminum powder filled epoxy adhesive. *J. Mater. Process. Technol.* **2008**, *205*, 183–189. [CrossRef]
18. Singh, R.; Zhang, M.; Chan, D. Toughening of a brittle thermosetting polymer: Effects of reinforcement particle size and volume fraction. *J. Mater. Sci.* **2002**, *37*, 781–788. [CrossRef]
19. Quan, D.; Carolan, D.; Rouge, C.; Murphy, N.; Ivankovic, A. Mechanical and fracture properties of epoxy adhesives modified with graphene nanoplatelets and rubber particles. *Int. J. Adhes. Adhes.* **2018**, *81*, 21–29. [CrossRef]
20. Zakiah, J.; Ansell, M.P.; Smedley, D.; Md Tahir, P. The Effect of Long Term Loading on Epoxy-Based Adhesive Reinforced with Nano-Particles for In Situ Timber Bonding. *Proc. Adv. Mater. Res.* **2012**, *545*, 111–118.
21. Vinay, M. Optimization of process parameters of aluminium 2024–T3 joints bonded using modified epoxy resin. *Mater. Today Proc.* **2022**, *54*, 325–329. [CrossRef]
22. Nemati Giv, A.; Ayatollahi, M.R.; Ghaffari, S.H.; da Silva, L.F. Effect of reinforcements at different scales on mechanical properties of epoxy adhesives and adhesive joints: A review. *J. Adhes.* **2018**, *94*, 1082–1121. [CrossRef]
23. Sanghvi, M.R.; Tambare, O.H.; More, A.P. Performance of various fillers in adhesives applications: A review. *Polym. Bull.* **2022**, *79*, 1–63. [CrossRef]
24. Suleiman, I.Y.; Kasim, A.; Mohammed, A.T.; Sirajo, M.Z. Evaluation of Mechanical, Microstructures and Wear Behaviours of Aluminium Alloy Reinforced with Mussel Shell Powder for Automobile Applications. *Stroj. Vestn.-J. Mech. Eng.* **2021**, *67*, 27–35. [CrossRef]
25. Udatha, P.; Babu, Y.N.; Satyadev, M.; Bhagavathi, L.R. Effect of natural fibers reinforcement on lap-shear strength of adhesive bonded joints. *Mater. Today Proc.* **2020**, *23*, 541–544. [CrossRef]
26. Mishra, A.; Singh, S.; Kumar, R.; Mital, A. Natural fiber based adhesive butt joints as a replacement to gas welded butt joints for thin tubes: An experimental study. *IOP Conf. Ser. Mater. Sci. Eng.* **2018**, *404*, 012024. [CrossRef]
27. Kumar, D.; Boopathy, S.R.; Sangeetha, D.; Bharathiraja, G. Investigation of mechanical properties of horn powder-filled epoxy composites/Raziskava mehanskih lastnosti epoksi kompozitov s polnilom iz rozevine v prahu. *Stroj. Vestn.-J. Mech. Eng.* **2017**, *63*, 138–148. [CrossRef]
28. Akhavan-Safar, A.; Delzendehrooy, F.; Ayatollahi, M.; da Silva, L.F.M. Influence of Date Palm Tree Fibers on the Tensile Fracture Energy of an Epoxy-based Adhesive. *J. Nat. Fibers* **2022**, *19*, 14379–14395. [CrossRef]
29. Delzendehrooy, F.; Ayatollahi, M.; Akhavan-Safar, A.; da Silva, L. Strength improvement of adhesively bonded single lap joints with date palm fibers: Effect of type, size, treatment method and density of fibers. *Compos. Part B Eng.* **2020**, *188*, 107874. [CrossRef]
30. Barbosa, A.; da Silva, L.; Öchsner, A. Effect of the amount of cork particles on the strength and glass transition temperature of a structural adhesive. *Proc. Inst. Mech. Eng. Part L J. Mater. Des. Appl.* **2014**, *228*, 323–333. [CrossRef]
31. Ri, J.-H.; Kim, M.-H.; Hong, H.-S. A mixed mode elasto-plastic damage model for prediction of failure in single lap joint. *Int. J. Adhes. Adhes.* **2022**, *116*, 103134. [CrossRef]
32. de Castro, A.L.P.; Serrano, R.O.P.; Pinto, M.A.; da Silva, G.H.T.Á.; de Andrade Ribeiro, L.; de Faria Viana, E.M.; Martinez, C.B. Case study: Abrasive capacity of *Limnoperna fortunei* (golden mussel) shells on the wear of 3 different steel types. *Wear* **2019**, *438*, 202999. [CrossRef]
33. Hamester, M.R.R.; Balzer, P.S.; Becker, D. Characterization of calcium carbonate obtained from oyster and mussel shells and incorporation in polypropylene. *Mater. Res.* **2012**, *15*, 204–208. [CrossRef]
34. Koçhan, C. An experimental investigation on mode-I fracture toughness of mussel shell/epoxy particle reinforced composites. *Pamukkale Univ. J. Eng. Sci.-Pamukkale Univ. Muhendis. Bilim. Derg.* **2020**, *26*, 599–604. [CrossRef]
35. Badawy, W.; Smetanska, I. Utilization of olive pomace as a source of bioactive compounds in quality improving of toast bread. *Egypt. J. Food Sci.* **2020**, *48*, 27–40.

36. Queirós, C.S.; Cardoso, S.; Lourenço, A.; Ferreira, J.; Miranda, I.; Lourenço, M.J.V.; Pereira, H. Characterization of walnut, almond, and pine nut shells regarding chemical composition and extract composition. *Biomass Convers. Biorefin.* **2020**, *10*, 175–188. [CrossRef]
37. Materials, C. Walnut Shells Applications and Uses. Available online: https://www.azom.com/article.aspx?ArticleID=10430 (accessed on 23 April 2021).
38. Deburring, A.B. Walnut Shell. Available online: https://abdeburr.com/corn-cob-walnut-shell/ (accessed on 12 August 2022).
39. *ASTM D 1002-10*; Standard Test Method for Apparent Shear Strength of Single-Lap-Joint Ahdesively Bonded Metal Specimens by Tension Loading (Metal-to-Metal). ASTM International: West Conshohocken, PA, USA, 2010.
40. Kavak, N.; Altan, E. A new hybrid bonding technique: Adhesive-soft soldered joints. *Proc. Inst. Mech. Eng. Part L J. Mater. Des. Appl.* **2014**, *228*, 137–143. [CrossRef]
41. Budhe, S.; Ghumatkar, A.; Birajdar, N.; Banea, M. Effect of surface roughness using different adherend materials on the adhesive bond strength. *Appl. Adhes. Sci.* **2015**, *3*, 20. [CrossRef]
42. Monteiro, J.; Salgado, R.; Rocha, T.; Pereira, G.; Marques, E.; Carbas, R.; da Silva, L. Effect of adhesive type and overlap length on the mechanical resistance of a simple overlap adhesive joint. *U. Porto J. Eng.* **2021**, *7*, 1–12. [CrossRef]
43. Jones, C.W.; Sun, W.; Boulter, H.; Brown, S. 3D roughness standard for performance verification of topography instruments for additively-manufactured surface inspection. *Meas. Sci. Technol.* **2022**, *33*, 084003. [CrossRef]
44. Kavak, N.; Altan, E. Influence of filler amount and content on the mechanical performance of joints bonded with metal powder filled adhesive. *Mater. Sci. Forum* **2014**, *773–774*, 226–233. [CrossRef]

Disclaimer/Publisher's Note: The statements, opinions and data contained in all publications are solely those of the individual author(s) and contributor(s) and not of MDPI and/or the editor(s). MDPI and/or the editor(s) disclaim responsibility for any injury to people or property resulting from any ideas, methods, instructions or products referred to in the content.

Article

A Laser Shock-Based Disassembly Process for Adhesively Bonded Ti/CFRP Parts

Panagiotis Kormpos [1], Selen Unaldi [2], Laurent Berthe [2] and Konstantinos Tserpes [1,*]

[1] Laboratory of Technology & Strength of Materials (LTSM), Department of Mechanical Engineering & Aeronautics, University of Patras, 26504 Patras, Greece
[2] PIMM, UMR8006 ENSAM, CNRS, CNAM, 151 bd de l'Hôpital, 75013 Paris, France
* Correspondence: kitserpes@upatras.gr

Abstract: The application of adhesively bonded joints in aerospace structural parts has increased significantly in recent years and the general advantages of their use are well-documented. One of the disadvantages of adhesive bonding is the relevant permanence, when compared to traditional mechanical fastening. End-of-life processes generally require the separation of the adherents for repair or recycling, and usually to achieve this, they combine large mechanical forces with a high temperature, thus damaging the adherents, while consuming large amounts of energy. In this work, a novel disassembly technique based on laser-induced shock waves is proposed for the disassembly of multi-material adhesively bonded structures. The laser shock technique can generate high tensile stresses that are able to break a joint, while being localized enough to avoid damaging the involved adherents. The process is applied to specimens made from a 3D-woven CFRP core bonded to a thin Ti layer, which is a common assembly used in state-of-the-art aircraft fan blades. The experimental process has been progressively developed. First, a single-sided shot is applied, while the particle velocity is measured at the back face of the material. This method proves ineffective for damage creation and led to a symmetric laser configuration, so that the tensile stress can be controlled and focused on the bond line. The symmetric approach is proved capable of generating a debonding between the Ti and the CFRP and propagating it by moving the laser spot. Qualitative assessment of the damage that is created during the symmetric experimental process indicates that the laser shock technique can be used as a material separation method.

Keywords: laser shock; disassembly and recycle; laser adhesion test; 3D-woven CFRP; bonded structures

1. Introduction

Recent aircraft construction has replaced nearly 50% of metallic parts with composite materials, as their specific properties are essential for weight reduction. While the weight decrease is accompanied by reduced fuel consumption, the broad application of composites has raised some different environmental challenges. The biggest environmental impact is the increasing generation of large amounts of waste and landfill material as more structures reach their end-of-life (EoL) [1].

Advancements in recycling have made it possible to recover carbon and glass fibers from composite materials that have reached their EoL, using strategies like pyrolysis [2] or solvolysis [3]. Furthermore, other strategies have been proposed to utilize shredding of the composite to be used as a filler in bio-based resins, thus delaying their landfill deposition [4]. To employ such strategies with sufficient efficiency, it is important that the different materials are well-separated and sorted.

General aircraft EoL processes have been designed for material separation before component recycling. Separation strategies are chosen based on a cost–benefit analysis, with the extreme cases being systematic disassembly and shredding. Systematic disassembly is the process of separating and sorting all the components based on material composition.

It is a labor-intensive process and yields the best material segregation. On the other end of the cost–benefit spectrum, shredding is the process of cutting pieces of the aircraft containing a multitude of materials, such as aluminum, titanium, composites, glass, etc. This process is concentrated on quantity over quality [5,6]. The two extremes are not usually desirable because of the excessive cost or the poor material quality, respectively. Intermediate strategies, on the other hand, use the mapping of the aircraft and specific combinations of shredding, cutting, and disassembly are utilized, according to material homogeneity, as illustrated in Figure 1.

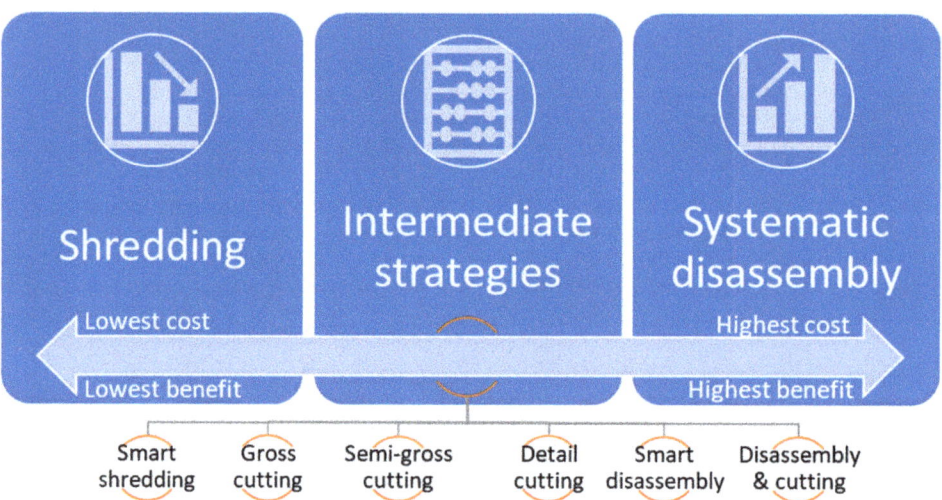

Figure 1. Current EOL strategies sorted based on their cost–benefit ratio.

Bonded metal, composite, and multi-material structures are a common design choice of state-of-the art aircraft structural parts. EoL planning for such structures is a major challenge [7], since separation of adherent materials, often of dissimilar nature (titanium or aluminum bonded with a composite), is usually achieved by applying large mechanical forces and extremely high temperatures, both of which can damage the adherents and lower the quality of the recovered materials [8]. Thermally induced disassembly approaches can be achieved through both thermal softening (exceeding the adhesive's T_g) and thermal decomposition (exceeding the temperature of flammability-in-air or auto-ignition point) [9]. The solutions provided by the literature are currently focused on the creation of reversible or dismantlable adhesives. These adhesives use thermally expandable particles and heat as an activation method of debonding [10,11]. While a joint design using such adhesives is promising, mechanical properties are inevitably decreasing [12], making the structures unappealing for aeronautical structural applications.

An interesting case of a multi-material bonded structure is modern aircraft fan blades, which consist of an advanced 3D-woven composite core bonded to a metallic leading edge, most commonly being titanium alloy. The described assembly has a complex end of life, since titanium is a valuable material that can be recycled, but not while it is bonded to the composite core. Additionally, the recycling process of the 3D-woven composite can recover the fibers to be reused. In the frame of the EC-funded project MORPHO, a novel disassembly process is developed for the separation of the two materials utilizing the laser-shock technique. The laser-shock technique can provide high precision in terms of tensile force application to separate bonded materials and, if it is calibrated correctly, it can avoid damaging the adherents. Additionally, minimum to no adhesive residue to at least one of the involved materials can be achieved when the process is optimized for adhesive failure of the bond. Moreover, in contrast to thermal methods for material separation, no harmful by-products are generated during the process.

The experimental work presented in this paper is composed of two parts. First, the single-sided configuration is used, while the particle velocity is measuredonthe free surface of the material. This configuration was not able to damage the bond between the Ti and CFRP. The second part uses a symmetric configuration; this approach creates two shockwaves that propagate in opposite directions, and their interference can be controlled by the time delay between them. The symmetric configuration was able to debond the Ti from the CFRP and propagate the debonding by moving the spot.

First, the laser-shock principle is described, as well as the different configurations of the technique to highlight the bases for the disassembly process. The materials and the different experimental procedures are then detailed to explain how the results were obtained. Finally, the results of the single-sided and symmetric configurations are presented and explained.

2. Materials and Methods

2.1. The Laser Shock Technique

Laser shock has been used in the industry for material processing, improving fatigue strength of metallic materials, by creating residual compressive stresses [13,14]. Recent research has been able to use the ability of the technique to generate high tensile stresses for non-destructive purposes in the form of an adhesion test called LASAT [15–18], and for the purpose of damage creation, as it is the case for selective paint stripping [19–21], and the formation of controlled delaminations in composites [22].

2.1.1. The Laser-Shock Principle

Laser shock is a technique based on laser–matter interaction. The plasma expansion that is obtained when a high-powered laser, with a duration in nanosecond range, is focused onto the surface of a target induces a pulsed pressure, which is the result of the recoil momentum of the ablated material [23]. If the plasma expansion occurs under a confinement regime, the pressure level and duration are increased significantly [24]. The confinement regime needs to be a dielectric material transparent to the laser, such as water, glass, or pliable polymer [14]. The pressure generated by the plasma expansion results in an elastic precursor shock followed by an elastic-plastic compression shock that propagates inside the material. After the plasma expansion, the surface is unloaded and a plastic-decompression shock alongside an elastic-plastic decompression shock begin to propagate and are described as a release wave. The interaction between the release wave and the elastic precursor shock wave develops high localized tensile stresses [25]. Figure 2 is an illustration of the laser-shock principle.

2.1.2. Laser-Shock Configurations

Depending on the configuration of the two beams, different set-ups are possible. To illustrate the wave interactions in the different configurations, it is common in the literature to use simplified space–time diagrams that present stresses, assuming a linear one-dimension propagation though the specimen thickness. Figure 3a demonstrates the standard single-sided shot configuration, where one beam is focused on the surface of the specimen; the high tensile stress area appears near the opposite side of the specimen caused by the interaction between the reflected release wave and the release wave resulting from surface unloading. This interaction is only dependent on specimen geometry and the pulse duration.

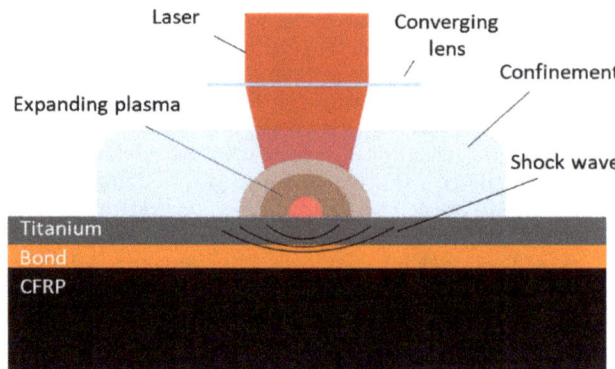

Figure 2. Schematic representation of laser-generated shockwave.

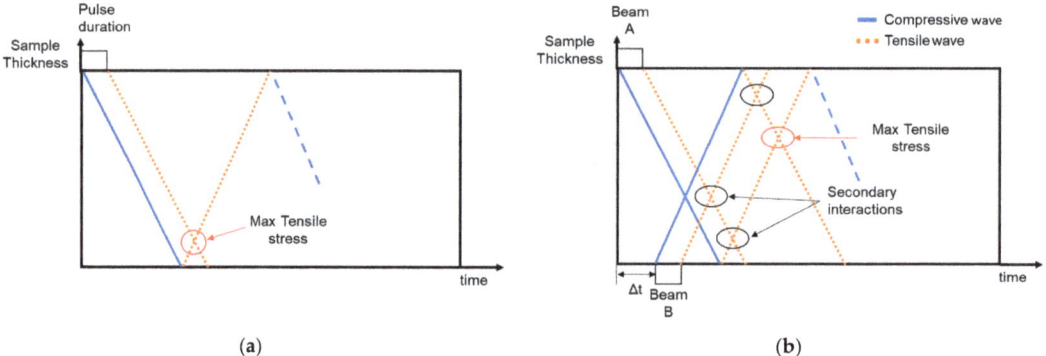

Figure 3. Space–time diagram for single shot (**a**) and symmetric (**b**) configuration.

Additionally, when both sides of the specimen are irradiated, two shock fronts are created and allowed to propagate in opposite directions [16]. To achieve the symmetric configuration illustrated in Figure 3b, two polarized beams are separated using a 90° polarizer and transported to each side of the specimen using optics. Utilizing the symmetric technique, the maximum tensile stress does not exclusively rely on pulse duration; instead, shifting the location is possible by applying time delay (Δt) between the pulses. Tensile stress areas are still created by the interactions of each individual shock propagating inside the material and their location cannot be controlled; however, the maximum tensile stress is produced by the interference of the two reflected release waves and its position at the material's thickness depends on the Δt.

2.2. Materials and Specimens

The specimens were provided by Safran in the frame of the MORPHO European project. The composition of the specimens is a 3D-woven CFRP core bonded to a thin Ti alloy edge using an adhesive film. Each specimen was cut by Safran from a single block and the final dimensions of the specimens are 100 mm × 40 mm with a thickness of 10.6 mm. The specimens tested are illustrated in Figures 4 and 5. During all experiments, the composite side of the specimens was covered by a thin aluminum tape. The reasons are twofold: during direct shots on the composite the aluminum acts as an ablation layer, providing higher pressure while protecting the composite from the ablation effects; additionally, aluminum provides a reflective surface that can be used for optical measurements during Ti side-shots.

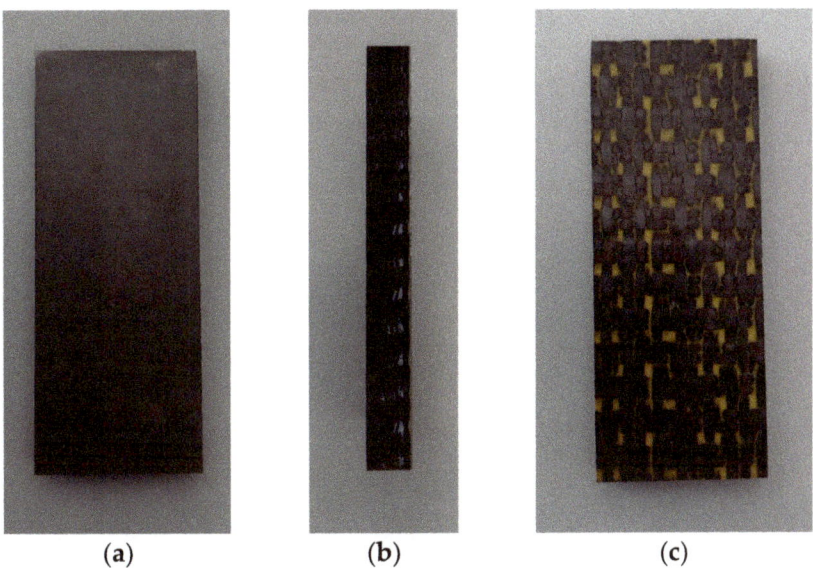

Figure 4. Specimen used for laser-shock experiments. (**a**) Top view; (**b**) side view; (**c**) bottom view.

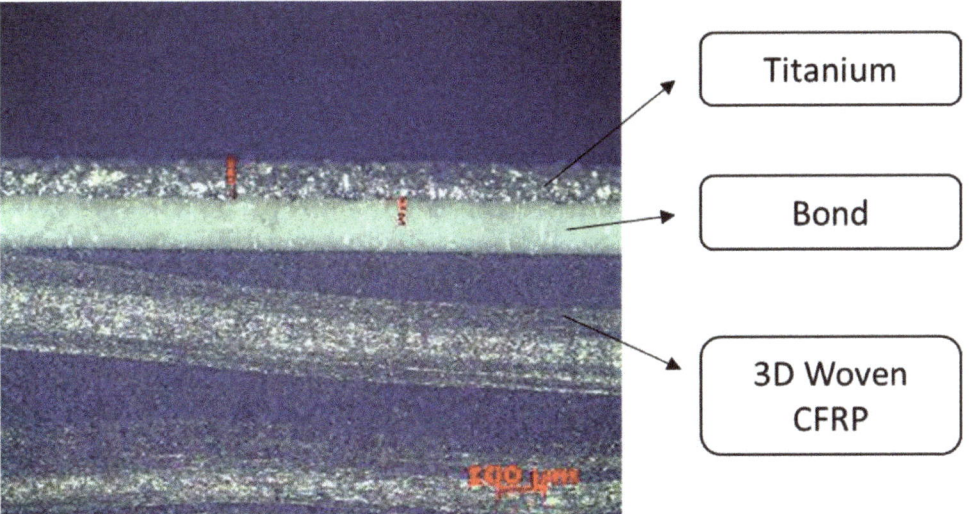

Figure 5. Microscope image of the interface between CFRP and Ti.

2.3. Laser Specifications

The experiments were conducted using the Hephaïstos facility located at the Laboratory for Processes and Engineering in Materials and Mechanics (PIMM), ENSAM, ParisTech. The facility uses two Nd:YAG (neodymium-doped yttrium aluminum garnet) Gaia HP lasers from THALES, which can emit synchronized or delayed pulses, at 532 nm with a repetition rate of 2 Hz. The pulse has a gaussian temporal profile with a duration of 7 ns and a maximum energy of 7 J each that can be combined into a 14 J pulse when both lasers are superposed. The beams are focused using an optical lens to control the focal diameter, with a range of 3 to 5 mm. After the focus, a diffractive optical element is used to obtain a uniform top-hat-shaped spatial profile.

2.4. Experimental

2.4.1. Single-Sided Shot Configuration

For the first part of the experiments, the single-sided shot configuration was used, as shown in Figure 6. During the experiments, the particle velocity of the free face of the material (opposite to the loading surface) is measured by an optical diagnostic tool called VISAR (Velocity Interferometer System for Any Reflector). VISAR is an interferometer that can measure the doppler shift of a 532 nm wavelength, low-power laser, when it is reflected by a free surface that is moving because of the arrival of a shock wave. The time resolution that the tool provides is 1 ns. The tool has been used in the study of shock waves in solids [18–20,23,26] as a method of damage identification during LASAT, or a way to validate numerical models for study and optimization.

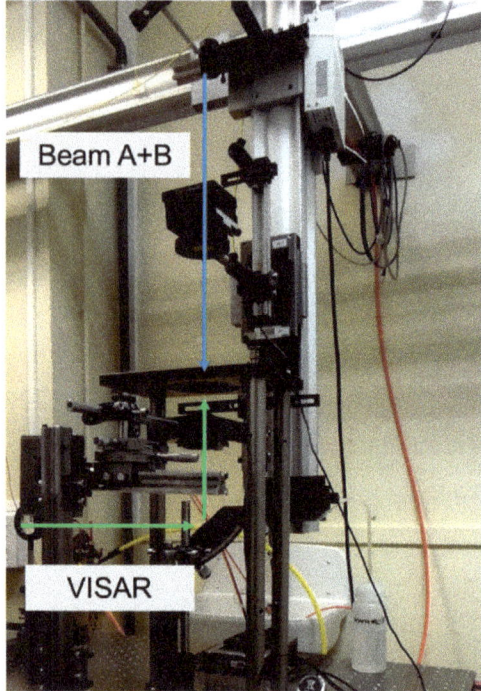

Figure 6. Single-sided laser experimental set-up.

This experimental series consists of single-sided shots on both the Ti and CFRP sides of the specimen with laser intensity varying from 1 GW/cm^2 to 6 GW/cm^2 using water as the confinement regime and a spot diameter of 4 mm. Due to the nature of the 3D-woven composite, it is expected that the VISAR measurements have an increased location dependent variability. This is attributed to the inhomogeneity of the material that results in local stiffness and thickness variations, influencing the particle velocity. To account for the variability of the measurements, each laser intensity experiment was repeated 10 times. Table 1 contains all the single-side shot experiments.

Table 1. Experiments with the single-sided laser configuration.

Experiment	Number of Specimens	Laser Intensity	Number of Shots
Titanium side shots with VISAR	4	1.2 GW/cm^2 3 GW/cm^2 4.5 GW/cm^2 6 GW/cm^2	40 (10/specimen)
CFRP side shots with VISAR	4	1.2 GW/cm^2 3 GW/cm^2 4.5 GW/cm^2 6 GW/cm^2	40 (10/specimen)

2.4.2. Symmetric Laser Configuration

The symmetric configuration uses the same two Nd:YAG lasers, splitting the beams using a polarizer to deliver one beam at each side of the specimen. The symmetric experimental set-up is shown in Figure 7. For this configuration to work, it is important that the two beams are perfectly aligned. To maintain the alignment, movement of the spot is achieved by placing the specimen on a robotic arm (Figure 6). The water confinement regime is challenging in this set-up, and thus, it was replaced by a solid pliable polymer. The shots were focused on the edge of the specimen so that the damage that is created can be visible through an electronic microscope. The energy of each beam was set to 100%, meaning 3.49 J and 5.14 J for beam A and B, respectively. Using a 3.2 mm spot diameter, the resulting laser intensity is 5.45 GW/cm^2 for beam A and 7.99 GW/cm^2 for beam B.

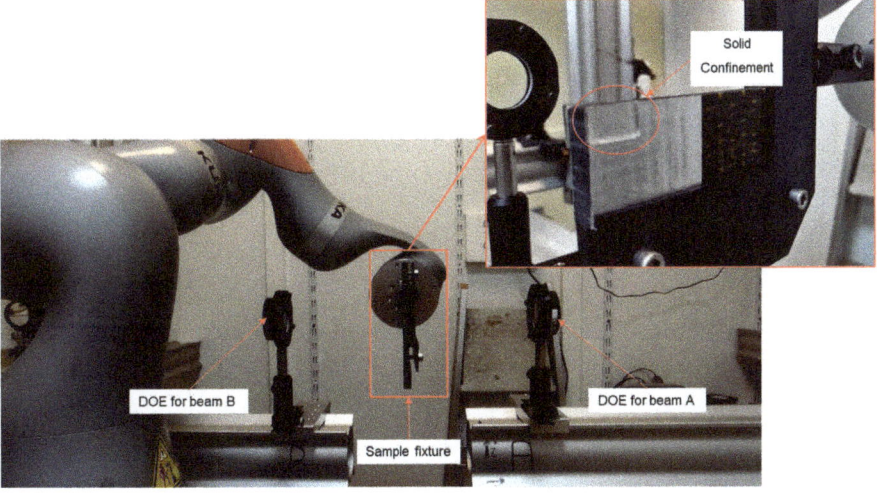

Figure 7. Symmetric experimental set-up.

During the experiments, the delay time can be set between beam A and B with an accuracy of 1 ns. The delay times that were calculated for the specimen are 3.45 µs and 3.55 µs when the tensile zone created by the shock-wave propagation aims to damage the interface between the bond line and the titanium or the bond line and the CFRP, respectively. Shots using a delay between the two mentioned values did not produce any damage; thus, the effort was focused on the creation of debonding caused by adhesive failure, aiming at the interface instead of cohesive failure. Figure 8 illustrates the wave propagation for each time delay.

Previous work [27,28] has shown that damage can be progressively created in a two-step loading. If the first shot provides enough energy to weaken the bond, then a second shot can initiate the debonding. Utilizing this strategy, each spot was shot two times, the first to weaken the bond and the second one to damage it.

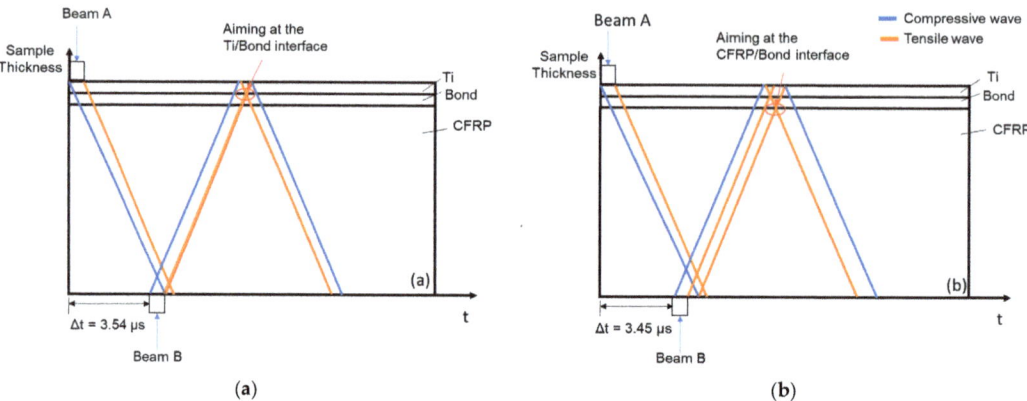

Figure 8. Space–time diagrams showcasing the wave propagation for delay times of (**a**) 3.54 µs and (**b**) 3.45 µs.

3. Results

3.1. Single-Sided Configuration Results

The shots conducted using the single-sided configuration were expected to show a shift in the back-face velocity measurements as the intensity increased, indicating damage at higher energies. Figures 9 and 10 show the back-face velocities of the specimens shot with 4.5 GW/cm^2 and 6 GW/cm^2 for the Ti side and composite side shots, respectively. Each curve corresponds to a different spot using the same intensity to observe the variability of the measurements. The response of the specimens was identical for lower and higher laser intensities, differentiating only on the peak value of the velocity; this is an indication that no damage was created. It is interesting to mention that Ti side shots have increased variability for the same experimental parameters. Measurements for the Ti side shots are conducted at the surface of the composite, so the inhomogeneous nature of the 3D-woven structure is more dominant for those measurements. On the other hand, the shots conducted at the surface of the CFRP show less variability because the measurements were taken at the surface of the titanium, which is homogenous.

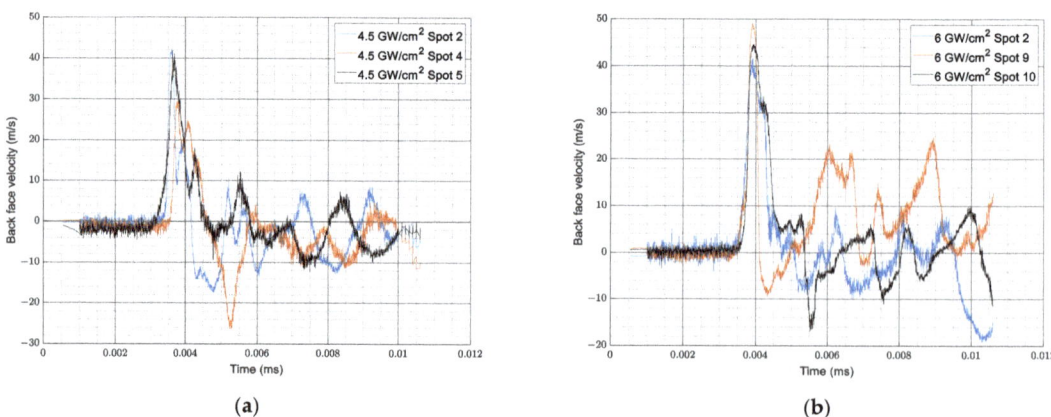

Figure 9. Back-face velocity measurements for Ti side shots for (**a**) 4.5 GW/cm^2 and (**b**) 6 GW/cm^2.

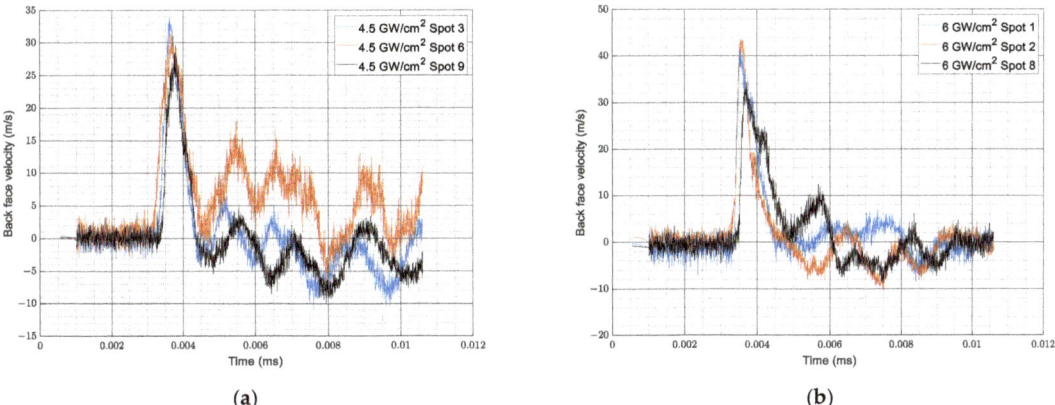

Figure 10. Back-face velocity measurements for CFRP side shots for (**a**) 4.5 GW/cm^2 and (**b**) 6 GW/cm^2.

To validate the absence of damage to the specimens, ultrasound tests were used in the form of a C-scan. Observing the C-scan of the 6 GW/cm^2, it is clear that no damage was created during the shots. Figure 11 shows the result for the Ti and composite side shots.

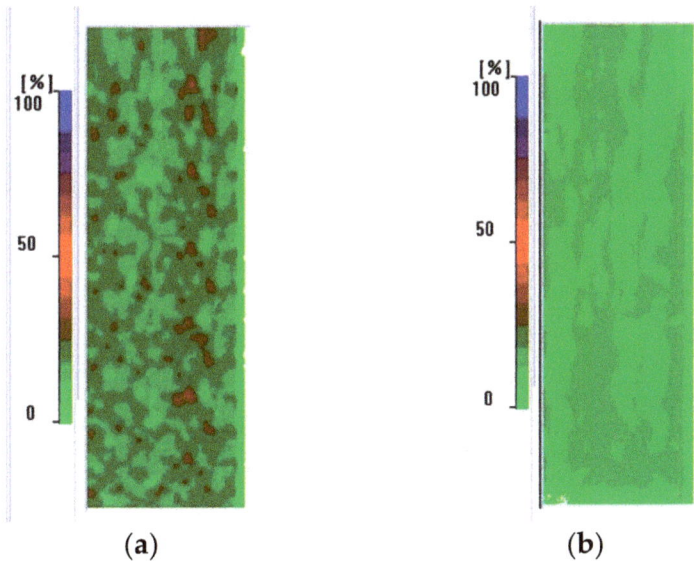

Figure 11. C-scan for the 6 GW/cm2 specimen: composite side (**a**) and Ti side (**b**).

3.2. Symmetric Configuration Results

3.2.1. Symmetric Shots Targeted at the Adhesive/CFRP Interface

The shots conducted with a delay between pulse A and B of 3.45 µs were aimed at the adhesive/CFRP interface. Figures 12 and 13 are images of an electronic microscope, where the Ti adhesive and CFRP layers are visible. Measurements of the adhesive's thickness outside and inside the spot area indicate an increase in thickness inside the spot area that is due to plastic deformation of the adhesive. The second shot at the same spot resulted in matrix damage in the composite material. This is visible in Figure 12b where the fibers have been exposed underneath the adhesive layer. The experiment was repeated with the same results in Figure 13.

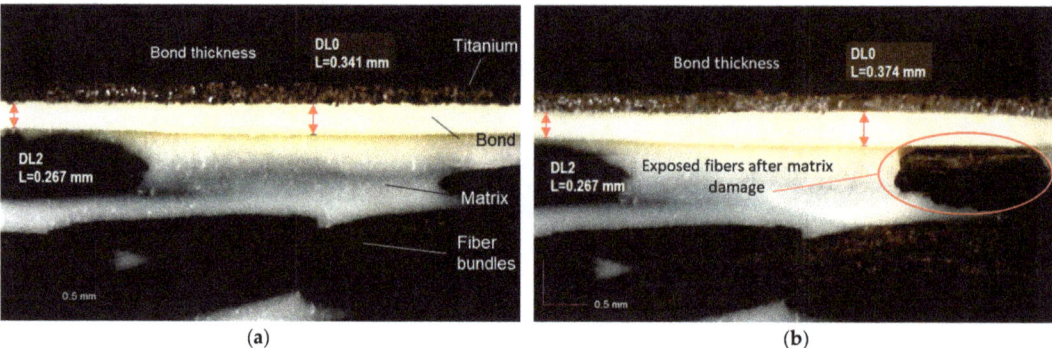

Figure 12. Symmetric shot aiming at the interface between CFRP and adhesive. (**a**) Plastic deformation caused by the first shot; (**b**) matrix damage caused by the second shot.

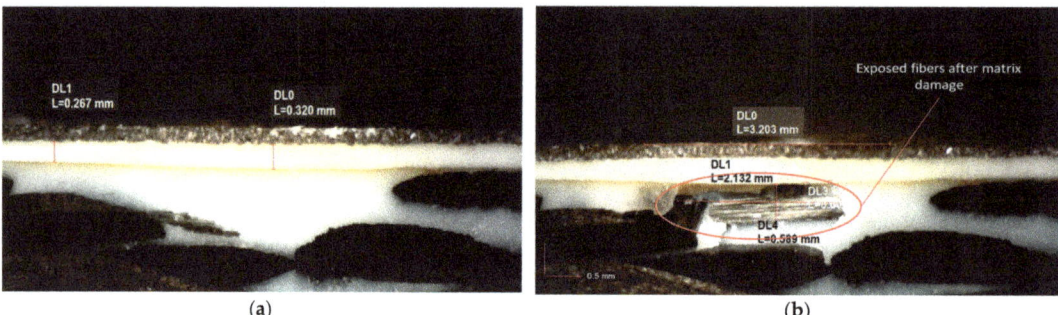

Figure 13. Repeated experiments for matrix damage. (**a**) Deformation after the first shot; (**b**) exposed fibers after the second shot.

3.2.2. Symmetric Shots Targeted at the Adhesive/Ti Interface

The shots that were conducted using the 3.54 µs delay time between beam A and B have led to adhesive failure at the adhesive/Ti interface, using the same two-shot methodology. After the first shot, the adhesive developed plastic deformation, like the experiments using delay time of 3.45 µs, and the second shot created the debonding, as shown in Figure 14. No other damage was visible other than the debonding.

After establishing the capability of the method to create the debonding, two propagation techniques were tested. The first experimental trial employed repeated shots at the same spot. Each consecutive shot increased the debonded area until it became equal to the spot diameter at the fourth shot. Figure 15 shows the propagation of the debonding after each shot. This method, although effective, produced fiber damage to the CFRP close to the free surface of the composite. Figure 16 is a microscope picture of the damaged CFRP.

The second approach for the propagation of the debonding is the moving spot. The sequence of shots during this trial is the following: the first two shots initiate the debonding; then, the spot is moved by 2 mm and two more shots are used to propagate the debonding. Finally, the spot is moved by another 2 mm in the same direction, and after two more shots, the final debonding was measured as 5.5 mm. The progression of damage is shown in Figure 17. Visual inspection of the specimen using the electronic microscope did not show any indication of damage to the CFRP.

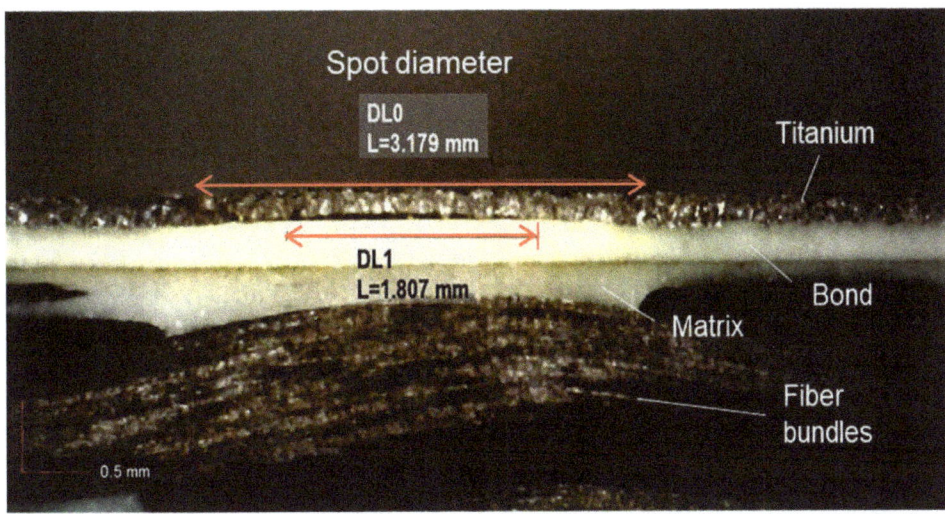

Figure 14. Debonding between titanium and adhesive after two symmetric shots with delay of 3.54 µs.

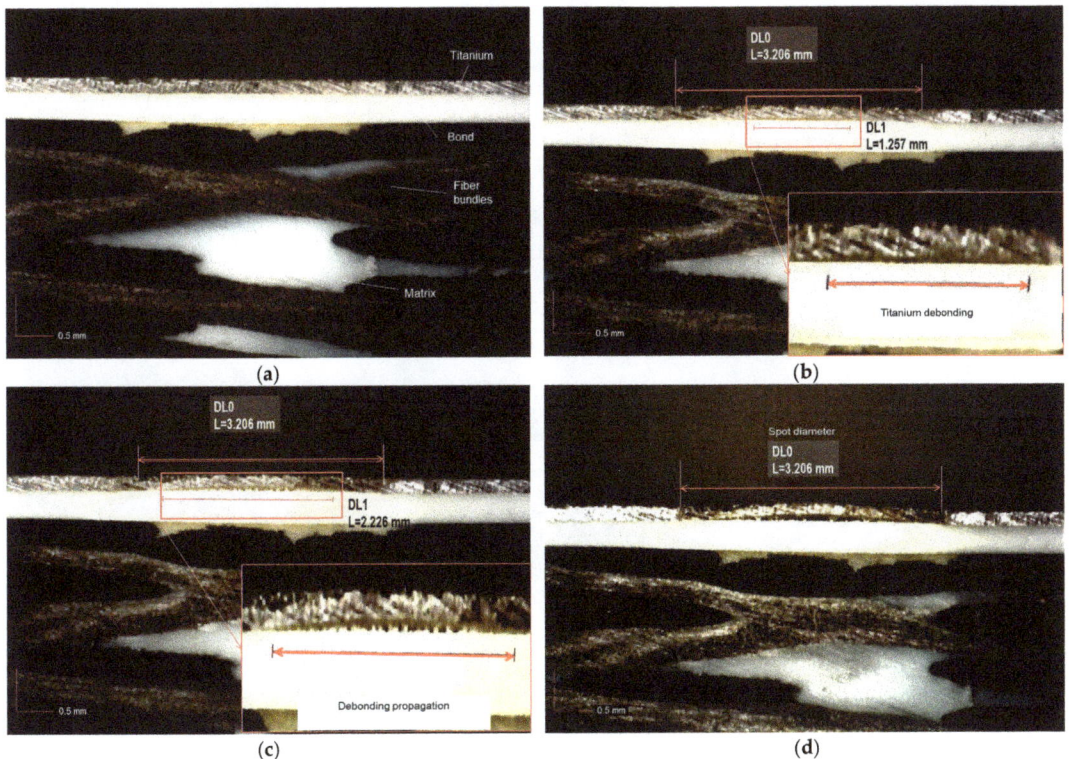

Figure 15. Consecutive shots at the same spot: (**a**) healthy specimen; (**b**) debonding initiation after two shots; (**c**) debonding propagation after the third shot; (**d**) debonding equal to the spot diameter.

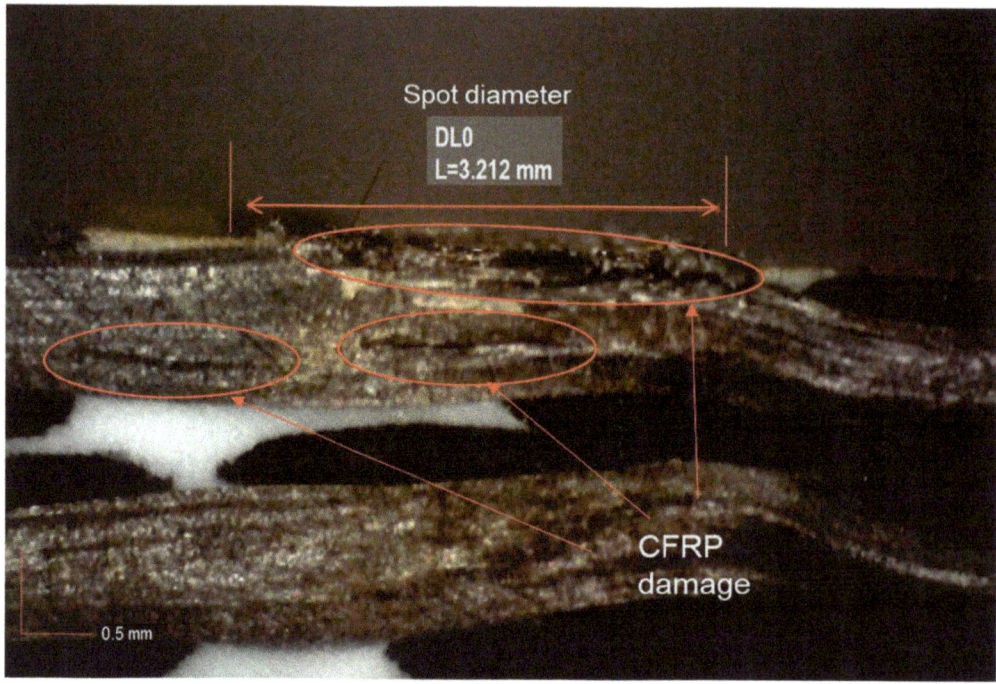

Figure 16. Fiber damage at the free surface of the composite after four consecutive shots at the same spot.

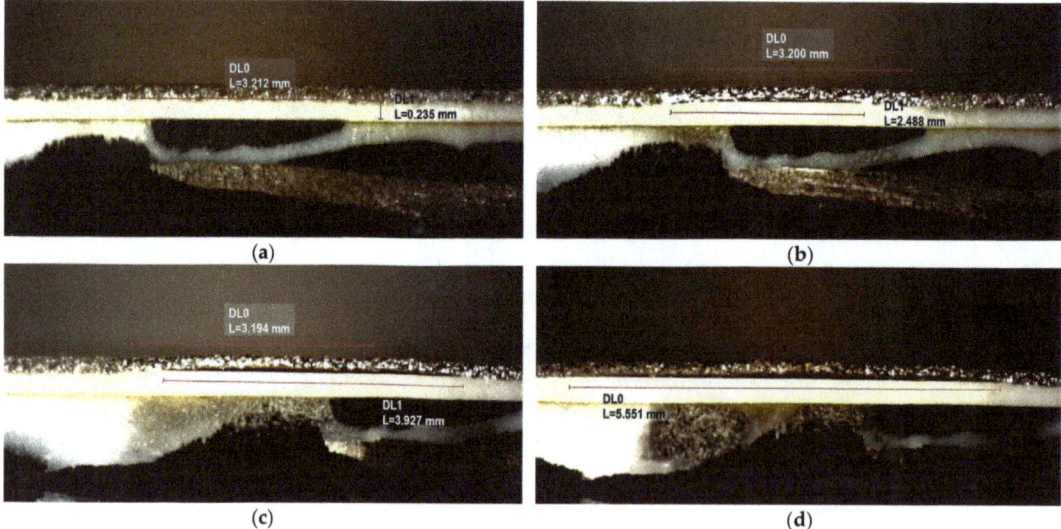

Figure 17. Damage propagation using a moving spot: (**a**) healthy specimen; (**b**) debonding initiation after two shots; (**c**) debonding propagation, after moving the spot by 2 mm and shooting two times; (**d**) final damage after repeating the previous step.

4. Discussion

The symmetric laser-shock configuration shows that it is possible to create a debonding between the Ti and CFRP by carefully calibrating the delay time of the two shots. This work proved experimentally the hypothesis that was investigated numerically in previous work [27,28] that the laser-shock technique can progressively create damage by shooting multiple times at the same spot. Additionally, the propagation of the debonding can be also achieved using repetitive shots at the same spot but also by moving the spot. The experiments indicated that repetitive shots at the same spot have an upper limit in their effectiveness because they can damage the involved adherents, as is the case of the first trial, which damaged the CFRP at the area where secondary tensile interactions are predicted to occur. Furthermore, this was the first time that a specimen of this thickness was tested, and although challenging, it is possible to disassemble even thick structures like an aircraft engine fan blade.

Although the first steps of the process development show promise, the implementation of the technique to a full-scale industrial disassembly application needs improvement. First, the symmetric process is tedious and should be streamlined for automation used in industrial-scale disassembly. This can be achieved using a continuous water stream as the confinement regime instead of the single-use solid polymer. In addition, the symmetric configuration, although effective, is restrictive in its use as both faces of the material should be accessible and taken into account during the implementation of the process. Furthermore, the shot sequence can be optimized for maximizing the debonded area. That, as well as the calculation of the optimal time delay, can be achieved by a simulation of the process using the data created by this experimental series to validate a digital twin of the specimen.

5. Conclusions

A novel disassembly process of adhesively bonded structures is being developed using the laser-shock technique. Two experimental trials were conducted to prove the method's ability to debond a thin titanium leading edge from a 3D-woven composite core. The single-shot configuration experiments revealed that it is not possible to create a debonding on the specimen. The method's inability is attributed to the fact that the required stress field near the bond line cannot be created by one shot. Nevertheless, the experimental series provided useful back-face velocity measurements that can be used to validate numerical models for further optimization of the method. The second experimental series that was conducted using the symmetric configuration succeeded to create and propagate a debonding between the composite and the titanium leading edge. Two methods of propagation were tested; on the one hand, subsequent shots at the same spot, although successful for propagation of the debonding, caused severe damage to the CFRP, something the process is aiming to avoid. On the other hand, moving the spot proved to be the most sufficient methodology for the debonding propagation. Additionally, using the symmetric configuration, it is possible to control with accuracy the concentration of tensile stresses by shifting the delay between the beams. This is showcased by the damage creation differences once the wave interference was aimed at the interface between the Ti/adhesive and CFRP/adhesive, respectively. Finally, although the results are promising in creating and propagating a debonding, to use the technique as a disassembly process both the damage initiation and the propagation need optimization.

Author Contributions: Conceptualization, L.B. and K.T.; methodology, P.K., S.U., K.T. and L.B.; investigation, P.K. and S.U.; resources, K.T. and L.B.; writing—original draft preparation, P.K. and K.T.; writing—review and editing, K.T., S.U. and L.B.; visualization, P.K. All authors have read and agreed to the published version of the manuscript.

Funding: The research leading to these results is part of the MORPHO project and has received funding from the European Union's Horizon 2020 research and innovation program under grant agreement No. 101006854.

Data Availability Statement: Not applicable.

Conflicts of Interest: The authors declare no conflict of interest.

References

1. Liu, W.; Huang, H.; Zhu, L.; Liu, Z. Integrating Carbon Fiber Reclamation and Additive Manufacturing for Recycling CFRP Waste. *Compos. Part B Eng.* **2021**, *215*, 108808. [CrossRef]
2. Meyer, L.O.; Schulte, K.; Grove-Nielsen, E. CFRP-Recycling Following a Pyrolysis Route: Process Optimization and Potentials. *J. Compos. Mater.* **2009**, *43*, 1121–1132. [CrossRef]
3. Dauguet, M.; Mantaux, O.; Perry, N.; Zhao, Y.F. Recycling of CFRP for High Value Applications: Effect of Sizing Removal and Environmental Analysis of the SuperCritical Fluid Solvolysis. *Procedia CIRP* **2015**, *29*, 734–739. [CrossRef]
4. Moslehi, A.; Ajji, A.; Heuzey, M.; Rahimizadeh, A.; Lessard, L. Polylactic Acid/Recycled Wind Turbine Glass Fiber Composites with Enhanced Mechanical Properties and Toughness. *J. Appl. Polym. Sci* **2022**, *139*, 51934. [CrossRef]
5. Sabaghi, M.; Cai, Y.; Mascle, C.; Baptiste, P. Sustainability Assessment of Dismantling Strategies for End-of-Life Aircraft Recycling. *Resour. Conserv. Recycl.* **2015**, *102*, 163–169. [CrossRef]
6. Zhao, X.; Verhagen, W.J.C.; Curran, R. Disposal and Recycle Economic Assessment for Aircraft and Engine End of Life Solution Evaluation. *Appl. Sci.* **2020**, *10*, 522. [CrossRef]
7. Tserpes, K. Adhesive Bonding of Aircraft Structures. In *Revolutionizing Aircraft Materials and Processes*; Pantelakis, S., Tserpes, K., Eds.; Springer: Cham, Switzerland, 2020; pp. 337–357. ISBN 978-3-030-35346-9.
8. Sato, C.; Carbas, R.J.C.; Marques, E.A.S.; Akhavan-Safar, A.; da Silva, L.F.M. Effect of Disassembly on Environmental and Recycling Issues in Bonded Joints. In *Adhesive Bonding*; Elsevier: Amsterdam, The Netherlands, 2021; pp. 407–436. ISBN 978-0-12-819954-1.
9. Hutchinson, A.; Liu, Y.; Lu, Y. Overview of Disbonding Technologies for Adhesive Bonded Joints. *J. Adhes.* **2017**, *93*, 737–755. [CrossRef]
10. Banea, M.D. Debonding on Demand of Adhesively Bonded Joints: A Critical Review. *Rev. Adhes Adhes.* **2019**, *7*, 33–50. [CrossRef]
11. Banea, M.D.; da Silva, L.F.M.; Carbas, R.J.C. Debonding on Command of Adhesive Joints for the Automotive Industry. *Int. J. Adhes. Adhes.* **2015**, *59*, 14–20. [CrossRef]
12. Piazza, G.; Burczyk, M.; Gerini-Romagnoli, M.; Belingardi, G.; Nassar, S.A. Effect of Thermally Expandable Particle Additives on the Mechanical and Reversibility Performance of Adhesive Joints. *J. Adv. Join. Process.* **2022**, *5*, 100088. [CrossRef]
13. Ivetic, G. Three-Dimensional FEM Analysis of Laser Shock Peening of Aluminium Alloy 2024-T351 Thin Sheets. *Surf. Eng.* **2011**, *27*, 445–453. [CrossRef]
14. Le Bras, C.; Rondepierre, A.; Seddik, R.; Scius-Bertrand, M.; Rouchausse, Y.; Videau, L.; Fayolle, B.; Gervais, M.; Morin, L.; Valadon, S.; et al. Laser Shock Peening: Toward the Use of Pliable Solid Polymers for Confinement. *Metals* **2019**, *9*, 793. [CrossRef]
15. Ecault, R.; Touchard, F.; Boustie, M.; Berthe, L.; Dominguez, N. Numerical Modeling of Laser-Induced Shock Experiments for the Development of the Adhesion Test for Bonded Composite Materials. *Compos. Struct.* **2016**, *152*, 382–394. [CrossRef]
16. Sagnard, M.; Ecault, R.; Touchard, F.; Boustie, M.; Berthe, L. Development of the Symmetrical Laser Shock Test for Weak Bond Inspection. *Opt. Laser Technol.* **2019**, *111*, 644–652. [CrossRef]
17. Arrigoni, M. Inputs of Numerical Simulation into the Development of Shock Adhesion Tests on Advanced Materials. *Int. J. Struct. Glass Adv. Mater. Res.* **2020**, *4*, 1–9. [CrossRef]
18. Ecault, R.; Touchard, F.; Berthe, L.; Boustie, M. Laser Shock Adhesion Test Numerical Optimization for Composite Bonding Assessment. *Compos. Struct.* **2020**, *247*, 112441. [CrossRef]
19. Tserpes, K.; Papadopoulos, K.; Unaldi, S.; Berthe, L. Development of a Numerical Model to Simulate Laser-Shock Paint Stripping on Aluminum Substrates. *Aerospace* **2021**, *8*, 233. [CrossRef]
20. Ünaldi, S.; Papadopoulos, K.; Rondepierre, A.; Rouchausse, Y.; Karanika, A.; Deliane, F.; Tserpes, K.; Floros, G.; Richaud, E.; Berthe, L. Towards Selective Laser Paint Stripping Using Shock Waves Produced by Laser-Plasma Interaction for Aeronautical Applications on AA 2024 Based Substrates. *Opt. Laser Technol.* **2021**, *141*, 107095. [CrossRef]
21. Papadopoulos, K.; Tserpes, K. Analytical and Numerical Modeling of Stress Field and Fracture in Aluminum/Epoxy Interface Subjected to Laser Shock Wave: Application to Paint Stripping. *Materials* **2022**, *15*, 3423. [CrossRef]
22. Ghrib, M.; Berthe, L.; Mechbal, N.; Rébillat, M.; Guskov, M.; Ecault, R.; Bedreddine, N. Generation of Controlled Delaminations in Composites Using Symmetrical Laser Shock Configuration. *Compos. Struct.* **2017**, *171*, 286–297. [CrossRef]
23. Fabbro, R.; Peyre, P.; Berthe, L.; Scherpereel, X. Physics and Applications of Laser-Shock Processing. *J. Laser Appl.* **1998**, *10*, 265–279. [CrossRef]
24. Fairand, B.P.; Clauer, A.H. Laser Generation of High-amplitude Stress Waves in Materials. *J. Appl. Phys.* **1979**, *50*, 1497–1502. [CrossRef]
25. Davison, L.W. *Fundamentals of Shock Wave Propagation in Solids*; Shock wave and high pressure phenomena; Springer: Berlin/Heidelberg, Germany, 2008; ISBN 978-3-540-74568-6.
26. Scius-Bertrand, M.; Videau, L.; Rondepierre, A.; Lescoute, E.; Rouchausse, Y.; Kaufman, J.; Rostohar, D.; Brajer, J.; Berthe, L. Laser Induced Plasma Characterization in Direct and Water Confined Regimes: New Advances in Experimental Studies and Numerical Modelling. *J. Phys. D Appl. Phys.* **2021**, *54*, 055204. [CrossRef]

27. Kormpos, P.; Tserpes, K.; Floros, G. Towards Simulation of Disassembly of Bonded Composite Parts Using the Laser Shock Technique. *IOP Conf. Ser. Mater. Sci. Eng.* **2022**, *1226*, 012081. [CrossRef]
28. Kormpos, P.; Unaldi, S.; Ayad, M.; Berthe, L.; Tserpes, K. Towards the Development of a Laser Shock-Based Disassembly Process for Adhesively Bonded Structural Parts: Experiments and Numerical Simulation. In Proceedings of the 20th European Conference on Composite Materials—Composites Meet Sustainability B, Lausanne, Switzerland, 26–30 June 2022; pp. 873–880. [CrossRef]

Disclaimer/Publisher's Note: The statements, opinions and data contained in all publications are solely those of the individual author(s) and contributor(s) and not of MDPI and/or the editor(s). MDPI and/or the editor(s) disclaim responsibility for any injury to people or property resulting from any ideas, methods, instructions or products referred to in the content.

Analysis of Hydrothermal Ageing on Mechanical Performances of Fibre Metal Laminates

Costanzo Bellini *, Vittorio Di Cocco, Francesco Iacoviello, Larisa Patricia Mocanu, Gianluca Parodo, Luca Sorrentino and Sandro Turchetta

Department of Civil and Mechanical Engineering, University of Cassino and Southern Lazio, Via G. Di Biasio 43, 03043 Cassino, Italy
* Correspondence: costanzo.bellini@unicas.it

Abstract: Fibre Metal Laminates (FMLs) are very interesting materials due to their light weight coupled with their high stiffness, high fatigue resistance, and high damage tolerance. However, the presence of the polymeric matrix in the composite layers and of polymeric adhesive at the metal/composite interface can constitute an Achille's heel for this class of materials, especially when exposed to a hot environment or water. Therefore, in the present article, aluminium/carbon fibre FML specimens were produced, aged by considering different hydrothermal conditions, and then, subjected to mechanical testing. The End-Notched Flexure (ENF) test was considered for this activity. It was found that the first ageing stage, consisting of submersion in saltwater, was very detrimental to the specimens, while the second stage, composed of high and low temperature cycles, showed an increase in the maximum load, probably due to a post-curing effect of the resin during the higher temperatures of the ageing cycles and to the dissolution of salt crystals during the subsequently ageing stages in distilled water.

Keywords: fibre metal laminates; hydrothermal ageing; end-notched flexure test

1. Introduction

Structural applications in advanced fields, such as aeronautics, demand innovative materials presenting high mechanical properties, low density, and resistance against ageing. In fact, the mechanical properties of structural parts have to remain unaltered throughout the entire lifecycle, even if the part is exposed to hazardous environments [1]. Fibre Metal Laminates (FMLs) are a class of materials able to meet the aforementioned properties. In fact, they are formed by metal sheets alternated with Fibre-Reinforced Polymer (FRP) layers, and this confers to the material the desired mechanical characteristics [2,3]. From a historical point of view, FMLs were developed to overcome the poor fatigue resistance of aluminium sheets [4,5]. Moreover, FRPs suffer a decrease in mechanical properties due to exposure to water/moisture environments, which are able to damage the matrix of the composite layers [6,7]. In fact, the matrix of FRP generally is hydrophilic. This leads to a tendency for the material to absorb moisture from the outside to the inside. Generally, the mechanisms by which this occurs are essentially two: volumetric and "interaction" [8]. In the first mechanism, the absorption of water in free volumes and microcavities that are present in the FRP such as micro-voids and pores due to gases generated during the polymerisation of the resin throughout the curing process is considered the cause. These gases could remain entrapped in the matrix, increasing the void content in the FRP. In the second mechanism, there is an interaction between the water molecules and the polar groups present in the molecular structure of the polymer. This interaction not only allows diffusion within the polymer, but also involves a plasticisation of the polymer itself as a weakening of the primary and secondary bonds of the molecular structure and distortions of the molecular chains. This phenomenon results in a change in the thermomechanical characteristics of the polymer, which are usually manifested by a reduction in the glass transition temperature

(Tg). On the other hand, thermal ageing is strongly linked to the Tg of the resin used for the production of FRP. Generally, when the temperature of the environment exceeds the Tg of the resin, its mechanical characteristics undergo a strong reduction, which can be recovered if the working temperature is reduced below the Tg. However, such temperature changes affect the molecular structure by generating configuration changes [9]. However, the effects of physical ageing related to the presence of moisture in the polymer or the presence of high-temperature environments can be recovered if chemical degradation of the molecular bonds does not occur. In fact, if a polymer is exposed for a certain time to a temperature close to its Tg, its aging history is lost. For epoxy resins, this phenomenon can also occur at temperatures lower than the Tg, definable as erasure temperatures [10].

The presence of the metal sheets on the exposed surface of the laminate is able to reduce the penetration of moisture into the FRP resin [11]. For this reason, FMLs can be employed in environments characterised by intense temperature variations and elevated humidity [12]. However, ageing can affect the mechanical properties of the FML, leading to dangerous delaminations in the laminate and, consequently, to its failure [13]. Moreover, the presence of an interface could accelerate the moisture uptake in the FML because it can be the most-critical area in terms of mechanical strength and temporal reliability [14].

There are several studies about the mechanical properties of FMLs, paying attention to the composite type [15], the thickness of the layers [16], the metal's surface preparation [17], and the fibre orientation [18]; however, there has not been much research performed on the hygrothermal effects on carbon-based FMLs, and it is still unclear how long layers will stay bonded, particularly in seawater environments or at high temperatures. Yu et al. [14] prepared titanium/Carbon-Fibre-Reinforced Polymer (CFRP) FMLs by anodising the titanium sheets and grafting the CFRP layers with multi-walled carbon nanotubes. Then, they compared the interlaminar fracture toughness of the produced laminates with that of an equivalent, but untreated, one, by considering both as-produced and aged materials. They found much better behaviour for the treated laminate. Instead, Wang et al. [19] studied the effect of graphene nanoplatelets on the impact resistance of FMLs subjected to different hygrothermal ageing conditions, finding that the addition of nanoplatelets decreased the water absorption and increased the impact resistance. Pan et al. [20] investigated the effects of aluminium sheet treatments on the mechanical performances of a hydrothermally aged CFRP FML. They found a greater decrease of the mechanical properties in the untreated laminates, due to the corrosion of the aluminium sheets. Hamill et al. [21] studied the effect of galvanic corrosion induced by ageing in saltwater on traditional aluminium/CFRP FML and an innovative bulk metallic glass/CFRP FML. They found a lower corrosion resistance in the former one, which was reflected in the tensile properties of the materials, while the flexural properties remained unaffected. To reduce the effects of environment-induced galvanic corrosion, Stoll et al. [1] added an elastomeric interlayer between the aluminium sheets and the carbon layers, suitable to reduce the corrosion thanks to its high electrical resistance. They found decreased mechanical properties in the laminate produced without the elastomer. Ali et al. [13] compared the effect of hydrothermally induced corrosion on the mechanical properties of titanium sheets, CFRP laminates, and titanium/CFRP FMLs. They found higher mechanical characteristics in the FML compared to the CFRP laminates, while the titanium sheets presented the lowest corrosion. Viandier et al. [22] studied the corrosion resistance of an FML based on CFRP and stainless steel, finding that the former behaved as a cathode, while the latter as an anode, and it was affected by pitting corrosion as well. Alia et al. [12] studied the effect of hydrothermal ageing on the adhesive layer used for bonding the metal with the composite in the FML, evaluating the diffusion of water throughout the adhesive thickness and finding both microstructural changes and chemical degradation in the latter. Hu et al. [23] compared the effect of moisture absorption on the mechanical behaviour of carbon-fibre-reinforced polyimide and polyimide–titanium-based FMLs. They subjected samples of these materials to a high-temperature and high-relative-humidity environment for different amounts of time, and then, they evaluated the interlaminar shear strength and the flexural strength of the aged specimens. A certain decrease in these properties was found, induced by ageing, as confirmed also by scanning electron micrographs and dynamic mechanical analysis tests. Hu et al. [24] investigated the effect of hydrothermal ageing on Ti/CF/PMR polyimide composite laminates conditioned in environments at different combinations of temperature and relative humidity. They found that the saturated moisture absorption rate depended on the relative humidity, while the diffusion rate of water in the composite depended on the temperature. Zhang et al. [25] proposed a reduced graphene oxide modified Ti/CFRP laminate to be used for intelligent de-icing in aeroplanes and tested both the de-icing performances and the mechanical properties of this laminate. They found that the mechanical properties improved after several de-icing cycles.

The aim of the present work is to investigate the effect of hydrothermal ageing on the mechanical performances of FMLs. Specimens made of CFRP and aluminium sheets were subjected to a sequence of different environments, such as saltwater, hot water, and ice, as will be better described in the "Materials and Methods" Section. The motivation behind this choice can be explained as the willingness of reproducing the possible environments an aeronautical part is subjected to. Therefore, in the present work, End-Notched Flexure (ENF) specimens were manufactured, aged under different conditions, and finally, tested through a three-point bending scheme in order to analyse the effect of hydrothermal ageing on the bonding interface between metal sheets and composite layers in an FML made of aluminium and CFRP.

2. Materials and Methods

FMLs made of carbon composite laminates and aluminium sheets were chosen for the investigation presented in this work. According to published research, galvanic corrosion affects FMLs made of CFRP and aluminium sheets because the standard electrode potentials of carbon and aluminium differ. Due to this peculiarity, the combination of these materials was chosen since it is extremely important when thinking about the issue of environment-induced corrosion, the goal of this work being to explore the impact of hydrothermal ageing on the mechanical properties. The FMLs considered in this work were manufactured using the prepreg vacuum bag process, as already described in [26]. The metal layers of the produced FMLs were made of EN AW 3105, a commercial aluminium alloy, while the layers of CFRP were made of M92/48%/220H4/AS4C/3K, a woven thermoset prepreg system. The aluminium sheets had a thickness of 0.8 mm, while the prepreg layers had a thickness of about 0.25 mm in the uncured state. The layup sequence is reported in Figure 1. The composite layer consisted of two plies of M92/48%/220H4/AS4C/3K, layered in the roll direction. The composite laminates were co-bonded to the aluminium sheets using the structural adhesive AF 163-2, manufactured by 3M. As the pretreatment for the bonding, the aluminium sheets were degreased using Methyl Ethyl Ketone (MEK). Specifically, the FMLs were made using the vacuum bag technique: the cure cycle consisted of a heat ramp of 2 °C/min up to the temperature of 125 °C and a dwell at this temperature of about 90 min. This thermal cycle was suitable for both the prepreg and adhesive, as reported in the respective technical sheets. Two notches were made using a very thin Polytetrafluoroethylene (PTFE) release film at the interface between the composite and the aluminium inner sheet in order to realise a notch length of about 30 mm in the finished specimens. In fact, once the cure was complete, the laminate was removed from the mould, and ENF specimens were made through a cutting operation using a diamond blade. In particular, the specimen dimensions were determined in accordance with ASTM D7905 [27], as shown in Figure 2. In total, 20 specimens were produced.

Ageing Treatments

To study the behaviour of the hybrid laminate in various types of possible working conditions, five specimens, which represented the reference for measuring the mechanical performance in unaged conditions, were stored in a chamber with a humidity of 30% and a temperature of 25 °C, while the other specimens were subjected to the hydrothermal ageing cycle reported in Figure 3. It is possible to subdivide the ageing cycle into three stages. Specifically, in the first stage, for a duration of 14 days, the specimens were immersed at room temperature in saltwater with a chemical composition according to ASTM D1141 [28]. For the preparation of saltwater, 2 stocks and a final container were used. In the first stock, 7 L of distilled water was placed and 3889 g of magnesium chloride hexahydrate, 405 g of calcium chloride anhydrous, and 15 g of strontium chloride hexahydrate were dissolved. In the second stock, 486 g of potassium chloride, 141 g of sodium bicarbonate, 70 g of potassium bromide, 19 g of boric acid, and 2 g of sodium fluoride were diluted in 7 L of distilled water. Finally, the preparation of the saltwater was made in the final container. To prepare it, 245 g of sodium chloride and 41 g of sodium sulphate were dissolved in 8 L of water. After, 200 mL of the solution in Stock 1 and 100 mL of the solution in Stock 2 were added to the container. Finally, the obtained solution was diluted with distilled water until a volume of 10 L was reached.

Figure 1. FML layup sequence adopted in this work.

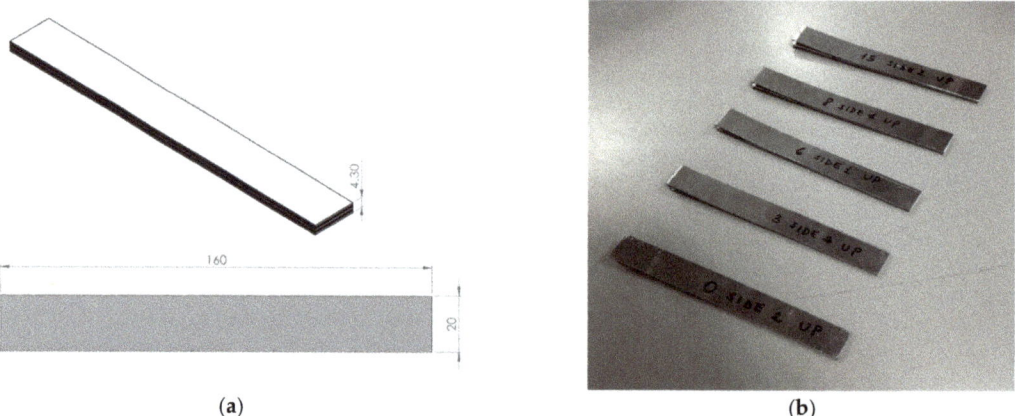

Figure 2. Manufactured ENF specimens: (**a**) geometry and dimensions in mm; (**b**) a photo of a group.

Subsequently, five specimens were extracted and stored in a dry environment at room temperature, while the others were immersed in distilled water and subjected to thermal shocks between −28 °C and 80 °C. Each ageing cycle lasted 7 days, and the thermal variations imposed are illustrated in Figure 4. The adopted ageing cycles simulated the most-critical environments an aeroplane could be exposed to: parking in the snow or ice, near the sealine, and high-altitude flight after a take off from a very hot and sunny location. It is worth pointing out that the edges of the specimens were not sealed in order to procure the most-critical condition for the specimens. In fact, the metal sheet, placed on the external surfaces, protects the FML from moisture and water absorption. By exposing the edges to the environment, the detrimental effects, due to the galvanic coupling between carbon fibres and aluminium sheets, are intensified.

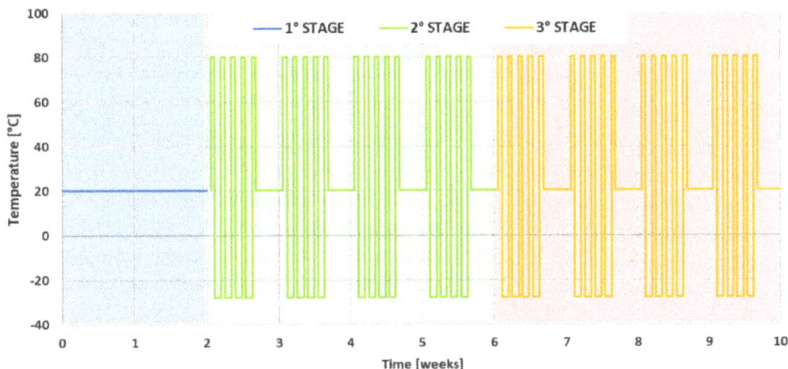

Figure 3. Hydrothermal ageing cycle adopted in this work.

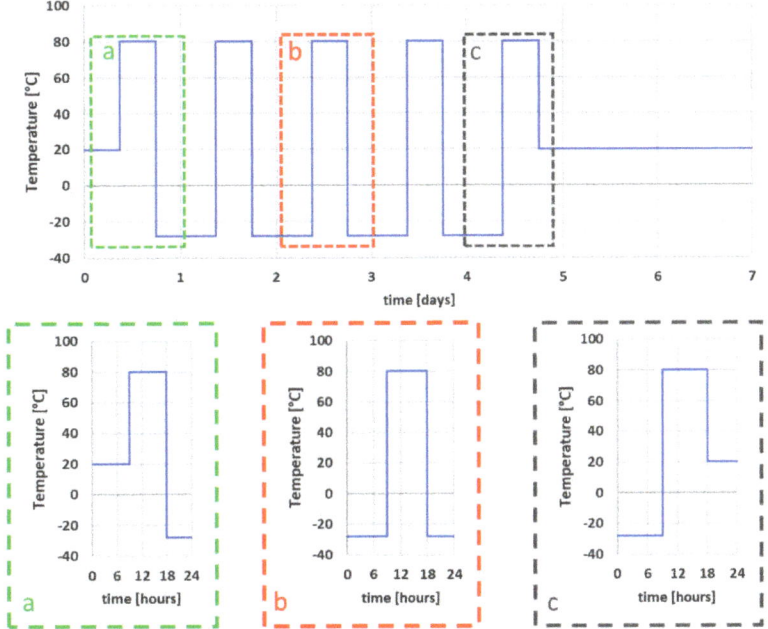

Figure 4. Temperature variations of specimens immersed in distilled water for one week: (**a**) first day of the week; (**b**) from the second to fourth day; (**c**) fifth day of the week.

After four cycles, five specimens were removed from the distilled water and stored in a dry environment at room temperature. At the end of each ageing stage, the specimens were dried and weighed using a precision balance. Weight variations were calculated only for samples that were subjected to the ageing stages. For each specimen, a coefficient of variation with respect to the unaged condition cwv_{Ref} (defined as the reference weight, which consisted of the weight of the specimen before ageing) was calculated using Equation (1):

$$cwv_{Ref} = \frac{P_{n\,stage} - P_{ref}}{P_{ref}} \qquad (1)$$

where $P_{n\,stage}$ is the weight of the specimen after the nth stage, while P_{ref} is the weight of the specimen in unaged conditions. Similarly, it is possible to calculate the coefficient of weight variation with respect to the previous ageing stage cwv_n according to Equation (2):

$$cwv_n = \frac{P_{n\,stage} - P_{n-1\,stage}}{P_{n-1\,stage}} \qquad (2)$$

where $P_{n-1\,stage}$ is the specimen weight after the previous ageing stage considered.

Once having completed the ageing treatments, the specimens were stored for two years at room temperature and subsequently subjected to flexural tests. These tests were performed using a universal testing machine and equipment produced to realise three-point bending tests. Specifically, the test configuration consisted of a three-point bending test with a span of about 100 mm (Figure 5). In particular, the diameters of the support were equal to 8 mm, while the punch diameter was fixed to 10 mm. For completeness, the imposed crosshead speed was equal to 2 mm/min during the tests.

Figure 5. The three-point bending test of a specimen during testing and its deflection/deformation.

3. Results

The first results obtained from the analyses were related to the weight variations of the specimens at the end of each ageing stage. The obtained values of cwv_{Ref} and cwv_n are reported in Figures 6 and 7, respectively. Observing Figure 6, it is possible to state that saltwater was uptaken into the specimens, allowing an increase in weight of about 8%. This increase in weight was also due to the formation of salt crystals on the free surfaces of the specimens. The subsequent ageing stages allowed a reduction in weight of about 10%, reducing the weight of each sample to a value lower than the reference one. This result could be due to the microcracking of the composite matrix, the corrosion of the aluminium in saltwater, and the subsequent solution of salt crystals and aluminium oxides in the distilled water.

During the three-point bending test, crack propagation between the adhesive and aluminium sheets was observed. The failure modes can be classified according to ASTM

D5573 [29]. Here, the failure modes can be subdivided into six types: adhesive failure, if the failure appeared at the adhesive–adherend interface; cohesive failure, if the separation appeared within the adhesive itself; thin layer cohesive failure, if the failure appeared very near the adhesive–adherend interface with the presence of traces of FRP adherends on the adhesive; fibre tear failure, if the rupture appeared only in the FRP matrix with the exposure of the fibres on the failure surface; light fibre tear failure, if the rupture was in the FRP matrix near the bonded surface; stock break failure, if the failure appeared outside the bonded region. Starting from the observation of all the failure surfaces, it is possible to classify the failure mode obtained from testing as an adhesive failure for all the ageing conditions. Specifically, failures appeared between the adhesive and the outer aluminium sheets. It is possible to state that the interface between the adhesive and CFRP was optimal, while the interface between the adhesive and aluminium was the most-critical. Figure 8 shows the representative load–displacement curves obtained from the experimental tests. It is possible to observe that all the ageing conditions caused a decrease in mechanical performance, but contrary to what one might think, the last and the second to last ageing stage involved an increase in the mechanical performance with respect to the first stage of the ageing treatment. This could be due to a possible effect of the high temperature used for the thermal shock, which could allow for a post-cure effect on the adhesive and the composite layer. The presence of a working environment with higher temperatures during the last two stages of the ageing cycle could have a positive effect on the mechanical resistance of the specimens by a revamping of the physical ageing accumulated during the first stage of the aging cycle [10]. Moreover, the immersion in distilled water during the last two ageing stages resulted in a reduction in the salt content deposited within the specimens. In fact, while the saltwater-aged samples showed, in addition to a noticeable increase in weight, also the presence of salt crystals spread over the entire free surface of the aluminium, the specimens that were subsequently subjected to immersion in distilled water showed, in addition to a weight reduction lower than the initial reference one, also a clear reduction of the salt crystals on the free surface of the aluminium.

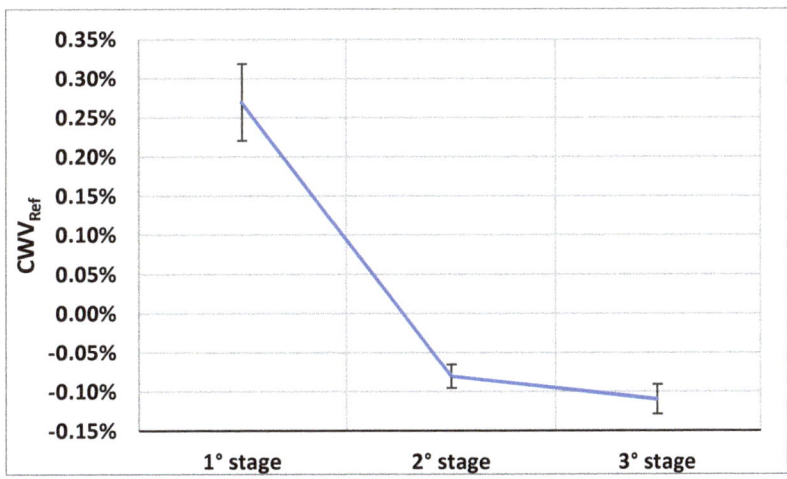

Figure 6. Coefficient of variations with respect to the unaged condition for each ageing stage.

Observing the maximum load obtained from the tests (Figure 9), it is possible to state that the average maximum load was lower, particularly for specimens aged only in saltwater at room temperature. Considering also the dispersion of the results, it is possible to state that the mechanical resistance variations were negligible compared to the other stages, despite the last stage showing a higher distribution. A decrease in the interfacial fracture energy in Ti-CFRP FMLs was also found by Yu et al. [14]. In particular, they aged in simulated seawater both common and pre-treated laminates and found a decrease of 67% of the interfacial energy in the former case, while it was of 62% and 43% in the other cases. This decrement of the mechanical properties was induced by the hydration of the metal oxide layer, which had poor bonding with the composite matrix, and by the penetration of water into the matrix. Alia et al. [12] determined a decrease in the mechanical properties of adhesives equal to 25% due to microstructural changes induced by the hydrolytic action of the water. Pan et al. [20] found a decrease in the interlaminar shear of an aluminium-based FML equal to 14% after 15 days of hygrothermal ageing in seawater and 26% after 90 days.

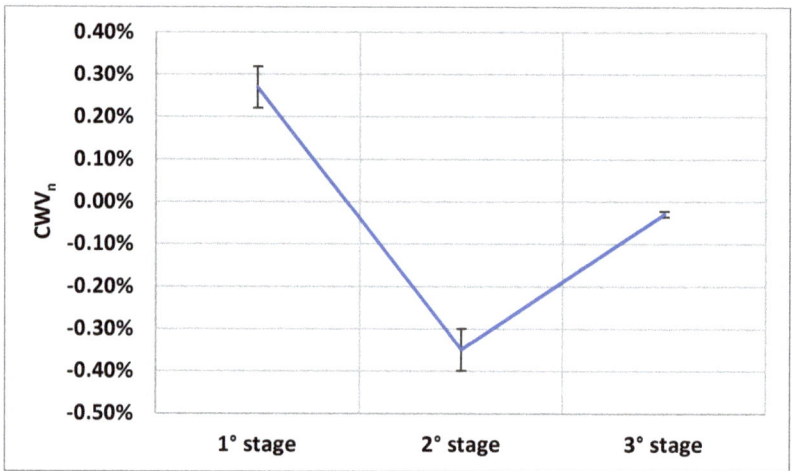

Figure 7. Coefficient of weight variation with respect to the previous ageing stage for each ageing stage.

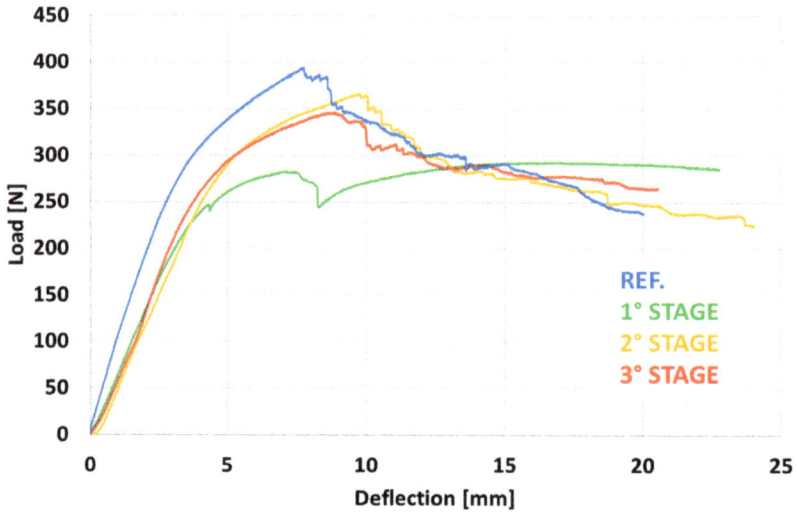

Figure 8. Average load–displacement curves obtained from the experimental tests.

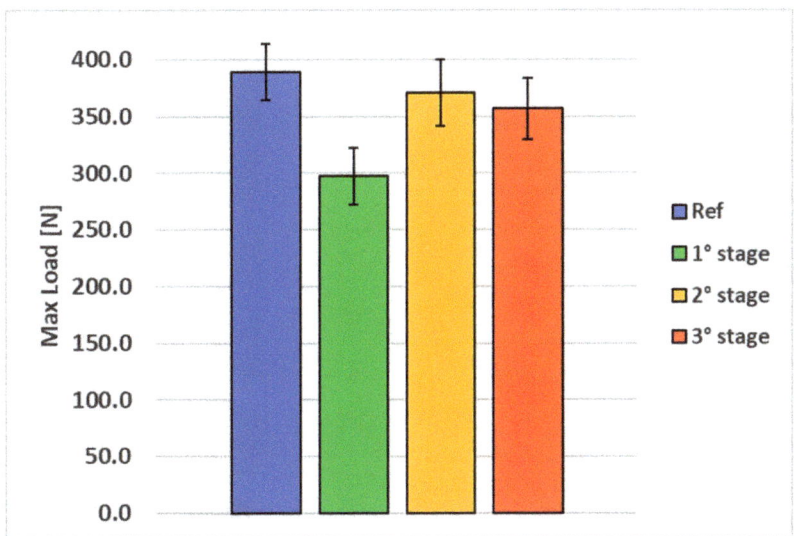

Figure 9. Maximum loads obtained from the experimental test (as a function of the ageing stage).

The observed variation in failure loads probably depended on the presence of salt crystals on the metal surfaces of the specimens. To understand this, the failure surfaces between the aluminium and composite were observed by optical microscope. Figure 10 shows the failure surfaces near the notch tip of a specimen aged only with the first stage, which was in saltwater. At the notch tip, it is possible to observe an extensive formation of salt crystals, which was not only concentrated at the notch apex, represented by the blue line, but developed within the bonding at the interface between the aluminium sheet and the adhesive. In fact, salt crystals at the notch tip were grown not only along the free surface of the aluminium, but also in the transverse direction. Such a phenomenon probably generated some peeling stress, which caused the crack propagation at the adhesive–aluminium interface. The free surface generated by this propagation led to a growth of salt crystals on the aluminium side (Figure 10c), which increased this phenomenon. The subsequent stages of ageing in distilled water involved a dissolution of salt crystals, therefore a decrease in the peeling stress at the crack tip and a consequent recovery of the mechanical performance of the FML specimens. It is likely that the surface pretreatment applied to the aluminium did not guarantee the performance needed to avoid the debonding between the aluminium and film adhesive. The analysis of the effect of different surface treatments on the aluminium sheets in these ageing conditions will be the subject of further work by the authors.

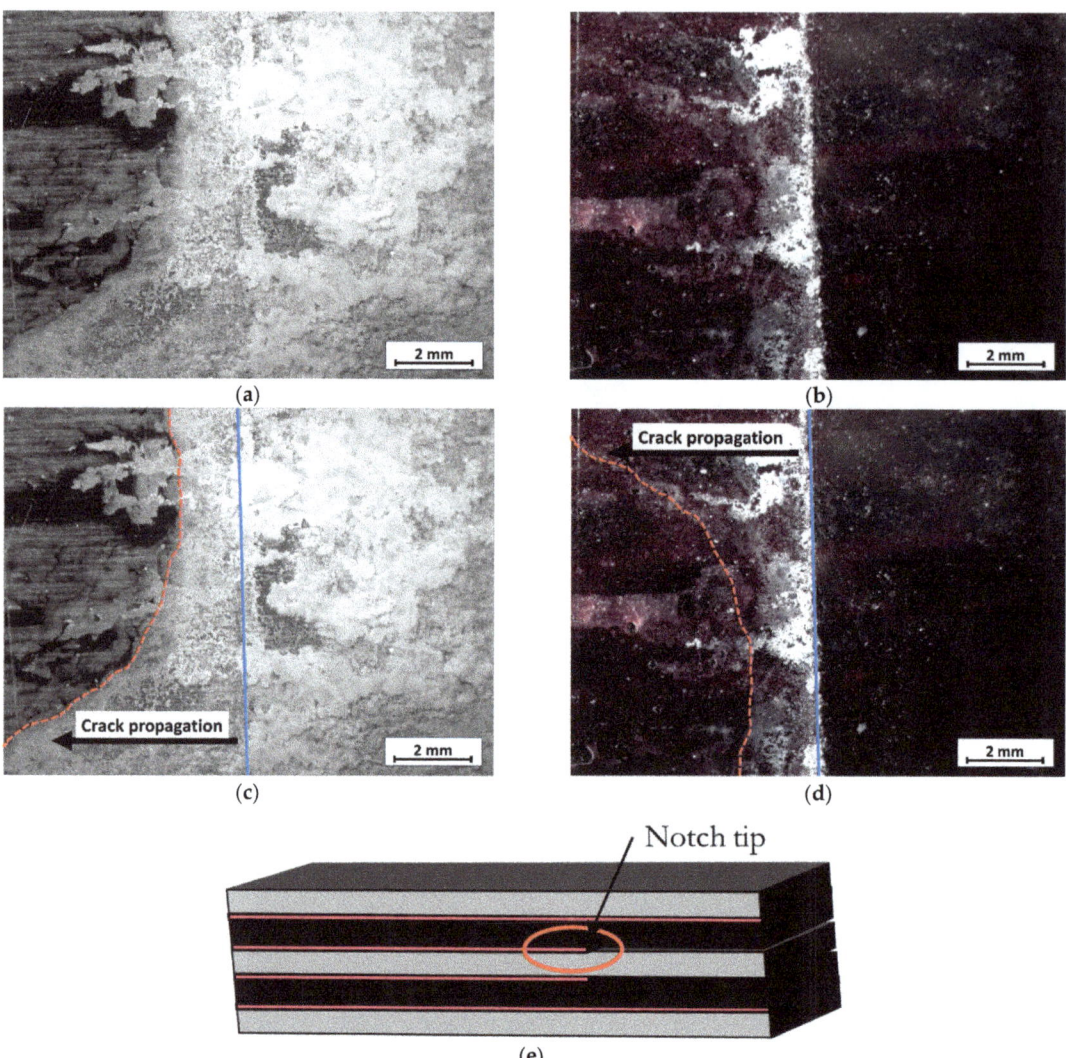

Figure 10. Failure surfaces of a specimen subjected to the first stage of ageing (saltwater): (**a**) salt crystals on the aluminium side; (**b**) salt crystal deposition on the adhesive side; (**c**) crack propagation on the aluminium side; (**d**) crack propagation on the adhesive side; (**e**) position of the region shown in the previous images.

4. Conclusions

In this work, the effect of ageing conditions on the mechanical performance of end-notched Fibre Metal Laminate (FML) specimens was investigated. The samples were manufactured through the vacuum bag technique and subjected to ageing cycles, which consisted of immersion in saltwater followed by thermal shocks in distilled water. At the end of each ageing cycle, the specimens were removed from the ageing environment, weighed, and tested through three-point bending tests. Specifically, the gravimetric analyses showed an increase in the weight of the specimens immersed in saltwater and a decrease in the weight in the subsequent ageing conditions in distilled water. This was due to the nucleation and growth of salt crystals on the free surfaces of the aluminium, which was severely limited by the dissolution of the crystals in the distilled water in the subsequent ageing stages. After the gravimetric analyses, mechanical testing was performed, and the results showed a decrease in the failure loads of about 30% for the specimens aged in saltwater, while the subsequent ageing in distilled water showed a recovery of mechanical performance. This was probably due to an imperfect pretreatment of the aluminium surface before bonding and to the formation and growth of salt crystals near the crack tip, which allowed the concentration of peel stress and was removed by the dissolution of the salt crystals during the subsequent ageing stage in distilled water. Moreover, the presence of a working environment with higher temperatures during the last two stages of the ageing cycles could have an effect on the resistance of the specimens by a revamping of the physical ageing accumulated during the first stage of aging. Future works will investigate the effect of different aluminium surface pretreatments on the reliability of FML specimens in the same ageing conditions.

Author Contributions: Conceptualization, C.B., V.D.C., F.I., L.P.M., G.P., L.S. and S.T.; Methodology, C.B., V.D.C., F.I., L.P.M., G.P., L.S. and S.T.; Investigation, C.B., V.D.C., F.I., L.P.M., G.P., L.S. and S.T.; Writing—original draft, C.B., V.D.C., F.I., L.P.M., G.P., L.S. and S.T. All authors have read and agreed to the published version of the manuscript.

Funding: This research received no external funding.

Institutional Review Board Statement: Not applicable.

Informed Consent Statement: Not applicable.

Data Availability Statement: The data presented in this study are available upon request from the corresponding author. The data are not publicly available due to a non-disclosure agreement.

Conflicts of Interest: The authors declare no conflict of interest.

References

1. Stoll, M.; Stemmer, F.; Ilinzeer, S.; Weidenmann, K.A. Optimization of Corrosive Properties of Carbon Fiber Reinforced Aluminum Laminates Due to Integration of an Elastomer Interlayer. *Key Eng. Mater.* **2017**, *742*, 287–293. [CrossRef]
2. Li, X.; Zhang, X.; Zhang, H.; Yang, J.; Nia, A.B.; Chai, G.B. Mechanical Behaviors of Ti/CFRP/Ti Laminates with Different Surface Treatments of Titanium Sheets. *Compos. Struct.* **2017**, *163*, 21–31. [CrossRef]
3. Bellini, C.; Di Cocco, V.; Iacoviello, F.; Sorrentino, L. Experimental Analysis of Aluminium/Carbon Epoxy Hybrid Laminates under Flexural Load. *Frat. Ed Integrità Strutt.* **2019**, *13*, 739–747. [CrossRef]
4. Campilho, R.D.S.G.; da Silva, L.F.M.; Banea, M.D. Adhesive Bonding of Polymer Composites to Lightweight Metals. In *Joining of Polymer-Metal Hybrid Structures: Principles and Applications*; Wiley: New York, NY, USA, 2018; pp. 29–59. ISBN 9781119429807.
5. Bellini, C.; Di Cocco, V.; Sorrentino, L. Interlaminar Shear Strength Study on CFRP/Al Hybrid Laminates with Different Properties. *Frat. Ed Integrità Strutt.* **2019**, *14*, 442–448. [CrossRef]
6. Botelho, E.C.; Costa, M.L.; Pardini, L.C.; Rezende, M.C. Processing and Hygrothermal Effects on Viscoelastic Behavior of Glass Fiber/Epoxy Composites. *J. Mater. Sci.* **2005**, *40*, 3615–3623. [CrossRef]
7. Bellini, C.; Parodo, G.; Sorrentino, L. Effect of Operating Temperature on Aged Single Lap Bonded Joints. *Def. Technol.* **2020**, *16*, 283–289. [CrossRef]
8. Gibhardt, D.; Buggisch, C.; Meyer, D.; Fiedler, B. Hygrothermal Aging History of Amine-Epoxy Resins: Effects on Thermo-Mechanical Properties. *Front. Mater.* **2022**, *9*, 826076. [CrossRef]
9. Odegard, G.M.; Bandyopadhyay, A. Physical Aging of Epoxy Polymers and Their Composites. *J. Polym. Sci. Part B Polym. Phys.* **2011**, *49*, 1695–1716. [CrossRef]
10. Lee, J.K.; Hwang, J.Y.; Gillham, J.K. Erasure below Glass-Transition Temperature of Effect of Isothermal Physical Aging in Fully Cured Epoxy/Amine Thermosetting System. *J. Appl. Polym. Sci.* **2001**, *81*, 396–404. [CrossRef]
11. Botelho, E.C.; Almeida, R.S.; Pardini, L.C.; Rezende, M.C. Elastic Properties of Hygrothermally Conditioned Glare Laminate. *Int. J. Eng. Sci.* **2007**, *45*, 163–172. [CrossRef]
12. Alia, C.; Biezma, M.V.; Pinilla, P.; Arenas, J.M.; Suárez, J.C. Degradation in Seawater of Structural Adhesives for Hybrid Fibre-Metal Laminated Materials. *Adv. Mater. Sci. Eng.* **2013**, *2013*, 869075. [CrossRef]
13. Ali, A.; Pan, L.; Duan, L.; Zheng, Z.; Sapkota, B. Characterization of Seawater Hygrothermal Conditioning Effects on the Properties of Titanium-Based Fiber-Metal Laminates for Marine Applications. *Compos. Struct.* **2016**, *158*, 199–207. [CrossRef]
14. Yu, B.; He, P.; Jiang, Z.; Yang, J. Interlaminar Fracture Properties of Surface Treated Ti-CFRP Hybrid Composites under Long-Term Hygrothermal Conditions. *Compos. Part A Appl. Sci. Manuf.* **2017**, *96*, 9–17. [CrossRef]

15. Ostapiuk, M.; Bieniaś, J.; Surowska, B. Analysis of the Bending and Failure of Fiber Metal Laminates Based on Glass and Carbon Fibers. *Sci. Eng. Compos. Mater.* **2018**, *25*, 1095–1106. [CrossRef]
16. Rajan, B.M.C.; Kumar, A.S. The Influence of the Thickness and Areal Density on the Mechanical Properties of Carbon Fibre Reinforced Aluminium Laminates (CARAL). *Trans. Indian Inst. Met.* **2018**, *71*, 2165–2171. [CrossRef]
17. Mamalis, D.; Obande, W.; Koutsos, V.; Blackford, J.R.; Brádaigh, C.M.Ó.; Ray, D. Novel Thermoplastic Fibre-Metal Laminates Manufactured by Vacuum Resin Infusion: The Effect of Surface Treatments on Interfacial Bonding. *Mater. Des.* **2019**, *162*, 331–344. [CrossRef]
18. Romli, N.K.; Rejab, M.R.M.; Bachtiar, D.; Siregar, J.; Rani, M.F.; Salleh, S.M.; Merzuki, M.N.M. Failure Behaviour of Aluminium/CFRP Laminates with Varying Fibre Orientation in Quasi-Static Indentation Test. *IOP Conf. Ser. Mater. Sci. Eng.* **2018**, *319*, 012029. [CrossRef]
19. Wang, S.; Cao, M.; Wang, G.; Cong, F.; Xue, H.; Meng, Q.; Uddin, M.; Ma, J. Effect of Graphene Nanoplatelets on Water Absorption and Impact Resistance of Fibre-Metal Laminates under Varying Environmental Conditions. *Compos. Struct.* **2022**, *281*, 114977. [CrossRef]
20. Pan, L.; Ali, A.; Wang, Y.; Zheng, Z.; Lv, Y. Characterization of Effects of Heat Treated Anodized Film on the Properties of Hygrothermally Aged AA5083-Based Fiber-Metal Laminates. *Compos. Struct.* **2017**, *167*, 112–122. [CrossRef]
21. Hamill, L.; Hofmann, D.C.; Nutt, S. Galvanic Corrosion and Mechanical Behavior of Fiber Metal Laminates of Metallic Glass and Carbon Fiber Composites. *Adv. Eng. Mater.* **2018**, *20*, 1700711. [CrossRef]
22. Viandier, A.; Cramer, J.; Stefaniak, D.; Schröder, D.; Krewer, U.; Hühne, C.; Sinapius, M. Degradation Analysis of Fibre-Metal Laminates under Service Conditions to Predict Their Durability. *UPB Sci. Bull. Ser. D Mech. Eng.* **2016**, *78*, 49–58.
23. Hu, Y.; Li, H.; Fu, X.; Zhang, X.; Tao, J.; Xu, J. Hygrothermal Characterization of Polyimide-Titanium-Based Fibre Metal Laminate. *Polym. Compos.* **2018**, *39*, 2819–2825. [CrossRef]
24. Hu, Y.; Liu, C.; Wang, C.; Fu, X.; Zhang, Y. Study on the Hygrothermal Aging Behavior and Diffusion Mechanism of Ti/CF/PMR Polyimide Composite Laminates. *Mater. Res. Express* **2020**, *7*, 076508. [CrossRef]
25. Zhang, Y.; Wei, J.; Liu, C.; Hu, Y.; She, F. Reduced Graphene Oxide Modified Ti/CFRP Structure-Function Integrated Laminates for Surface Joule Heating and Deicing. *Compos. Part A Appl. Sci. Manuf.* **2023**, *166*, 107377. [CrossRef]
26. Bellini, C.; Di Cocco, V.; Iacoviello, F.; Sorrentino, L. Numerical Model Development to Predict the Process-Induced Residual Stresses in Fibre Metal Laminates. *Forces Mech.* **2021**, *3*, 100017. [CrossRef]
27. *ASTM D7905/D7905M-19e1*; Standard Test Method for Determination of the Mode II Interlaminar Fracture Toughness of Unidirectional Fiber-Reinforced Polymer Matrix Composites. ASTM International: West Conshohocken, PA, USA, 2019. [CrossRef]
28. *ASTM D1141-98*; Standard Practice for Preparation of Substitute Ocean Water. ASTM International: West Conshohocken, PA, USA, 2021. [CrossRef]
29. *ASTM D5573-99*; Standard Practice for Classifying Failure Modes in Fiber-Reinforced-Plastic (FRP) Joints. ASTM International: West Conshohocken, PA, USA, 2019. [CrossRef]

Disclaimer/Publisher's Note: The statements, opinions and data contained in all publications are solely those of the individual author(s) and contributor(s) and not of MDPI and/or the editor(s). MDPI and/or the editor(s) disclaim responsibility for any injury to people or property resulting from any ideas, methods, instructions or products referred to in the content.

Article

Adhesive Thickness and Ageing Effects on the Mechanical Behaviour of Similar and Dissimilar Single Lap Joints Used in the Automotive Industry

Raffaele Ciardiello [1,2,*], Carlo Boursier Niutta [1] and Luca Goglio [1,2,*]

[1] Department of Mechanical and Aerospace Engineering, Politecnico di Torino (IT), 10129 Turin, Italy
[2] J-TECH@POLITO, Advanced Joining Technologies, Politecnico di Torino (IT), 10129 Turin, Italy
* Correspondence: raffaele.ciardiello@polito.it (R.C.); luca.goglio@polito.it (L.G.)

Abstract: The effects of the adhesive thickness and overlap of a polyurethane adhesive have been studied by using different substrate configurations. Single lap joint (SLJ) specimens have been tested with homologous substrates, carbon fibre-reinforced plastics and painted metal substrates. Furthermore, a configuration with dissimilar substrates has been included in the experimental campaign. Both types of these adhesive and substrates are used in the automotive industry. The bonding procedure has been carried out without a surface treatment in order to quantify the shear strength and stiffness when surface treatments are not used on the substrates, reproducing typical mass production conditions. Three different ageing cycles have been used to evaluate the effects on SLJ specimens. A finite element model that uses cohesive modelling has been built and optimised to assess the differences between the different adopted SLJ configurations.

Keywords: single lap joints; polyurethane adhesive; finite element model; cohesive model; ageing cycles

1. Introduction

In recent years, the use of composite materials is also spreading in mass production vehicles, especially in luxury cars where the requests for safety and comfort equipment are increasing drastically. Of course, this contributes negatively to the total weight of the vehicles, and thus, increases fuel consumption and vehicle emission. This tendency is promoting lightweight design even in the mass production of vehicles by replacing many components with reinforced plastics, such as crash absorbers, spoilers, side mouldings and roofs [1–5]. A drawback of the adoption of composite materials is that traditional mechanical fasteners (i.e., bolts, rivets and screws) are not easily adaptable, since holes are detrimental for the mechanical properties of composite materials and the presence of the holes in the composite should be properly designed or integrated to avoid premature failure [6]. For these reasons, adhesive bonding is preferred when composite materials have to be joined [7].

In recent years, the use of polyurethane adhesives as structural adhesives has been widely increased due to their resistance to dynamic load and their capacity to withstand larger deformations [7]. Nowadays, polyurethane adhesive is the most used adhesive, together with acrylics and epoxies, due to the increased resistance of the new proposed formulations [8].

Furthermore, polyurethane adhesives present a larger viscosity before curing that allows to assemble components or specimens with larger clearances, since the adhesive does not pour without proper tool or the application of pressure. Thus, once cured, these adhesives present sealant properties that are needed for materials that can present large clearances, such as thermoplastic composite components [9]. Although they present the aforementioned advantages, many substrate materials need to be pretreated in order to establish strong bonds with the adhesives. Pereira et al. [10] studied the effect of five different surface pretreatments (two different etching with sodium dichromate–sulphuric acid and caustic solution, Tucker's reagent, abrasive polishing and wiping solvent) on the mechanical properties of SLJ tests. The analysed pretreatments allowed to obtain surface roughness between 18.6 and 5.6 µm. The decrease in the surface roughness led to an increase in the shear strength. The etching with sodium dichromate–sulphuric acid led to a decrease of the surface roughness of 5.6 µm and the highest ultimate shear load (~7800 N). Prolongo et al. [11] studied the effects of mechanical abrasive cleaning, alkaline cleaning and two complex sulfuric acid-based solutions on the ultimate shear strength of SLJ prepared with aluminium alloy. They showed that the etch with sulfuric acid-based solution led to the highest increase in the shear strength. Stammen et al. [12] and Ciardiello et al. [13] proposed a methodology to use plasma treatment to adhesively bond polypropylene-based materials with a polyurethane adhesive. Stammen et al. [12] showed that by using air and pyrosil as gas carriers to plasma-treat the aluminium substrates, the maximum shear strength can be increased by at least 2.5 times than the adhesive joints prepared by simply degreasing the substrate surface. Ciardiello et al. [13] showed that by using nitrogen as a gas carrier, a polyurethane adhesive can be used to adhesively bond polypropylene substrates without pretreatment. Zain et al. [14] showed that a decrease in the contact angle of aluminium substrates can be achieved by using an alkaline etching, dipping in warm water followed by treating with silane solution. The tests carried out on adhesive joints prepared with polyurethane adhesive showed that the shear strength can be increased by at least five times by using the surface treatment compared to joints bonded with untreated specimens. Although surface treatments can increase the mechanical performances of adhesive joints, these treatments cannot be easily adopted for specific applications where the assemblies are made along the production line in mass production due to the time production. In fact, many adhesive producers are studying specific formulations that can be used without pretreatments.

In this work, a polyurethane adhesive is used to prepare adhesive joints made of carbon fibre-reinforced plastic (CFRP) specimens, painted steel specimens and the relative dissimilar joints (CFRP/Steel). An extensive experimental campaign was carried out to assess the mechanical properties of SLJ specimens made with similar and dissimilar substrates considering three different thicknesses and two overlaps. Three different ageing cycles [13,15] used in the automotive industries were adopted to assess the effect of extreme environmental conditions on the adhesive joints. A finite element model (FEM) that uses cohesive zone modelling has been calibrated to find the cohesive parameters based on the obtained experimental results.

2. Materials and Methods

2.1. Materials and Experimental Methodology

Steel and CFRP substrates are both adopted in the automotive industry. A DD11 steel was used in this work as a metal substrate. The substrates were painted with a cataphoresis cycle designed by the automotive industry for this material. The composite substrates were obtained from a laminate that was fabricated with a specific stacking sequence optimised for painting the composite laminates without aesthetic defects. For this reason, the composite laminate is stacked with four layers of prepreg provided by Impregnatex Compositi (Italy) with different tow sizes. The prepregs are balanced twill fabrics that present different areal weights and fibres within a tow. They are laminated with the following sequence: GG630T (12 K, 630 gsm), GG204T (3 K, 204 gsm), DYF15 180P (15 K, 180 gsm) and finally GG204T (3 K, 204 gsm). The mechanical properties of the two substrates are reported in Table 1. The thickness of the steel and CFRP substrates are, respectively, 2.2 mm and 1.3 mm. The substrates present a length of 100 mm and a width of 20 mm. The size of the substrates guarantees no plastic deformation in the substrates during SLJ tests.

Table 1. Mechanical properties of the specimens.

	Steel	CFRP	Polyurethane Adhesive
Tensile strength [MPa]	440	730	8.2
Young's Modulus [Mpa]	207×10^3	60×10^3	20
Maximum elongation [%]	24.0	1.2	114

The substrates were adhesively bon"ed b' using a bi-component polyurethane adhesive, Betaforce 2850L by Du Pont (Wilmington, DE, USA). The adhesive properties are also presented in Table 1 and have been assessed by using a Zwick Roell-Z005 (Ulm, Germany) testing machine in displacement control, 2 mm/min. On the other hand, the mechanical properties of the substrates are provided by the datasheet.

The SLJ tests were performed at a speed of 2 mm/min with an Instron (United States) 8801 testing machine. Tabs of different thicknesses were adopted to geometrically avoid the misalignment with the grips of the testing machine. Both similar and dissimilar SLJ were prepared and tested with only metal steel substrates (named here MS) and CFRP substrates (named CS) and their combinations. The SLJ tests were prepared with the following configurations CS-CS, MS-MS, and MS-CS. For each material pair configuration, three adhesive thicknesses were adopted, 1.5 mm (advised by the producer), 3 mm and 4 mm, and two different overlaps, 12 and 24 mm. The adoption of the larger thickness aims to understand the drop of the shear properties when this adhesive is used for larger clearances. At least three replications were carried out for each joint configuration.

SLJ specimens prepared by using a thickness of 1.5 mm and the two overlaps, 12 and 24 mm, for the three adopted configurations, CS-CS, MS-MS and CS-MS, have been aged with three different ageing cycles. These ageing cycles were also used by Ciardiello et al. [13,15] in previous works, and as reported in [16], they are used to study the effects of long exposure to extreme environmental conditions on the mechanical properties of adhesive joints. The following ageing cycles are carried out:

Cycle A: Exposure at 90 °C without control of the relative humidity (RH) for 500 h.
Cycle B: Exposure at 40 °C with RH set at 98% for 500 h.
Cycle C: Exposure at 80 °C without RH for 24 h; exposure at 40°C with RH set at 98% for 24 h; exposure at -40 °C for 24 h.

Aging cycles are carried out by using two different chambers (Votsch VT4020 and Votsch Heraeus HC0020). The aim of the ageing treatment is to assess whether cycles A, B or C can significantly affect the mechanical properties of the adhesive joints. As reported by Belingardi et al. [16], mechanical tests after ageing are always carried out in the automotive industry on adhesive joints since they can modify the mechanical behaviour of the adhesive joints in some cases. In the present work, the ageing cycles have been carried out since the surface of the substrates was not pretreated. Thus, a possible effect of the ageing on the surfaces had to be considered.

2.2. Finite Element Model

The mechanical models of SLJ with two different overlaps, 12 and 24 mm, have been simulated in the configuration with an adhesive thickness of 1.5 mm. The software used for the simulations is LS-Dyna. The numerical activity aimed to assess the mechanical properties of the SLJ and to study the drop of mechanical properties for the SLJ prepared with composite substrates which is illustrated in Section 3.1. The substrates have been modelled as four-nodes Belytshcko-Tsai shell elements. Eight-node solid elements are used to model the adhesive. The integration points of MS and CS substrates through the adhesive thickness are three and four respectively (as the number of layers of the composite laminate). The cohesive formulation of the adhesive solid element uses four integration points that are placed at the midpoint of the element surface. The substrates present a mesh of 2 mm that is refined to 1 mm approaching the overlap area of the SLJ specimens. Figure 1a displays the FEM model of the SLJ specimen. Figure 1b,c display the cohesive material model that has been adopted and the six points that have been used to normalise the force-displacement curve based on the experimental trends that have been observed. Geometrically, one extremity of the SLJ specimen was constrained and a motion law was applied to the other substrate. This motion law is set as an initial ramp followed by a constant value of the speed, as in [17]. The MS substrates are modelled as elastic, while the CS substrates are modelled with an orthotropic model, as in [1]. Due to the intrinsic nature of SLJ specimens, a mixed mode (mode I and mode II) failure is induced. For this reason, a cohesive material model that takes into account both failure modes is chosen, namely *MAT_GENERAL_COHESIVE [18,19]. The main peculiarities of this material model are shear and peel stress and their relative energy release rate that are defined with user-defined points; the shear stress and the energy release rate can be handled as design variables. Six normalised points on the normalised force-displacement experimental curve were chosen to replicate the mechanical behaviour of the adhesive joint: the origin, the load at 20%, the load at 80%, the load at 100%, the load at 50% of the drop after the maximum peak load was reached and the ultimate displacement.

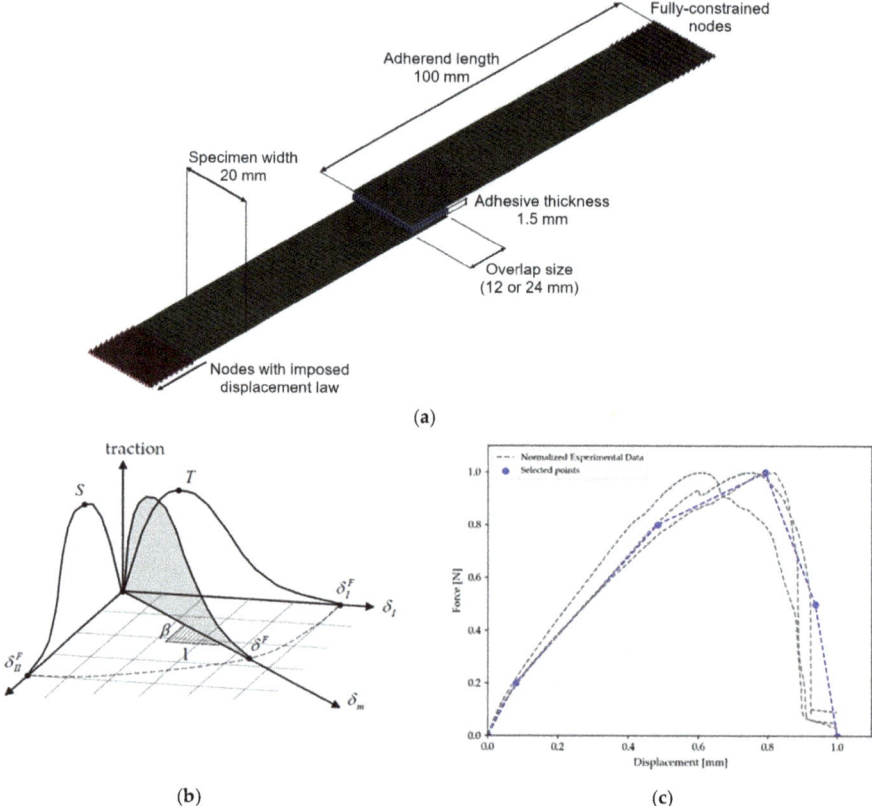

Figure 1. (a) SLJ in LS-Dyna environment (12 mm overlap); (b) material model; (c) normalised point chosen for the simulations.

Using a methodology already used by the authors [1] and the main experimental results obtained by Banea et al. [20] and Leal et al. [21] that found that the ratio between the energy release rate in mode II, G_{IIC}, is in the order of four times the energy release rate in mode I, G_{IC}, for these types of adhesives, the following strategy was adopted. The experimental force-displacement curves of the tests carried out on SLJ made with 12 mm overlap were adopted to impose the material behaviour. Together with the ratio between G_{IC} and G_{IIC}, the shear stress S (Figure 1b) and the G_{IIC} were considered to approximate the mechanical behaviour of the adhesive layer. On the other hand, the peel stress T is assumed to be double the maximum shear stress, in agreement with the Tresca criterion.

A surrogate model optimisation has been carried out on SLJ specimens produced with 12 mm overlap. The optimisation procedure can be summarised as reported below:

$$f(\underline{x})$$

$$\text{such that } g_1 = \frac{F_{max,num}(\underline{x})}{F_{max,\,exp}} - 1 \leq 0 \qquad (1)$$

$$\text{where } f = |EN_{num} - EN_{exp}| \qquad (2)$$

$$\underline{x} = [S, G_{IIc}]$$

Then, the same parameters were adopted to simulate the SLJ prepared with the 24 mm overlap in order to understand whether the mechanical model is able to replicate the experimental behaviour of the 24 mm overlap joints as well by using the same cohesive parameters. The optimisation model works on the experimental and numerical results related to the force (Equation (1)) and absorbed energy (Equation (2)). The optimisation aims to minimise the difference in absorbed energy of SLJ test between the experimental (E_{Nexp}) and numerical (E_{Nnum}) simulations. Furthermore, a constraint between the maximum experimental ($F_{max,exp}$) and numerical force ($F_{max,num}$) was adopted. Both maximum force and absorbed energy were surrogated using the approximation method of Kriging [22]. The surrogated surfaces were constructed by considering 20 samples that are stochastically disposed in the design domain in addition to the four corners samples. The optimisation is run with the algorithm COBYALA [23]. Table 2 reports the limit domains for the three different adhesive joint configurations MS-MS, CS-CS and MS-CS.

Table 2. Lower and upper limits for the three different configurations.

	Lower Bound	Upper Bound	Unit
	MS-MS		
S	8.0	13.0	MPa
G_{IIC}	16.0	26.0	N/mm
	CS-CS		
S	5.0	11.0	MPa
G_{IIC}	10.0	22.0	N/mm
	MS-CS		
S	5.5	11.5	MPa
G_{IIC}	10.0	22.0	N/mm

3. Results and Discussions

The results of the mechanical tests are reported in this section. Since different graphs are presented, the following nomenclature will be used in the present work: MS refers to the steel metal substrates, while CS refers to the CFRP substrates. Furthermore, the value of the overlap and thickness is reported in the nomenclature. For example, MS-MS_24_1.5 refers to the adhesive joints prepared with only steel substrate with an overlap length of 24 mm and a thickness of 1.5 mm. The letter A, B or C will be added at the end of the label for the adhesive joints exposed at the ageing cycles A, B or C, as illustrated in Section 2.

3.1. Single Lap Joint Tests

The load-displacement curves obtained by using SLJ tests are reported in Figure 2. In particular, Figure 2a reports the load-displacement curves of configuration that uses metal substrates for the two different overlap lengths and the three different adhesive thicknesses. Figure 2b,c report the curves obtained by using the same sizes of the SLJ (two overlap lengths and three thicknesses), but they are related to the configurations that use only composite substrates and metal-composite substrates respectively. Figure 2a–c show that the highest loads and displacements are obtained by testing SLJ prepared with metal substrates. On the other hand, the lowest loads and displacements are obtained by the tests of SLJ prepared with the composite substrates. Intermediate load values are obtained for the SLJ prepared with dissimilar, metal and composite, substrates for all the considered configurations. In general, Figure 2a–c illustrate that the highest loads are obtained for the configurations that use an adhesive thickness of 1.5 mm, while the SLJ prepared with 3 and 4 mm present very close maximum loads and similar displacements. However, the loads of the SLJ prepared with 3 mm are slightly higher than those prepared with 4 mm thickness.

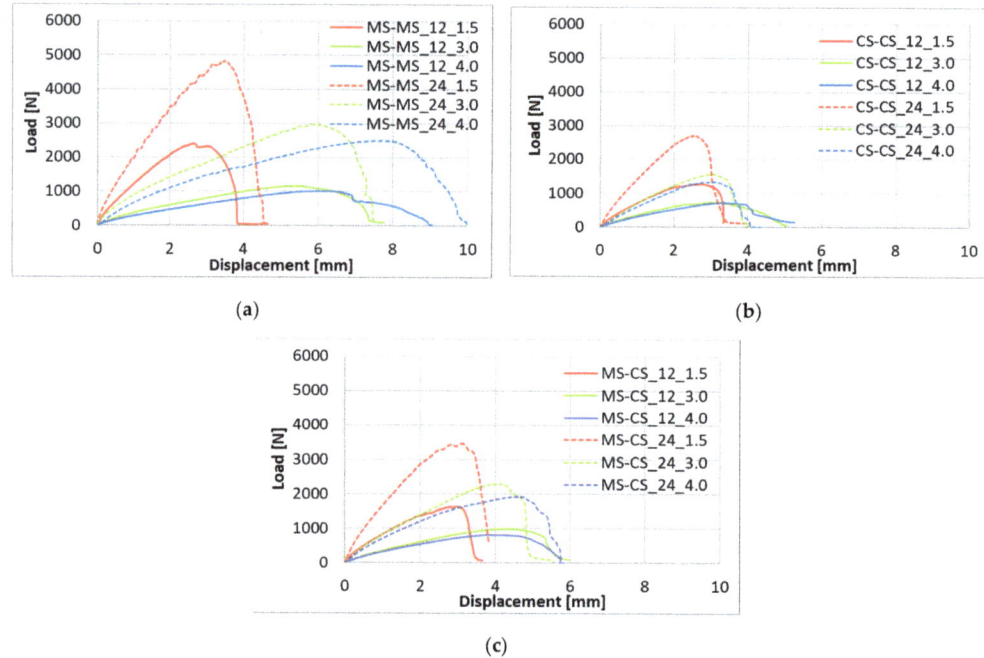

Figure 2. Load-displacement curves for the configurations MS-MS (**a**), CS-CS (**b**) and MS-CS (**c**).

Figure 3 reports the summary of the shear strengths and stiffnesses obtained by SLJ tests for all the configurations presented in Figure 2. The MS-MS configuration presents the highest strength as shown in Figure 2 for the three adhesive thicknesses. The CS-CS configuration presents the lowest shear strengths while the dissimilar configuration, MS-CS, presents intermediate values of shear strengths for all three adhesive thicknesses. The shear strength values shown in Figure 3a–c show that there is no significant difference between the two adopted overlaps, except for the configuration MS-CS for the SLJ specimens prepared with 3 and 4 mm adhesive thicknesses. However, the error bars show that there is no significant difference. Furthermore, Figure 3a,b reveal a drop in the shear strength by doubling the thickness of the SLJ specimens. The drop in shear strength doubling the adhesive thickness from 1.5 to 3 mm is about 46% for the SLJ prepared with 12 mm overlap. Of course, the 24 mm overlap has a similar trend since the values are quite superimposed. In contrast with this described drop, there is no significant difference between the shear strengths obtained by the SLJ tests prepared with 3 and 4 mm. Only the ultimate displacement is significantly changing between SLJ prepared with 3 and 4 mm adhesive thickness, as illustrated in Figure 1b,c. Although the two substrates present two different Young's moduli, the drop in the shear strength is not justified by the different stiffnesses of the two specimens. The reason for the drop can be found in the mixed adhesive and cohesive failure modes, which is shown in Section 3.2 at the interface of CS substrates. Da Silva et al. [24] reported a decrease of the maximum shear strength between 17% and 26% percent by increasing the thickness of the adhesive layers for three different epoxy adhesives from 0.2 to 1 mm (five times the initial thickness). On the other hand, the values obtained in the present work for the SLJ specimens prepared with 12 mm overlap report a decrease of about 45% for both MS-MS, CS-CS and MS-CS configurations. Thus, a decrease of 45% is obtained by increasing the adhesive layer thickness from 1.5 to 4 mm (2.7 times the initial thickness). Figure 3 reports the values of the stiffness as well. The stiffness has been computed by using the tendency line in the first linear trend of the load-displacement curve. Figure 3a reports that the joint stiffnesses vary significantly with the three different adopted configurations. This is due to the different stiffnesses of the substrate configuration and adhesive thicknesses, as shown in the graphs presented in Figure 3. Figure 3a–c illustrate that a reduction of stiffness is obtained for the configuration CS-CS and MS-CS compared to the configuration MS-MS for a specific adhesive thickness. The drop in stiffness of the CS-CS and MS-CS configuration compared to the baseline, MS-MS configuration, is reported in Figure 3d. Figure 3d shows that SLJ prepared with 12 and 24 mm overlaps present similar trends for all the SLJ prepared with different adhesive thicknesses. SLJ prepared with 1.5 mm thickness presents a drop close to 50% for the configuration CS-CS and 37% for the configuration MS-CS, which means higher stiffness for the configuration prepared with the dissimilar materials. Figure 3d illustrates a drop of 38% (CS-CS) and 22% (MS-CS) compared to the configuration MS-MS for the SLJ prepared with an adhesive thickness of 3 mm. Finally, a drop of 36% (CS-CS) and 12% (MS-CS) compared to the configuration MS-MS is shown for SLJ prepared with 4 mm thickness. This means that by increasing the adhesive thickness of SLJ, the stiffness decreases significantly as reported, while the drop of stiffness for higher thickness is lower compared to SLJ prepared with an adhesive thickness of 1.5 mm. A combination table of the reported values was built to show the configuration that presents similar results. Tables 3 and 4 present a summary of the configurations that show similar shear strength and stiffness, respectively. The values were considered similar when both shear strength and stiffness present a value that is at most ±10% from the considered value.

3.2. Fracture Surfaces

Figure 4a–c illustrate the representative failure surfaces obtained for the different configurations, a higher magnification of the CS-CS failure surface and an optical microscope image that shows the adhesive spots that have been depicted on CS specimens, respectively. Figure 4a shows that MS-MS configuration exhibits a fully cohesive failure (the SLJ samples fail through the adhesive) for both adhesive joints prepared with both 12 and 24 mm overlaps. The adhesive joints prepared with composite specimens and with dissimilar substrates macroscopically present a cohesive failure as well. However, a mixed adhesive/cohesive failure mode can be detected in Figure 4b,c by observing the small brighter spots using a microscope. In particular, Figure 4b presents some clearer areas that are zones of the visible surface of the CFRP substrate. As shown in Section 3.1, this led to a lower value of the shear strength. Figure 4c shows also that the adhesive failure spots are not always uniformly spread on the whole surface but are limited to the right part of the substrate in this specific case. However, the inspection of all the substrates involved in the experimental campaign showed that the spots can be present in different parts of the substrates. As a proof of this behaviour, CS-CS_12 mm and CS_CS_24 mm shown in Figure 4a present these spots on the left side and the top part. On the other hand, MS-CS_12 mm and MS-CS_24 mm show that these areas are on the right and top part of the specimens, respectively.

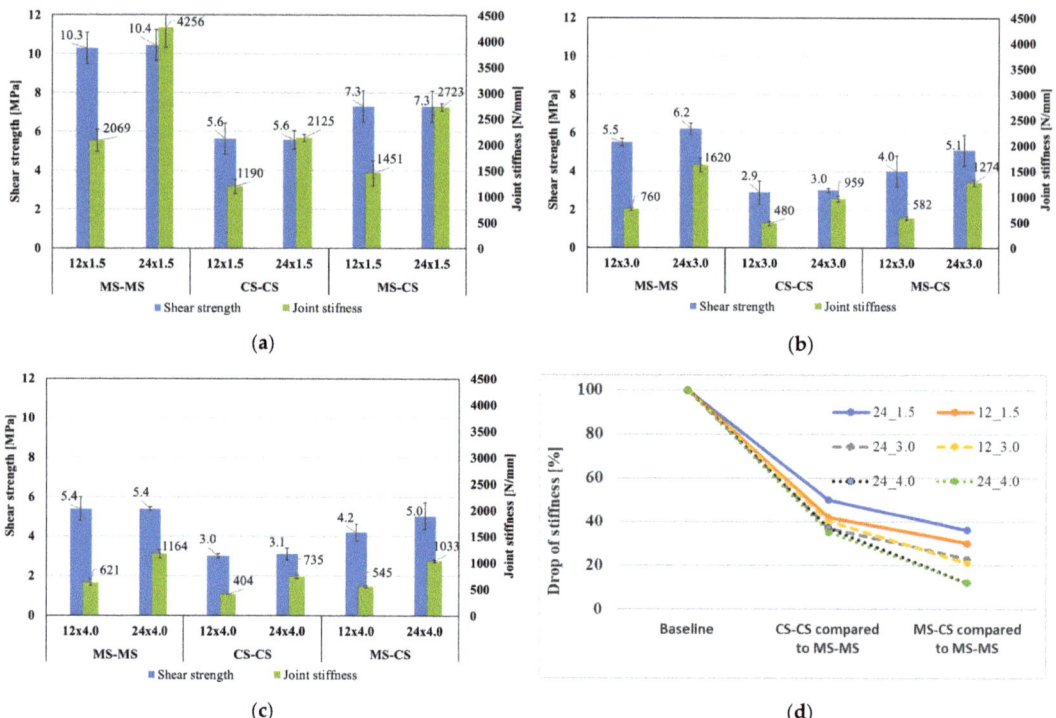

Figure 3. Shear strengths and stiffnesses for the configurations MS-MS (**a**), CS-CS (**b**) and MS-CS (**c**); decrease of stiffness for the different adhesive configurations (**d**).

Table 3. Configurations that present similar strength.

		12 mm	24 mm
1.5 mm	MS-MS	▼	▼
	CS-CS		●
	MS-CS		
3.0 mm	MS-MS	●	
	CS-CS	▤	▤
	MS-CS	▲	●
4.0 mm	MS-MS	●	●
	CS-CS	▤	▤
	MS-CS	▲	●

The same symbol indicates that the configurations present similar shear strength with 10% of variation included.

Table 4. Configurations that present similar stiffness.

		12 mm	24 mm
1.5 mm	MS-MS	● (blue circle)	
	CS-CS	▲ (red triangle)	● (blue circle)
	MS-CS		
3.0 mm	MS-MS	▬ (green square)	
	CS-CS		
	MS-CS	▽ (inv. triangle)	▲ (red triangle)
4.0 mm	MS-MS	▽ (inv. triangle)	▲ (red triangle)
	CS-CS		▬ (green square)
	MS-CS	▽ (inv. triangle)	

The same symbol indicates that the configurations present similar stiffness with 10% of variation included.

3.3. Numerical Model

In this section, the results of the optimisation process are used to simulate the mechanical behaviour of the adhesive joints. Figure 5a shows the result of the optimisation process obtained on the MS-MS specimens made with 12 mm overlap. Figure 5b displays the three experimental curves obtained for the MS-MS configuration (12 mm overlap) and the good agreement with the numerical curve. Figure 5c shows the experimental curves of SLJ tests carried out on MS-MS configuration that uses an overlap of 24 mm and the relative numerical curve that has been obtained by using the optimised cohesive parameters obtained from the 12 mm overlap configuration. Figure 5c shows that there is a very good agreement with the initial trend and with the maximum force. The model can also detect the change in the slope observed at 1000 N. On the other hand, the ultimate load is slightly underestimated. The table of Figure 5d reports the values obtained in numerical and experimental results for the MS-MS curve made with 24 mm overlap that shows that the absorbed energy is underestimated as well, mainly due to the lower ultimate displacement as well as the slightly lower maximum force that is obtained from the simulation compared to the experimental test.

Figure 6 shows the results obtained on the SLJ configuration made with composite substrates CS-CS. Figure 6a displays a table with the results of the optimisation procedure that has been used to simulate the SLJ made with 12 mm overlap. It is worth noticing that the value of the shear strength for the CS-CS configuration is reduced by 45%, similar to the experimental results. Figure 6b shows the comparison between experimental and numerical curves related to the CS-CS configuration made with 12 mm overlap. A very good agreement is found for the initial trend, the maximum load, slightly higher in the numerical simulation, and the final displacement. The results illustrated in Figure 6c display that a very good agreement is also found for 24 mm overlap configuration. Finally, the table reported in Figure 6d shows that the numerical and experimental results of the absorbed energy and maximum load, are very close to each other, confirming the good agreement of the FEM analysis.

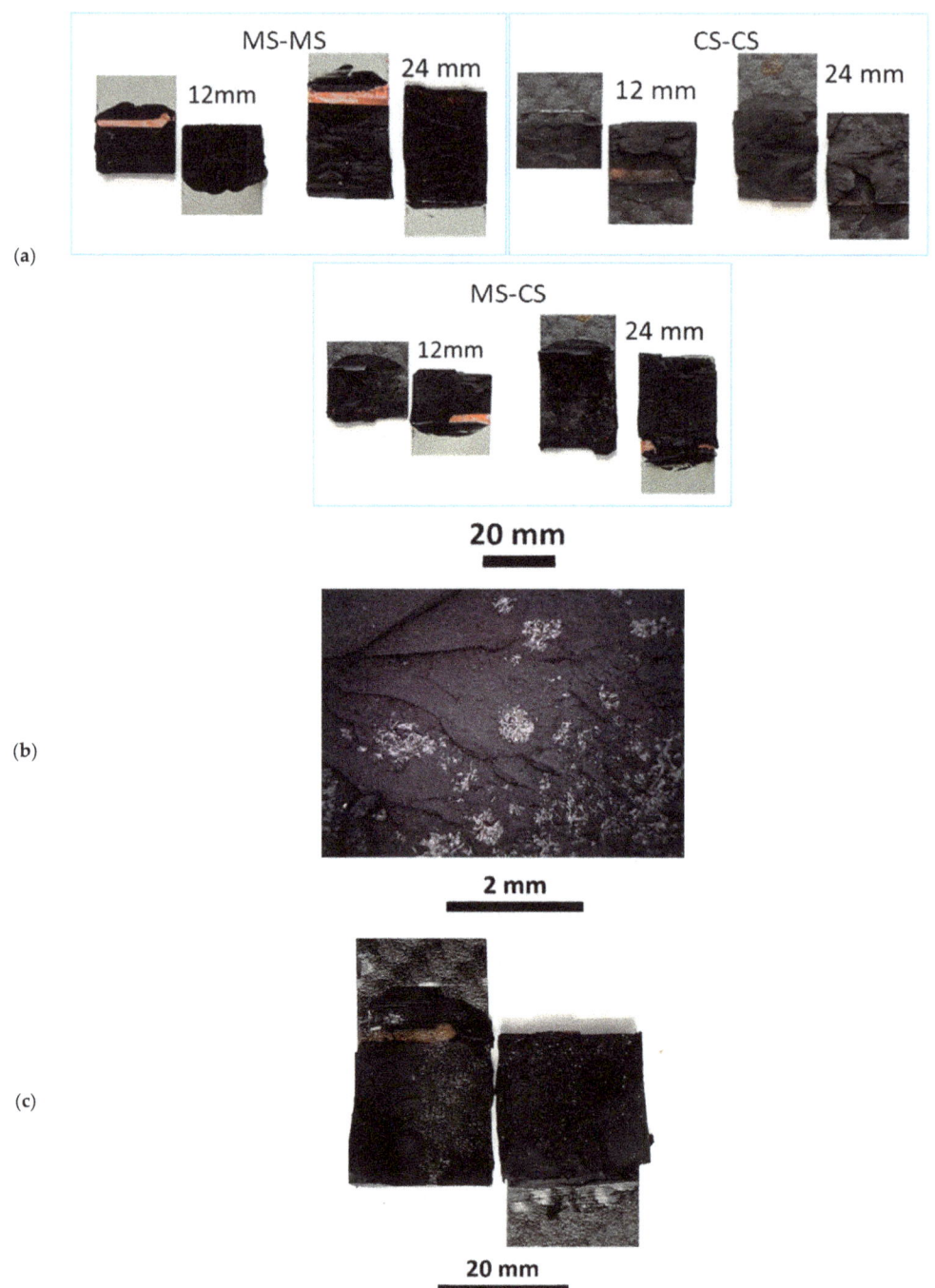

Figure 4. (a) Representative failure surfaces of the MS-MS, CS-CS and MS-CS SLJ configurations; (b) higher magnification of the CS-CS substrate; (c) representative adhesive failures of the CS substrate.

(a)

	Optimal results	Experimental results	Unit
S	10.0	-	MPa
G_{IIC}	23.9	-	N/mm
EN	5.75	6.13	J
F_{max}	2408	2408	N

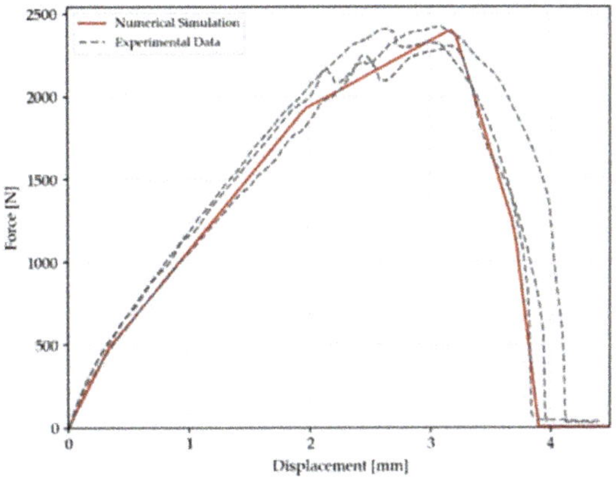

(b)

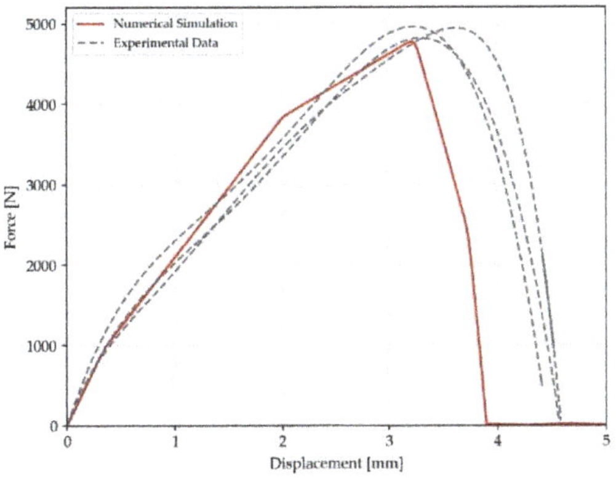

(c)

(d)

	Numerical results	Experimental results	Unit
EN	11.50	13.98	J
F_{max}	4772	4918	N

Figure 5. MS-MS configuration: (**a**) Optimised parameter of SLJ test with 12 mm overlap; (**b**) Comparison between experimental and numerical load-displacement curves (12 mm overlap); (**c**) Comparison between experimental and numerical load-displacement curves (24 mm overlap); (**d**) Comparison between experimental and numerical parameters (24 mm overlap).

Similar results have been obtained for the configuration MS-CS. The table reported in Figure 7a shows the optimised parameters for the configuration MS-CS. It is worth noting that the value of the shear stress is an intermediate value between the configurations MS-MS and CS-CS, closer to the CS-CS configuration due to the similar obtained failure surface. Figure 7b displays a good agreement between numerical and experimental curves for the configuration made with 12 mm overlap. A good agreement is still presented in Figure 7c for the CS-CS configuration made with 24 mm overlap. The use of the 12 mm overlap parameters led to correct numerical curves both for the initial trend, force and maximum displacements as well as for the SLJ made with 24 mm overlap. The maximum obtained load from numerical simulations and experiments led to a very similar value, while the absorbed energy is slightly lower for the numerical model, as shown in the Table of Figure 7d.

	Optimal results	Experimental results	Unit
S	5.5	-	MPa
G_{IIC}	12.0	-	N/mm
EN	2.92	2.92	J
F_{max}	1312	1266	N

(a)

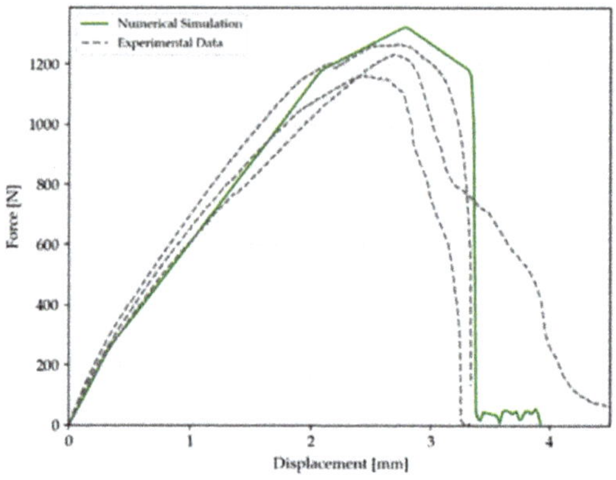

(b)

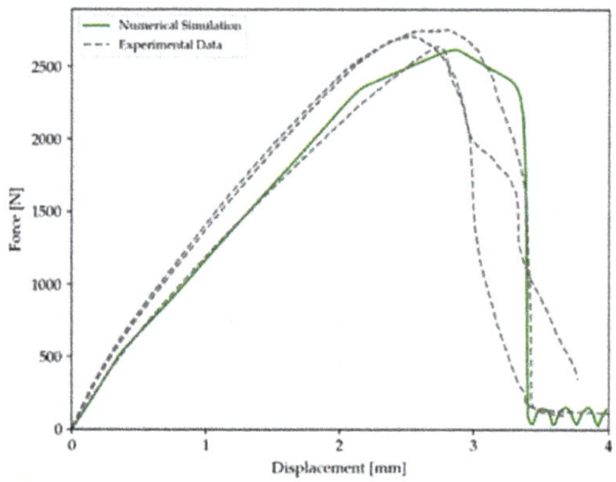

(c)

	Numerical results	Experimental results	Unit
EN	5.78	5.80	J
F_{max}	2616	2622	N

(d)

Figure 6. CS-CS configuration: (**a**) Optimised parameter of SLJ test with 12 mm overlap; (**b**) Comparison between experimental and numerical load-displacement curves (12 mm overlap); (**c**) Comparison between experimental and numerical load-displacement curves (24 mm overlap); (**d**) Comparison between experimental and numerical parameters (24 mm overlap).

The numerical modelling activity showed that the cohesive parameters that fits well with the SLJ test results for the specific configurations varies significantly among the three different configurations MS-MS, CS-CS and MS-CS. The numerical activity has been used to demonstrate that the mixed adhesive failure obtained with the SLJ specimens made with composite laminates led to a significant drop in the cohesive parameters. The drop of the cohesive parameters has been depicted for the SLJ configurations prepared with the composite materials, CS-CS and MS-CS, which have been compared to the MS-MS configuration since the failure surfaces exhibit a cohesive type. The drop of the parameters S, GIIC, EN and Fmax for the CS-CS configuration compared to MS-MS is, respectively, 45%, 50%, 50% and 45%, similar to the experimental mechanical result. On the other hand, the MS-CS configuration displays a drop of S, GIIC, EN and Fmax of 32%, 34%, 36% and 34%, respectively, compared to MS-MS configuration. Again, these results are very close to the experimental results. The experimental activity together with the numerical simulation proves that the mixed adhesive/cohesive failure led to lower values of the cohesive properties.

	Optimal results	Experimental results	Unit
S	6.8	-	MPa
G_{IIC}	15.75	-	N/mm
EN	3.69	3.70	J
F_{max}	1585	1645	N

(a)

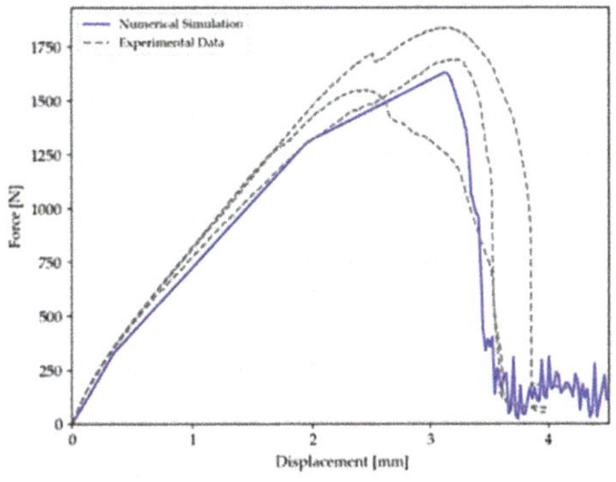

(b)

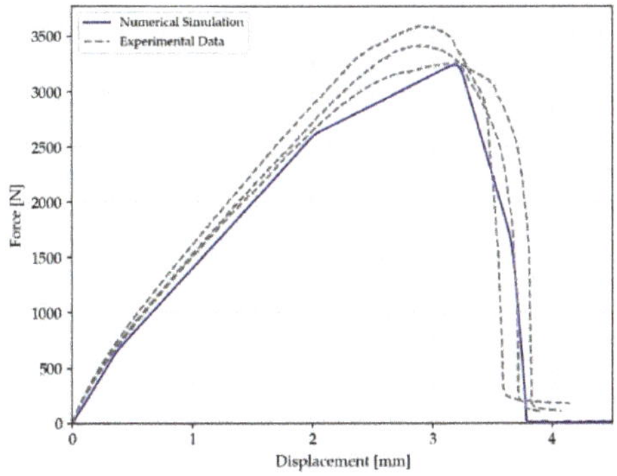

(c)

	Numerical results	Experimental results	Unit
EN	7.56	8.31	J
F_{max}	3236	3418	N

(d)

Figure 7. MS-CS configuration: (**a**) Optimised parameter of SLJ test with 12 mm overlap; (**b**) Comparison between experimental and numerical load-displacement curves (12 mm overlap); (**c**) Comparison between experimental and numerical load-displacement curves (24 mm overlap); (**d**) Comparison between experimental and numerical parameters (24 mm overlap).

3.4. Ageing Cycles

The results of the ageing cycles presented in Section 2.1 are reported in this section. As reported in Section 2.1, the SLJ tests have been carried out for three different ageing cycles on the configuration MS-CS prepared with both 12 and 24 mm overlaps and with a thickness of 1.5 mm, which is the advised datasheet thickness. Figure 8 presents the load-displacement curves of the MS-CS configuration with an overlap of 12 mm before and after the ageing cycle (a), the load-displacement curves for the configuration with an overlap of 24 mm before and after the ageing cycle (b) and the summary of the results with the shear strengths and stiffnesses (c). Figure 8a,b show that the load-displacement curve of SLJ prepared with both 12 and 24 mm overlaps present the same initial trend. For both graphs, the maximum load is slightly higher for the SLJ tests after the ageing A, which is the cycle that conditioned the SLJ at 90 °C. On the other hand, the maximum load illustrates that the SLJ tests after the ageing B led to slightly lower loads while cycle C is not influencing the load-displacement curve. Figure 8c reports the summary of the results related to the ageing cycles. The comparison between the unaged SLJ and SLJ aged with cycle A show that there is a slight increase in the shear strength although it is not significant by looking at the presented error bars. On the other hand, SLJ conditioned with cycle B exhibit a slight decrease in the shear strength. This slight drop was also reported by

Ciardiello et al. [13] for the same adhesive and an ageing cycle that presents high relative humidity. The drop is 25% for the SLJ prepared with 12 mm overlap and 15% for the SLJ prepared with 24 mm overlap. Finally, the stiffnesses reported in Figure 8c show that their values do not vary after the ageing cycles.

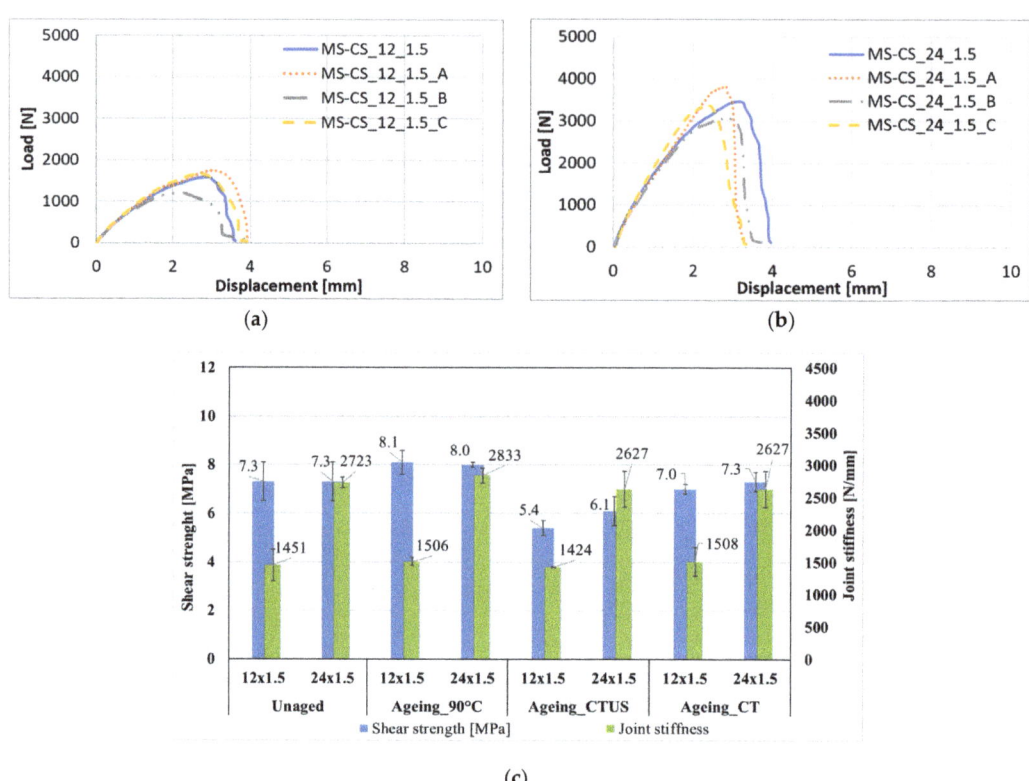

Figure 8. (a) Load-displacement with and without ageing, 12 mm overlap; (b) Load-displacement with and without ageing, 24 mm overlap; (c) Summary of the results.

4. Conclusions

The effects of 12 and 24 mm overlap together with three different adhesive thick-nesses 1.5, 3.0 and 4.0 mm have been studied with regard to different substrate configurations. These configurations are made only with painted steel and composite substrates. Furthermore, a dissimilar configuration with both steel and composite material has been studied. The SLJ specimens have been prepared by not using surface pretreatment in order to assess the mechanical performances of the adhesive joints without the treatment that is usually required when polyurethane adhesive is used. The following main conclusion has been reported and discussed in Section 3:

1. The SLJ showed that the adhesive joints with the composite substrate presented a mixed adhesive/cohesive failure surface that led to a detrimental effect on the mechanical properties of the SLJ. The shear strength presents a drop of 45% and 30%, respectively, for CS-CS and MS-CS configurations compared to MS-MS SLJ.
2. The analysis carried out by FEM modelling showed that the different failure surfaces led to different cohesive parameters. Thus, the drop in shear strength does not depend on the different substrate stiffnesses but is related to the different failure modes. The stiffnesses vary with the adhesive overlap and thickness.

3. Lap shear tests carried out on the aged adhesive joints showed that the adhesive presents a good mechanical response to both the hot cycle (ageing A) and mixed cycle (ageing C). On the other hand, the mechanical properties after the humid cycle (ageing B) are lower. This drop in shear strength is related to a decrease in the mechanical properties of the adhesive itself, since the failure surfaces do not change after ageing.

Author Contributions: Conceptualisation, R.C. and L.G.; methodology, R.C. and C.B.N.; software, C.B.N.; validation, R.C., C.B.N. and L.G.; investigation, R.C. and C.B.N.; data curation, R.C.; writing—original draft preparation, R.C.; writing—review and editing, C.B.N. and L.G.; supervision, L.G. All authors have read and agreed to the published version of the manuscript.

Funding: This research received no external funding.

Data Availability Statement: Data will be available on request.

Conflicts of Interest: The authors declare no conflict of interest.

References

1. Ciardiello, R.; Boursier Niutta, C.; Di Sciullo, F.; Goglio, L. Single-lap joints of similar and dissimilar adherends bonded with a polyurethane adhesive used in the automotive industry. *IOP Conf. Ser. Mater. Sci. Eng.* **2021**, *1038*, 12031. [CrossRef]
2. Ciampaglia, A.; Mastropietro, A.; De Gregorio, A.; Vaccarino, F.; Belingardi, G.; Busto, E. Artificial Intelligence for Damage Detection in Automotive Composite Parts: A Use Case. *SAE Int. J. Adv. Curr. Prac. Mob.* **2021**, *3*, 2936–2945.
3. Ciampaglia, A.; Santini, A.; Belingardi, G. Design and analysis of automotive lightweight materials suspension based on finite element analysis. *Proc. Inst. Mech. Eng. C J. Mech. Eng. Sci.* **2021**, *235*, 1501–1511. [CrossRef]
4. Ciampaglia, A.; Fiumarella, D.; Boursier Niutta, C.; Ciardiello, R.; Belingardi, G. Impact response of an origami-shaped composite crash box: Experimental analysis and numerical optimization. *Compos. Struct.* **2021**, *256*, 113093. [CrossRef]
5. Boursier Niutta, C.; Ciardiello, R.; Tridello, A. Experimental and Numerical Investigation of a Lattice Structure for Energy Absorption: Application to the Design of an Automotive Crash Absorber. *Polymers* **2022**, *14*, 1116. [CrossRef] [PubMed]
6. Chang, F.; Scott, R.A.; Springer, G.S. Failure of Composite Laminates Containing Pin Loaded Holes—Method of Solution. *J. Comp. Mater.* **1989**, *18*, 255–278. [CrossRef]
7. Somarathna, H.M.C.C.; Raman, S.N.; Mohotti, D.; Mutalib, A.A.; Badri, K.H. The use of polyurethane for structural and infrastructural engineering applications: A state-of-the-art review. *Constr. Build. Mater.* **2018**, *190*, 995–1014. [CrossRef]
8. Technical Industrial Report. In *Global Automotive Adhesives & Sealants Market-Size, Share, COVID-19 Impact & Forecasts up to 2028*; GlobeNewswire: Los Angeles, CA, USA, 2021.
9. Segura, D.M.; Nurse, A.D.; McCourt, A.; Phelps, R.; Segura, A. Chapter 3 Chemistry of Polyurethane Adhesives and Sealants. In *Handbook of Adhesives and Sealants*, 1st ed.; Cognard, P., Ed.; Elsevier Science Ltd.: Amsterdam, The Netherland, 2005; Volume 1, pp. 101–162.
10. Pereira, A.M.; Ferreira, J.M.; Antunes, F.V.; Bartolo, B.J. Analysis of manufacturing parameters on the shear strength of aluminium adhesive single-lap joints. *J. Mater. Process. Technol.* **2010**, *210*, 610–617. [CrossRef]
11. Prolongo, S.; Ureña, A. Effect of surface pre-treatment on the adhesive strength of epoxy–aluminium joints. *Int. J. Adhes. Adhes.* **2009**, *29*, 23–31. [CrossRef]
12. Stammen, E.; Dilger, K.; Böhm, S.; Hose, R. Surface Modification with Laser: Pretreatment of Aluminium Alloys for Adhesive Bonding. *Plasma Process. Polym.* **2007**, *4*, 39–43. [CrossRef]
13. Ciardiello, R.; D'Angelo, D.; Cagna, L.; Croce, A.; Paolino, D. Effects of plasma treatments of polypropylene adhesive joints used in the automotive industry. *Proc. Inst. Mech. Eng. C J. Mech. Eng. Sci.* **2022**, *236*, 6204–6218. [CrossRef]
14. Zain, N.M.; Ahmmad, S.H.; Ali, E.S. Green Polyurethane Adhesive Bonding of Aluminum: Effect of Surface Treatment. *Appl. Mech. Mater.* **2013**, *393*, 51–56. [CrossRef]
15. Ciardiello, R.; Belingardi, G.; Martorana, B.; Brunella, V. Effect of accelerated ageing cycles on the physical and mechanical properties of a reversible thermoplastic adhesive. *J. Adhes.* **2020**, *96*, 1003–1026. [CrossRef]
16. Belingardi, G.; Brunella, V.; Ciardiello, R.; Martorana, B. Thermoplastic Adhesive for Automotive Applications. In *Adhesives-Applications and Properties*, 1st ed.; Rudawska, A., Ed.; IntechOpen: London, UK, 2016; Volume 1, pp. 341–362.
17. Boursier Niutta, C.; Ciardiello, R.; Belingardi, G.; Scattina, A. Experimental and numerical analysis of a pristine and a nano-modified thermoplastic adhesive. In Proceedings of the PVP®Pressure & Vessels, Prague, Czech Republic, 15–20 July 2018.
18. Gleich, D.M.; Van Tooren, M.J.L.; Beukers, A. Analysis and evaluation of bondline thickness effects on failure load in adhesively bonded structures. *J. Adhes. Sci. Technol.* **2001**, *15*, 1091–1101. [CrossRef]
19. LSTC. *2019 LS-DYNA Keyword Manual Volume II*; Livermore Software Technology Corporation (LSTC): San Francisco, CA, USA, 2019.
20. Banea, M.D.; da Silva, L.F.M.; Campilho, R.D.G. The effect of adhesive thickness on the mechanical behavior of a structural polyurethane adhesive. *J. Adhes.* **2015**, *91*, 331–346. [CrossRef]

21. Leal, A.J.S.; Campilho, R.D.S.; Silva, S.; Silva, F.J.G.; Moreira, F.J.P. Comparison of different test configurations for the shear fracture toughness evaluation of a ductile adhesive. *Procedia Manuf.* **2019**, *38*, 940–947. [CrossRef]
22. Forrester, A.I.J.; Sóbester, A.; Keane, A.J. *Engineering Design via Surrogate Modelling: A Practical Guide*; John Wiley & Sons, Ltd.: Hoboken, NJ, USA, 2008; pp. 60–100.
23. Powell, M.J.D. *Advances in Optimization and Numerical Analysis*, 1st ed.; Kluwer Academic: Dordrecht, The Netherland, 1994; pp. 51–67.
24. da Silva, L.F.M.; Rodrigues, T.N.S.S.; Figueiredo, M.A.V.; de Moura, M.F.S.F.; Chousal, J.A.G. Effect of Adhesive Type and Thickness on the Lap Shear Strength. *J. Adhes.* **2006**, *82*, 1091–1115. [CrossRef]

Disclaimer/Publisher's Note: The statements, opinions and data contained in all publications are solely those of the individual author(s) and contributor(s) and not of MDPI and/or the editor(s). MDPI and/or the editor(s) disclaim responsibility for any injury to people or property resulting from any ideas, methods, instructions or products referred to in the content.

Article

Coupled Excitation Strategy for Crack Initiation at the Adhesive Interface of Large-Sized Ultra-Thin Chips

Tao Wu *, Xin Chen, Shiju Wen, Fangsong Liu and Shengping Li

College of Engineering, Shantou University, Shantou 515063, China; 21xchen@stu.edu.cn (X.C.); 22sjwen@stu.edu.cn (S.W.); 22fsliu@stu.edu.cn (F.L.); spli@stu.edu.cn (S.L.)
* Correspondence: taowu@stu.edu.cn

Citation: Wu, T.; Chen, X.; Wen, S.; Liu, F.; Li, S. Coupled Excitation Strategy for Crack Initiation at the Adhesive Interface of Large-Sized Ultra-Thin Chips. *Processes* **2023**, *11*, 1637. https://doi.org/10.3390/pr11061637

Academic Editor: Raul D. S. G. Campilho

Received: 25 April 2023
Revised: 19 May 2023
Accepted: 22 May 2023
Published: 26 May 2023

Copyright: © 2023 by the authors. Licensee MDPI, Basel, Switzerland. This article is an open access article distributed under the terms and conditions of the Creative Commons Attribution (CC BY) license (https:// creativecommons.org/licenses/by/ 4.0/).

Abstract: The initial excitation of interface crack of large-size ultra-thin chips is one of the most complicated technical challenges. To address this issue, the reversible fracture characteristics of a silicon-based chip (chip size: 1.025 mm × 0.4 mm × 0.15 mm) adhesive layer interface was examined by scanning electron microscope (SEM) tests, and the characteristics of a cohesive zone model (CZM) unit were obtained through peel testing. The fitting curve of the elastic bilinear model was in high agreement with the experimental data, with a correlation coefficient of 0.98. The maximum energy release rate required for stripping was G_C = 10.3567 N/m. Subsequently, a cohesive mechanical model of large-size ultra-thin chip peeling was established, and the mechanical characteristics of crack initial excitation were analyzed. The findings revealed that the larger deflection peeling angle in the peeling process resulted in a smaller peeling force and energy release rate (ERR), which made the initial crack formation difficult. To mitigate this, a coupling control method of structure and force surface was proposed. In this method, through structural coupling, the change in chip deflection was greatly reduced through the surface coupling force, and the peeling angle was greatly improved. It changed the local stiffness of the laminated structure, made the action point of fracture force migrate from the center of the chip to near the edge of the chip, the peeling angle was increased, and the energy release rate was locally improved. Finally, combined with mechanical analysis and numerical simulation of the peeling process, the mechanical characteristics of peeling were analyzed in detail. The results indicated that during the initial crack germination process, the ERR of the peel interface is significantly increased, the maximum stress value borne by the chip is significantly reduced, and the peel safety and reliability are greatly improved.

Keywords: cohesion zone model; peeling angle; non-destructive chip peeling of large-size ultra-thin chip; coupling excitation of crack initiation

1. Introduction

A chip is an integrated circuit manufactured on the surface of a semiconductor chip, which plays an important role in physics, military applications, science and technology, the chemical industry, medicine, and other fields [1–4]. Chip peeling and transfer are widely used for the packaging and manufacturing of high-performance devices, such as CPU, DSP, LED, RFID, and MEMERY, and represent the key to enhancing the performance and reliability of electronic devices. At present, chips are constantly developing in the direction of thinness, high performance, and low power consumption [5–7]. The industry IC chip thickness has been reduced from 120 μm to less than 40 μm, and those used in laboratory applications have reached the level of 10~20 μm. The chip transfer adopts a chip-based film laminate structure, which is characterized by a 5–10 μm adhesive layer between the chip and the base film (made of polyvinyl chloride) [8]. Chip peeling the separation of the interface to realize the separation of the chip and the substrate. As the chip size becomes larger and the thickness becomes thinner, the chip is easily bent and deformed together with the base film during the peeling process, resulting in chip damage or fracture. This problem is common in the field of high-end electronic packaging and testing equipment, and it is a technical bottleneck restricting the development of related processes, which has resulted in extensive research. Delamination is also a common problem in many other domains such as the aerospace industry when dealing with strain measurement instrumentation or composite delamination, even in rotating parts [9].

The combination of mechanical analysis, finite element simulation, and experimental analysis is a common method to analyze similar problems. Using finite element analysis and building a cohesive zone model, Zou et al. clarified the technical feasibility of an artificial barrier to control crack height [10]. Neves and Khan et al. used the finite element method and the cohesive zone model to study the adhesive joint. The simulation proved to have a good correlation with the experimental results, making it possible to model and design a hybrid model of the adhesive joint [11,12]. Hong et al. proposed the substrate dynamic release layer (DRL) chip structure and established the finite element and bond zone model (CZM) to study the evolution of ultra-thin chip peeling. The results of the model showed that a longer laser irradiation time can produce larger maximum vapor pressure and chip transfer speed [13]. Hong et al. also put forward a spring-buffered chip stripping technology, which can ensure chip stripping and inhibit chip cracking under large ejector pin force [14]. Using linear fracture theory, Yin, Peng, and Liu established the fracture mechanical model of interfacial peeling under impact load and calculated the stress intensity factor at the crack tip, the energy release rate during expansion, and the influencing factors [15,16]. The inaccuracy of the linear analysis in the situation, and the dual-standard competitive failure criterion of chip fragmentation and bending damage, as well as the flexible needle, multi-needle scheme, vacuum, and flexible suction control strategies were proposed. Cheng of Northwestern University in the United States analyzed the transfer process of stamp/ink-type chips, revealing the phenomenon of an initial fracture value at low speeds [17]. Jeon proposed a blowing peeling scheme, where the chip is peeled off by combining an air blowing force under the substrate ($\leq$90 kpa) and the appropriate speed of the upper suction nozzle ($\leq$50 mm/min) [18]. Behler implemented a step-difference needle to realize peeling of large-size ultra-thin memory chips through the coupling of structure and rate [7]. It can be found that, generally, the simplified beam model and the macroscopic fracture mechanics linear small deformation theory are used to analyze the peeling process. As the size of the chip becomes larger, the thickness becomes thinner, the flexibility of the chip increases, and the deflection during the peeling process becomes larger; the linear small deformation theory still cannot correctly analyze this process, and thus a new mechanical model is needed.

The SEM test in the next section shows that the chip peeling process is essentially the crack tip extension process between the chip and the adhesive layer interface, which can be analyzed by the adhesive interface separation mechanical model. Williams studied the stress and pre-strain in the process of interfacial fracture and proposed a general criterion of initial peeling from the perspective of energy [19]. Molinari focused on the introduction of the cohesive zone model (CZM) in the crack tip region and identified that the size of the cohesion zone is related to the peeling angle [20,21]. Kovalchick and Yang explained that the ERR and peeling rate showed a power-law relationship [22]. Other relevant studies focused on the peeling under rigid substrates, which lay a theoretical foundation for the research of peeling excitation under large deflection and large deformation of flexible substrates [23,24].

This paper is a new attempt to apply the adhesive peel model and CZM unit to chip peeling analysis. In this paper, an adhesive peeling model was established to analyze the initial peeling process. Combined with simulations under a large deformation, the impact of changes in the peeling angle on initial peeling was revealed. On this basis, structural coupling and force surface coupling strategies were proposed, dividing the chip peeling process into two stages; the initial peeling stage and the crack propagation stage, to effectively improve the peeling speed and reduce the stress–strain state of chips.

In this paper, our research is organized as follows:

Section 1 of the research discusses the importance of chip delamination and the proposed approach for analyzing the process. It focuses on the use of adhesive delamination models and CZM elements, along with structural and force-coupling strategies to divide the chip removal process into initial detachment and crack expansion stages.

Section 2 establishes a cohesive fracture mechanics model and analyzes the cohesive force unit model, determining that a bilinear-CZM model is appropriate for describing chip detachment behavior. Furthermore, a mechanical analysis of chip bonding and delamination is conducted to establish the relationship between lifting rate and the energy required for delamination.

Section 3 details the simulation experiments based on the findings from Section 2 and discusses the effects of factors such as chip size and lifting speed on crack initiation and stress during chip removal.

Section 4 proposes a coupling fracture initiation model for adhesive interface cracks, with a focus on the mechanical characteristics of this model.

In Section 5, we simulate the characteristics of crack initiation in large-sized ultra-thin chips and discuss how the delamination angle condition can be improved during crack initiation to reduce the local stress of chip removal.

Finally, Section 6 provides conclusions and outlines plans for further research.

2. Mechanical Modeling and Fracture Analysis of Chip Adhesion Interface

2.1. Adhesion and Peeling Model

The traditional research on chip peeling mainly focuses on the chip–adhesive–substrate structure, using linear small deformation theory and macroscopic fracture mechanics methods to analyze the adhesive layer fracture [25–28].

When the chip size is small and the thickness is large, and the chip deformation can be almost negligible, this model is somewhat representative and can reflect the fracture problem at the interface to a certain extent. As the chip size increases and the thickness decreases, the flexibility of the chip increases, and large deflection deformation occurs during the peeling process. To understand the essential characteristics of adhesive fracture, SEM experiments were conducted. Figure 1 shows the results after multiple round of adhesion and peeling. In Figure 1a, the surface of the epitaxial sheet was smooth with almost no adhesive residue, and the entire peeling process was reversible. On the other hand, Figure 1b presents a clear filament phenomenon at the crack tip region, which undergoes initiation, propagation, and detachment, and the crack tip region presented a common extension distribution.

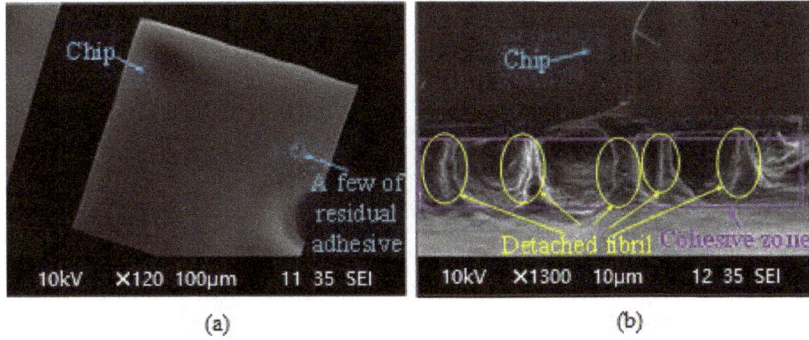

Figure 1. Electron microscope scanning test. (a) Chip epitaxy after multiple peeling; (b) interface separation crack tip.

The results show that the chip peeling process is consistent with the extension process of the crack tip between the chip bonding layer interface, which can be simulated using the CZM model. In reality, the chip peeling is shown in Figure 2, where the central hole on the upper surface of the needle cover is vacuum-absorbed to fix the chip, and the outer ring hole is vacuum-absorbed to fix the blue film around the chip. During peeling, the chip is pushed up by the needle, and the membrane deformation produces a peel force. Once the peeling force at the junction of the chip and the adhesive layer exceeds the adhesive force, the initial fracture begins to occur.

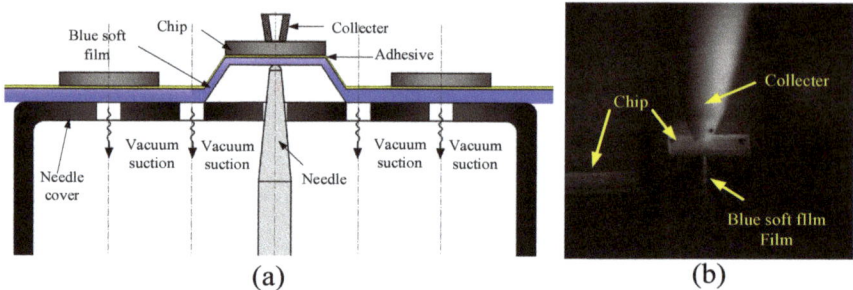

Figure 2. Traditional chip peeling method. (**a**) Actual peeling process; (**b**) schematic diagram.

The edge of the suction hole closest to the chip is regarded as the chip peeling fixed support boundary for chip peeling, and the peeling model is illustrated in Figure 3a. During peeling, the chip exhibits symmetrical deformation with the needle tip serving as the fulcrum, as shown in the simplified model depicted in Figure 3b.

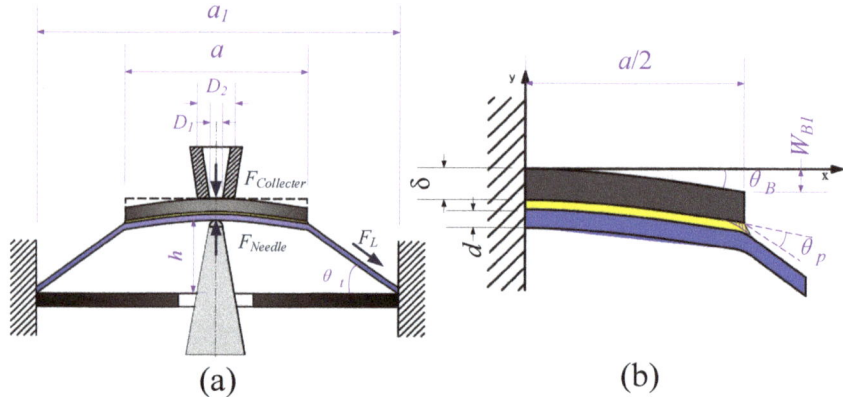

Figure 3. Chip Interface Crack Germination Model. (**a**) Simplified model; (**b**) cantilever model.

The figure illustrates various parameters and variables: a_1 is the length of the fixed support boundary, a is the chip length, D_1 is the inner diameter of the contact surface between the needle and the chip, δ is the thickness of the chip, D_2 is the outer diameter of the contact surface between the needle and the chip, d is the thickness of the blue film, and W_{B1} is the largest chip vertical deflection, θ_t is the deflection angle of the soft film, θ_B is the maximum turning angle of the chip, θ_P is the peeling angle, h is the lifting height of the chip, F_{needle} is the force exerted by the ejector pin, $F_{collector}$ represents the downward pressure applied by the collector, and F_L is the blue film pulling force.

2.2. Test for CZM UNIT

Adhesion and peeling are realized through the initiation and extension of the crack tip area. In order to describe the adhesive characteristics more accurately, a CZM cohesion unit was established through experimental measurements to determine the relationship between deformation and peeling force. Considering that the crack during initiation is mainly an opening mode crack, the probe test method was adopted in the test [29].

The instrument used in the test is the multifunctional push–pull force meter shown in Figure 4. The column is attached to an *XY* micro-movement translation stage (resolution: 0.002 mm). The Z-direction platform module (resolution: 0.001 mm, effective stroke 75 mm) is integrated with a load cell (accuracy: 0.001 g) at the end. The device samples through sensors and observes the experimental process using a microscope. The comprehensive accuracy error of this platform is within 0.01%, with a minimum speed of 0.01 mm/s and an average resolution of 2 mm.

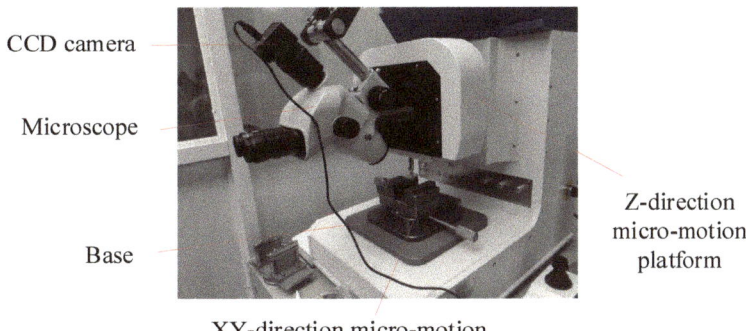

Figure 4. Test instruments.

The experiment was set up as shown in Figure 5. First, clean the glass plate with distilled water, wipe it with an acetone solution and keep it dry until use. Next, cut off part of the blue film (surface protective film, spv-224srb, Nitto, Osaka, Japan) and use ethyl cyanoacrylate to dry it quickly. The adhesive adheres the adhesive surface of the blue film to the substrate. After standing for 3 h, clamp the glass plate on the micro-motion platform of the push–pull tester. Finally, fix the single-sided polished silicon wafer (nanoscale, thickness 650 ± 20 μm, crystalline phase P<100>, growth method CZ, resistivity 0–20 Ωcm) on the push–pull knife that pushes the Z-axis. The test steps are as follows:

(1) Move the Z-axis downward at a constant speed of 0.05 mm/s until the square piece on the broach contacts the blue film and maintain the contact pressure at 10 N;
(2) Set the holding time to 120 s;
(3) Control the Z axis to lift at a speed of 1 mm/s;
(4) Repeat the above steps cyclically to obtain multiple sets of test data.

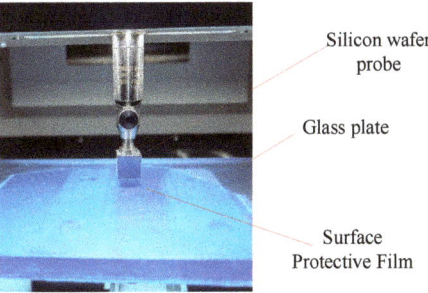

Figure 5. Test specimen and process.

The experiments were conducted at a test environment temperature of 21.2 °C, and the Levenberg–Marquardt iterative algorithm was utilized to fit the stress–displacement curve data acquired during the separation of the blue film from the polished silicon wafer. The fitting curves of the three CZMs are shown in Figure 6. The results show that the fitting curve of the elastic bilinear model (Bilinear-CZM) had the highest agreement with the experimental data, and the correlation coefficient was as high as 0.98. Therefore, the bilinear model (Bilinear-CZM) can be used to describe the chip peeling process. The cohesion model parameters obtained are provided in Table 1.

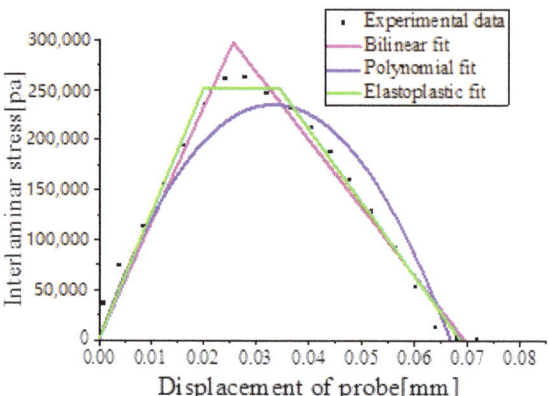

Figure 6. Fitting curves of CZM and experimental data.

Table 1. Fitting parameters of CZM.

CZM Model	σ_{max}/Pa	$\delta 0$/mm	$\delta 1$/mm	$\delta 2$/mm	δc/mm	R^2
Bilinear	297,821.664	0.02576			0.06955	0.97783
Polynomial	235,543.553	0.03341			0.06683	0.88844
Elastoplastic	251,316.990		0.02014	0.03464	0.06877	0.97076

2.3. Mechanical Criterion of Crack Initiation at Adhesive Interface

The mechanical model of chip peeling is similar to Kendall's soft tape peeling model [30,31], as presented in Figure 7. The experimental model uses rigid substrate soft film peeling mechanics. Among them, P_f is the peeling force, θ_0 is the peeling angle, Δc is the length of the cohesive zone, d is the thickness of the soft film, G_c is the unit surface peeling energy, and E_{film} is the elastic modulus of the soft film, E_{chip} is the elastic modulus of the chip. Assuming that the surface bonding energy of the bonding interface per unit area is U_S, when the peeling rate is constant, the adhesive layer absorbs the energy generated by the peeling force W_p and converts it into the cohesive energy of the adhesive layer in the CZM cohesive zone $\Delta \Pi$. As the interface separates, the solid–solid interface between the chip and the adhesive layer is transformed into the solid–gas interface between the chip air and the colloid air, and energy is accumulated within the adhesive layer $\Delta \Pi$. Overcoming the surface energy of bonding interfaces in $\Delta \Pi$, the effective work performed by ΔU_S is called the unit energy release rate G, and the dissipated energy is converted into internal energy, thermal energy, and other energies of the material.

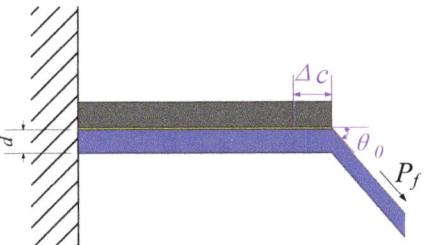

Figure 7. Rigid substrate–soft film peeling mechanical test model.

The relationship between the peeling energy release rate, peeling force, and peeling angle is:

$$G = \frac{P_f^2}{2E_{film}b^2d} + \frac{P_f}{b}(1 - \cos\theta_0) \quad (1)$$

In Equation (1), the first term on the right side of the equation is the deformation energy of the soft film, and the second term is the work done by the peeling force. Assuming that the unit surface peeling energy is G_c, the criterion for crack initiation by interfacial peeling is:

$$G = G_c \quad (2)$$

With a constant G_c, there is a negative correlation between the peeling angle θ_0 and the peeling force P_f. Essentially, as the peeling angle θ_0 decreases, the peeling force P_f increases, and as the peeling angle θ_0 increases, the peeling force P_f reduces. In addition, there is a positive correlation between the energy release rate G and the peeling angle θ_0, meaning that as the peeling angle θ_0 decreases, the G also decreases and as the peeling angle θ_0 increases, the G also increases.

When the chip is peeling, the peeling force P_f is affected by the geometric relationship between the peeling angle θ_p and the bending deflection of the chip. As shown in Figure 3b, let b be the adhesive width, and the relationship is as follows:

$$\begin{cases} F_L = E_{film}db\left(\dfrac{1}{\cos\theta_t} - 1\right) \\ \theta_t = \theta_{B1} + \theta_p = \dfrac{3\sin\theta_t F_L a^2}{2E_{chip}b\delta_{chip}^3} + \theta_p \end{cases} \quad (3)$$

From Equations (1) and (3), it can be seen that the peeling energy release rate G in the process of chip peeling is only related to the peeling angle θ_p and the geometric relationship, and the relationship is as follows:

$$G = \frac{1}{2}E_{film}d\left(\frac{1}{\cos\theta_t} - 1\right)^2 + E_{film}d\left(\frac{1}{\cos\theta_t} - 1\right)(1 - \cos\theta_p) \quad (4)$$

Therefore, it can be concluded that the peeling energy release rate G is primarily dependent on the peeling angle θ_p and the geometry during crack initiation. The peeling angle θ_p can therefore serve as an important parameter of the peeling energy release rate during the interface separation process. In addition, the peeling angle θ_p is easy to observe and measure during the peeling process, making it a key parameter in the analysis of chip peeling.

2.4. Mechanical Analysis of Crack Initiation Process

If the chip is thicker and more rigid, the deformation during peeling is small. As shown in Figure 8, the relationship between the peeling angle and the length of the chip and the fixed support boundary is as follows:

$$\tan\theta_p = \tan\theta_t = \frac{2h}{a_1 - a} \tag{5}$$

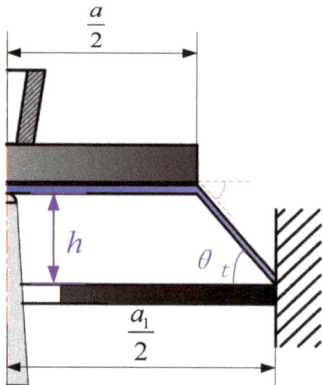

Figure 8. Diagram of the relationship between stripping Angle and chip and fixed support boundary size.

When the chip becomes larger and thinner, the flexibility becomes stronger, and the deformation is larger during peeling. As shown in Figure 9, the actual peeling angle θ_p is computed as the difference between the deflection angle θ_t of the soft film and the maximum rotation angle θ_B of the chip:

$$\theta_p = \theta_t - \theta_B \approx \arctan(\frac{2(h - w_{B1})}{a_1 - a}) \tag{6}$$

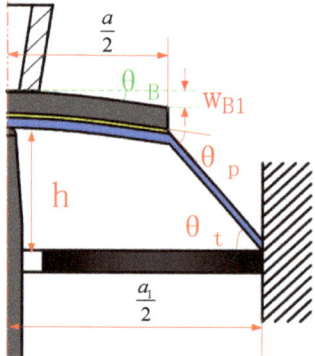

Figure 9. Mechanical model of crack initiation.

A comparison between the two indicates that the peel angle decreases as the bending deflection of the chip increases. This occurs because the film tension is harder to convert into an effective peel force, leading to a decreased energy release rate during peeling and a reduced likelihood of interface crack initiation.

Furthermore, Kovalchick has discovered that the crack initiation and propagation are also related to the peeling rate. The energy G_E required for peeling varies with the interface peeling rate v_c, and the two show a power-law relationship as shown in Equation (7):

$$G_E(v) = G_0 \left(\frac{v}{v_0}\right)^\varepsilon \tag{7}$$

where v_0 is the peeling rate, G_0 is the energy required for peeling, v is the actual peeling rate, and ε is a power-law constant (determined by the properties of the chip and the adhesive material.

The adhesion peeling initial crack is shown in Figure 10. The length of the cohesive zone is Δc and the length is about 10 μm. If the peeling angle remains approximately unchanged during the crack initiation, the relationship between needle speed v_{needle} and the interface peeling rate v_c approximately satisfies Equation (8).

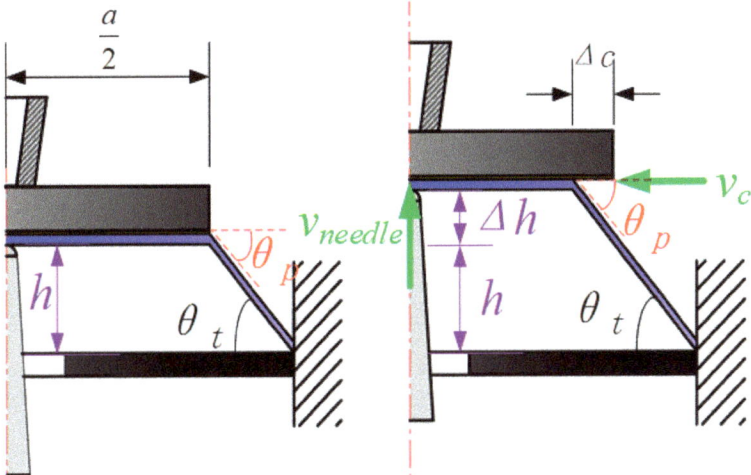

Figure 10. The process of crack initiation at the rigid die bond-peel interface during peeling.

$$v_c = \frac{v_{needle}}{\tan \theta_p} = \frac{v_{needle}}{\tan \theta_t} \tag{8}$$

Equation (8) shows that the required peeling energy is positively related to the needle lifting rate, which is deduced as Equation (9).

$$G_c = G_E(v) \propto v_c^n = G_E(v) \propto v_{needle}^n \tag{9}$$

where n is a constant related to base film material, substrate material, size specification, type of adhesive, etc.

When the chip becomes larger and thinner, its flexibility increases. If the needle exerts force as shown in Figure 11, the bending deformation of the chip increases, thus compressing the peeling angle θ_p in the mechanical model to θ_{p1}. Consequently, the interface peeling rate compared to the case of a rigid chip v_c is raised to v_{c1}, satisfying Equation (10).

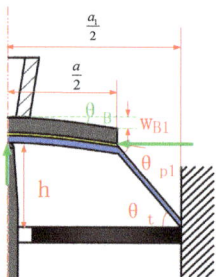

Figure 11. Initial crack initiation at the bond–peel interface of flexible chips during peeling.

$$v_{c1} = \frac{v_{needle}}{\tan \theta_{p1}} = \frac{v_{needle}}{\tan(\theta_{t1} - \theta_B)} > v_c \qquad (10)$$

Equation (10) shows that the required peel energy is increased compared to rigid chips, which is expressed as:

$$G_c = G_E(v) \propto v_{c1}^n > G_E(v) \propto v_c^n \qquad (11)$$

To enhance the peeling performance, the pushing velocity v_{needle} is continuously increased. This leads to an increase in the interface peeling energy G_c that needs to be overcome, necessitating a higher energy release rate for successful peeling. Consequently, crack initiation becomes challenging.

3. Results and Discussion

As discussed earlier, the size, thickness, and peeling speed of the chip all impact the initial interface crack generation process. These effects can be correlated and characterized by the peeling angle to a certain extent. In order to better understand this relationship, a three-dimensional finite element model of the peeling system was constructed in combination with the CZM unit to simulate the peeling process.

The chip, collector, and needle materials are shown in Table 2. In the chip peeling experiment, the thickness of the adhesive layer was only 5 µm, which is much smaller than the thickness of the blue film and the chip. It was no longer set separately in the geometric model and was replaced by the cohesive zone unit embedded in the zero-thickness layer by the finite element calculation platform.

Table 2. Simulation parameter settings.

	Materials	Elastic Modulus	Poisson's Ratio
Chip	Silicon substrate	129 GPa	0.28
Blue film	PVC	148 MPa	0.3
Needle	Structural steel	200 GPa	0.3
Collector	Structural steel	200 GPa	0.3

The downward pressure of the collector was 0.08 N, and the peeling process was simulated over a time of 1×10^{-4} s. The tip radius of the needle was 0.025 mm. The thickness of the blue film was 0.075 mm, and the width of the chip was 0.2 mm.

3.1. Influence of Chip Size on Crack Initiation

We set the chip thickness to δ = 0.03 mm and the needle pushing speed v_{needle} = 0.2 mm/5 ms. Two cases of a = 0.4 mm, a_1 = 1.2 mm and a = 2 mm, a_1 = 2.8 mm were selected for the simulation, and the influence of the chip size on crack initiation was analyzed.

The change process of peeling angle and peeling force in the two cases is shown in Figure 12. When a = 0.4 mm, the chip did not bend significantly during the whole crack initiation process, and the peeling angle was approximately equal to the deflection angle of the blue film. The initial crack was produced at a lifting height of 0.0993 mm, with the peeling angle and peeling force reaching 12.54788° and 0.02325 N, respectively. In contrast, when a = 2 mm, the crack did not initiate even when the lifting height was 0.2 mm, and the peeling angle and peeling force were considerably smaller than the former. This indicates that as the chip size increases, its deformation increases, leading to a smaller peeling angle at the same lifting height, and a more difficult crack initiation. These results align with the analytical findings.

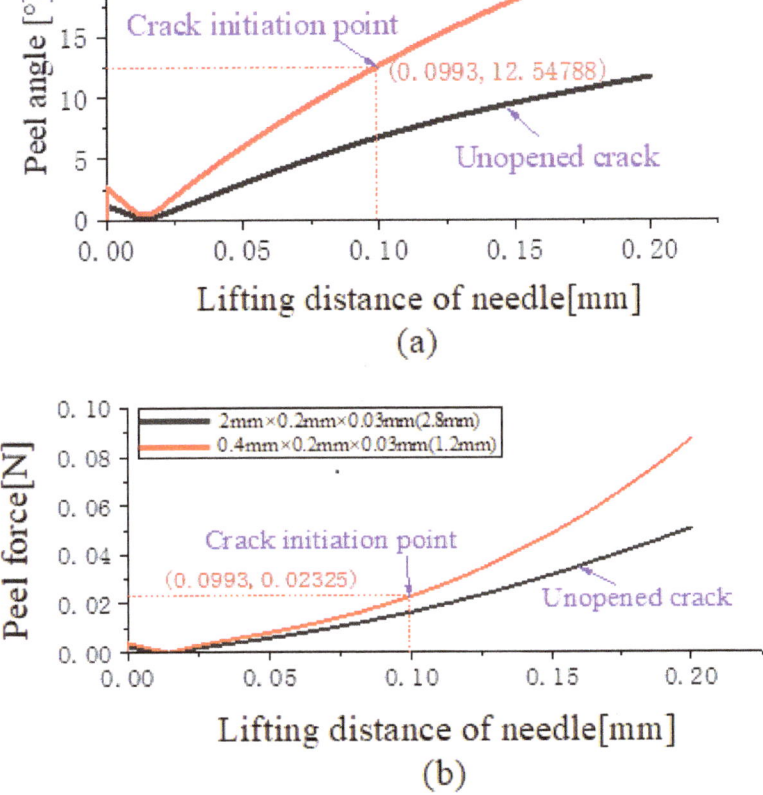

Figure 12. Comparison of changing chip length when v_{needle} = 0.2 mm/5 ms. (**a**) Peeling angle change curve; (**b**) Peel force change curve.

3.2. Influence of Chip Thickness on Crack Initiation

We set a = 2 mm, a_1 = 2.8 mm, needle lifting speed v_{needle} = 0.2 mm/5 ms, and chip thickness δ = 0.03 mm or δ = 0.1 mm. The influence of the chip thickness on crack initiation is analyzed.

The changes in peeling angle and peeling force were analyzed in two cases, as shown in Figure 13. In the first case, when δ = 0.1 mm, the crack initiation was completed when the chip was lifted to 0.1815 mm, and the peeling angle and peeling force reached 16.60008° and 0.03493 N, respectively. In the second case, when δ = 0.03 mm, the crack was not initiated even if the chip was lifted to 0.2 mm; the peeling angle was only 11.6672° and the peeling force reached 0.05047 N.

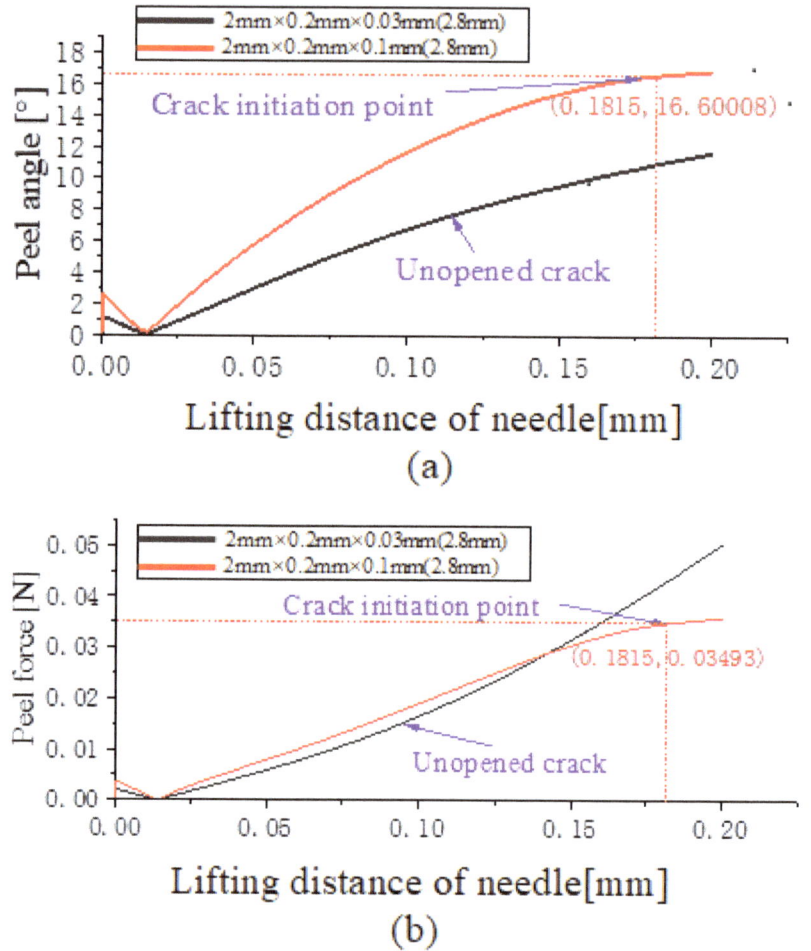

Figure 13. Comparison of changing chip thickness when v_{needle} = 0.2 mm/5 ms. (**a**) Peeling angle change curve; (**b**) Peel force change curve.

3.3. Influence of Lifting Speed on Crack Initiation

We set δ = 0.03 mm, and a = 0.4 mm, a_1 = 1.2 mm or a = 2 mm, a_1 = 2.8 mm. In both cases, the needle lift speed was set to v_{needle} = 0.2 mm/5 ms and v_{needle} = 0.2 mm/500 ms, and influence of lifting speed on crack initiation was analyzed.

The comparison of the peeling angle and peeling force in the two cases is shown in Figure 14. Both cases achieved initial crack germination at a lift of 0.0993 mm. The peeling angle and peeling force at 5 ms and 500 ms were 12.54788° and 0.02325 N and 12.53796° and 0.02266 N, respectively; the former increased by about 0.079% and 2.6%, indicating that the increase in the lifting rate leads to an increase in the energy required for initial peeling, which is consistent with the analytical results.

The comparison between the peeling angle and peeling force in the two cases is shown in Figure 15. The data results are consistent and indicate that the traditional peeling method is unable to initiate the crack, even if the lifting time is slowed down to an unacceptable level of 5 ms to 500 ms. This highlights the difficulty of crack initiation through traditional peeling techniques.

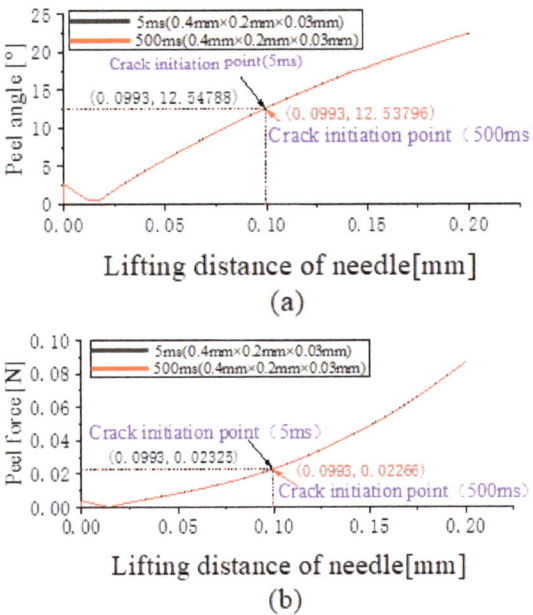

Figure 14. Comparison of changing lifting rate when $a = 0.4$ mm, $a_1 = 1.2$ mm, $\delta = 0.03$ mm. (**a**) Peeling angle change curve; (**b**) peel force change curve.

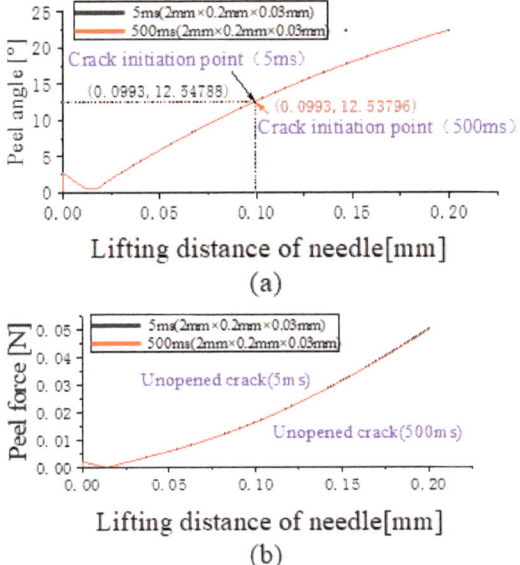

Figure 15. Comparison of changing lifting rate when $a = 2$ mm, $a_1 = 2.8$ mm, $\delta = 0.03$ mm. (**a**) Peeling angle change curve; (**b**) peel force change curve.

3.4. Stress of Chip during Crack Initiation

We set the lifting velocity v_{needle} = 0.2 mm/5 ms, δ = 0.03 mm, a = 0.4 mm, a_1 = 1.2 mm, δ = 0.03 mm or a = 2 mm, a_1 = 2.8 mm, δ = 0.1 mm, or a = 2 mm, a_1 = 2.8 mm. The changes in chip stress during the crack initiation under the three conditions are shown in Figure 16. In the first two cases, the chip stress remained within safe levels (the limit value of 1% probability fracture damage strength of silicon substrate, which is about 71 MPa when δ = 100 μm and about 345 MPa when δ = 30 μm), indicating that this peeling method is effective for chips with a smaller size or larger thickness, and can meet the requirements of crack initiation. However, in the third case, the chip stress exceeded the safety limit before the crack was initiated during the lifting process, resulting in damage or even fracture. Even when a = 2 mm, a_1 = 2.8 mm, δ = 0.03 mm, and the lifting velocity was reduced to 1%, the initial crack still could not be initiated. This suggests that as the chip becomes larger and thinner, reducing the lifting rate is not a feasible way to achieve crack initiation with the traditional peeling method.

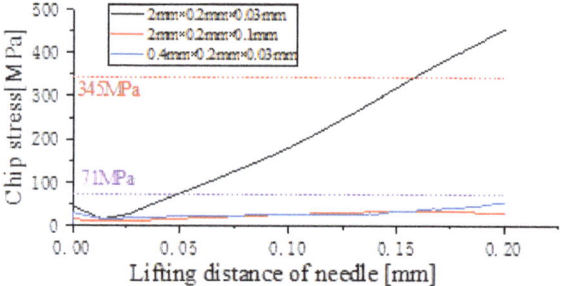

Figure 16. The maximum equivalent stress change curve of the chip (v_{needle} = 0.2 mm/5 ms).

The influence of lifting velocity on the maximum equivalent stress of the chip is shown in Figure 17. When a = 2 mm, a_1 = 2.8 mm, and δ = 0.03 mm, compared v_{needle} = 0.2 mm/500 ms with v_{needle} = 0.2 mm/5 ms, the ultimate stress of the chip was reduced from 455.17 MPa to 451.32 MPa, indicating a reduction of about 8.46‰. If a = 0.4 mm, a_1 = 1.2 mm, and δ = 0.03 mm, the peeling process became easier with a slower lifting velocity, but it only changed the equivalent maximum stress of the chip from 53.043 MPa to 53.215 MPa, showing a difference of only 3.24‰. Even if the lifting rate was reduced by 100 times, the variation range of the equivalent maximum stress of the chip was different by a few thousandths of a MPa. This indicates that the effect of reducing the lifting velocity on reducing the equivalent maximum stress of the chip is not obvious.

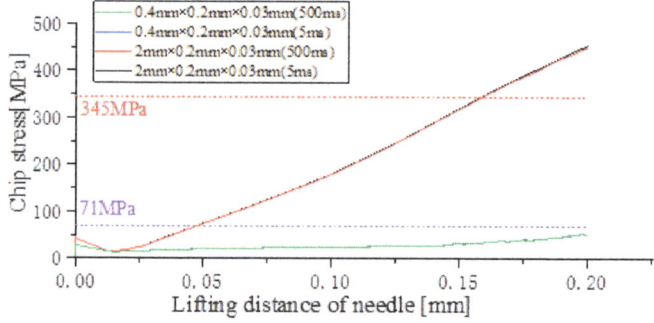

Figure 17. Change curve of chip maximum equivalent stress under changing jacking rate (v_{needle} = 0.2 mm/5 ms and v_{needle} = 0.2 mm/500 ms).

4. Coupled Initiation Mode of Adhesive Interfacial Crack

4.1. Principle of Coupling Initiation

The analysis above shows that with the increase of chip size and thickness, the probability of chip damage and fragmentation increases, and the initial stripping becomes more and more difficult. Additionally, the importance of the peel angle in crack initiation and the way in which the geometric structure influences the peel angle suggest that structural optimization may offer a new solution for increasing the peel angle during crack excitation. Therefore, the peeling strategy including two features, structural coupling and force surface coupling, was proposed and is shown in Figure 18.

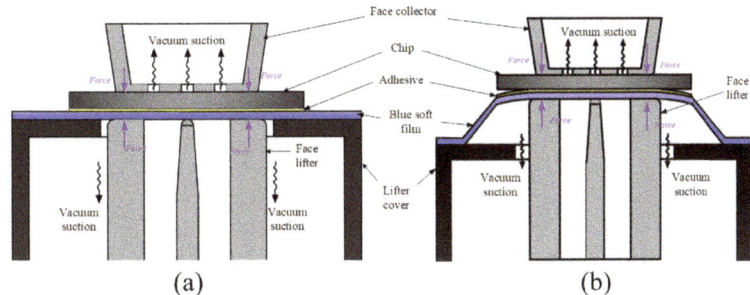

Figure 18. Schematic diagram of the crack coupling initiation strategy. (**a**) Initial state of face collector down pressure and face lifter coupling; (**b**) Initial state of face collector down pressure and face lifter coupling.

Firstly, the needle was redesigned as a face lifter embedded with the needle, and the displacement load is applied by the surface load (Figure 18a).

Secondly, the single hole suction nozzle (collector) was redesigned into a face collector with multi holes. At the beginning of peeling, the suction nozzle and the lifter surface are coupled with each other, and the coupling force surface is constructed by "lower pushing—upper pressing" to suppress the bending deformation of the chip during the crack initiation. Through structural coupling, the bending deformation of the chip is reduced, the change of peeling angle is controllable, and the effective peeling force is improved. Through force surface coupling, chip local stiffness is also improved, leading to a reduction in deformation and stress.

After the completion of the interface crack initiation, the interface crack propagation process begins. The phased control strategy process for crack propagation is shown in Figure 19.

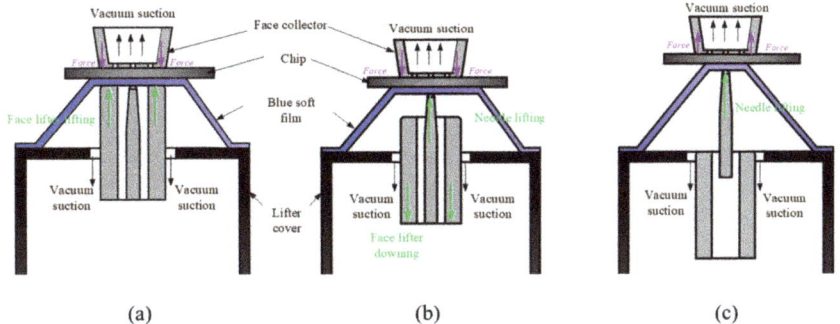

Figure 19. Staged control strategy for crack propagation. (**a**) Face-lifter lifting crack propagation stage; (**b**) needle lifting crack propagation stage; (**c**) crack propagation end stage.

(1) After the completion of the coupled strategy for interface crack initiation, the lifting surface continues to lift, and the crack begins to propagate. The contact area between the chip and the blue film interface gradually decreases until the blue film, which has completed the expansion, is about to contact the lifted structure of the surface. The crack propagation will be hindered by the lifted structure of the surface.

(2) Due to the fact that the lifting surface will hinder the propagation of the crack, the top needle is chosen to replace the lifting surface to continue lifting, in order to avoid the situation where the lifting surface affects the peel angle and interferes with the initiation of the peel energy release rate, which ultimately leads to a slowdown of the crack propagation rate. Therefore, it is necessary to control the rapid descent of the lifting surface.

(3) The top needle continues to lift until the crack propagates to around 70–80% of the original interface contact area, and the crack propagation is completed. Finally, only the swing arm is needed to control the suction cup to lift and continue to operate to complete the transfer of the chip.

Under the coupling action of the structure and force plane, only the part of the chip outside the coupling area is involved in the peeling, which reduces the length of the chip involved in the peeling and reduces the length thickness ratio of the chip. In fact, the step-difference needle implemented by Swiss Bessie, the gas needle proposed by Jeon of Korea, and the multi-thimble needle presented by Peng and Yin of Huazhong University of Science and Technology all imply the idea of structural coupling excitation. In essence, they are all specific applications of this strategy.

4.2. Optimization of Mechanical Properties

By comparing the mechanical characteristics before and after the coupling method, the optimization effect was analyzed. The simplified model of the chip in the traditional mode is shown in Figure 20, and the maximum deflection W_{B1} of the bending part is:

$$W_{B1} = \frac{F_L \sin\theta_t (\frac{a}{2})^3}{3 E_{chip} I_{chip}} = \frac{F_L \sin\theta_t}{2 E_{chip} b} \left(\frac{a}{\delta}\right)^3 \tag{12}$$

The maximum rotation angle θ_{B1} is:

$$\theta_{B1} = \frac{3 F_L \sin\theta_t (a)^2}{2 E_{chip} b \delta^3} = \frac{3 F_L \sin\theta_t}{2 E_{chip} b \delta} \left(\frac{a}{\delta}\right)^2 \tag{13}$$

The maximum stress $\sigma'/_{max}$ of the chip is:

$$\sigma'_{max} = \frac{3 F_L \sin\theta_t (a)}{b \delta^2} \frac{F_L \cos\theta_t}{b \delta} \tag{14}$$

The peeling angle θ_p is:

$$\theta_p = \arctan\left(\frac{2(h - W_{B1})}{a_1 - a}\right) - \theta_{B1} \tag{15}$$

The mechanical model under the coupling mode is shown in Figure 21. The deformation area of the chip is limited to the part between the edge of the coupling surface and the edge of the chip, and the chip bends with the edge of the coupling surface as the origin and is symmetrical on the left and right.

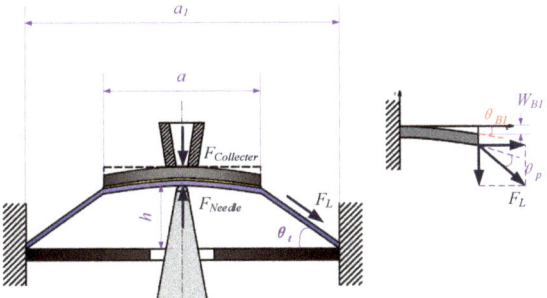

Figure 20. Simplified model of chip peeling and adhesion using traditional strategy.

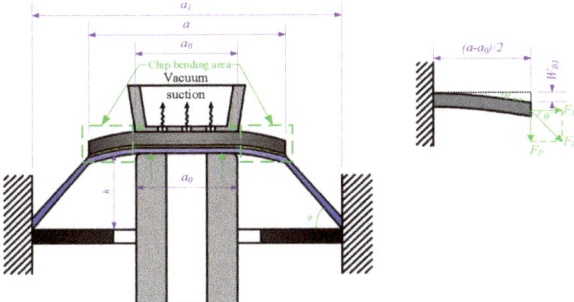

Figure 21. Simplified model of chip peel-off adhesion using coupling strategy.

The maximum deflection W_{B3} of the bending part is:

$$W_{B3} = \frac{F_L \sin\theta_t \left(\frac{a-a_0}{2}\right)^3}{3E_{chip}I_{chip}} = \frac{F_L \sin\theta_t}{2E_{chip}b}\left(\frac{a-a_0}{\delta}\right)^3 \tag{16}$$

The maximum rotation angle θ_{B3} is:

$$\theta_{B3} = \frac{3F_L \sin\theta_t (a-a_0)^2}{2E_{chip}b\delta^3} = \frac{3F_L \sin\theta_t}{2E_{chip}b\delta}\left(\frac{a-a_0}{\delta}\right)^2 \tag{17}$$

The maximum stress σ_{max} of the chip is:

$$\sigma_{max} = \frac{3F_L \sin\theta_t (a-a_0)}{b\delta^2}\frac{F_L \cos\theta_t}{b\delta} \tag{18}$$

The peeling angle θ''_p is:

$$\theta'_p = \arctan\left(\frac{2(h-W_{B3})}{a_1-a}\right) - \theta_{B3} \tag{19}$$

Under the action of the coupling force, the effective length thickness ratio involved in the peeling process changes from a/δ_{chip} to $(a_1-a)/\delta_{chip}$, and the length thickness ratio of the chip involved in crack initiation decreases. With this change, the length thickness ratio of the chip involved in crack initiation decreases and the peeling angle θ_p ($\theta_p = \theta_t - \theta_B$) is much larger compared to the traditional mode. The essence of the coupling action is to transform the peeling of large-scale ultra-thin chips into a small-size thick chip peeling. The size design of the coupling force surface should satisfy the following relationship:

$$\sigma_{\max} = \frac{3F_L \sin\theta_t (a - a_0)}{b\delta^2} + \frac{F_L \cos\theta_t}{b\delta} \leq [\sigma] \tag{20}$$

where $[\sigma]$ is the damage limit stress of the chip, and the size a_0 of the coupling surface should meet the following requirements:

$$a > a_0 \geq a - \frac{b\delta^2}{3F_L \sin\theta_t}[\sigma] + \frac{\delta}{3\tan\theta_t} \tag{21}$$

5. Simulation Analysis on Crack Initiation Characteristics

As mentioned earlier, the face lifter is a surface structure with a needle embedded in the middle. Similarly, the face collector also has a surface structure. In theory, their surfaces can have different shapes. For simplicity, it is assumed that they have the same surface shape. Crack initiation characteristics can be analyzed with simulations. The basic parameters of the geometry and materials are shown in Table 3.

Table 3. Simulation parameter settings.

	Materials	Elastic Modulus	Poisson's Ratio
Chip	Silicon substrate	129 GPa	0.28
Blue film	PVC	148 MPa	0.3
Face lifter	Structural steel	200 GPa	0.3
Face collector	Structural steel	200 GPa	0.3

The chip specifications of the simulation model were a = 2 mm, a_1 = 2.8 mm, and δ = 0.03 mm. The thickness of the coupling structure (simplified face lifter and collector) was 0.05 mm and the width was 0.18 mm. In the crack initiation, the surface suction nozzle was contacted and pressed down (the time required for the down pressure to be 0.08 N was set to 1×10^{-4} s), and then the face lifter was pushed up.

5.1. Peeling Angle and Chip Stress in Coupling Mode

The lifting velocity was set to v_{Face} = 0.2 mm/5 ms, and the coupling size was a_0 = 1.2 mm. The comparative analysis between the coupling mode and the traditional mode is shown in Figure 22. In the coupling mode, the chip was successfully lifted up to 0.04444 mm, with a peeling angle of 6.42462° and a peeling force of 0.00986 N. In contrast, the traditional mode could not lift the chip at 5 ms/0.2 mm, and the crack could only be initiated at a higher speed. Additionally, the peeling angle in the coupling mode was significantly larger, and the peeling force was effectively improved in comparison to the traditional mode.

The chip stress in the coupling mode is shown in Figure 23; the maximum stress of the chip was 98.907 MPa, which is significantly lower than that in the traditional mode, which exceeded the safety limit of 345 MPa.

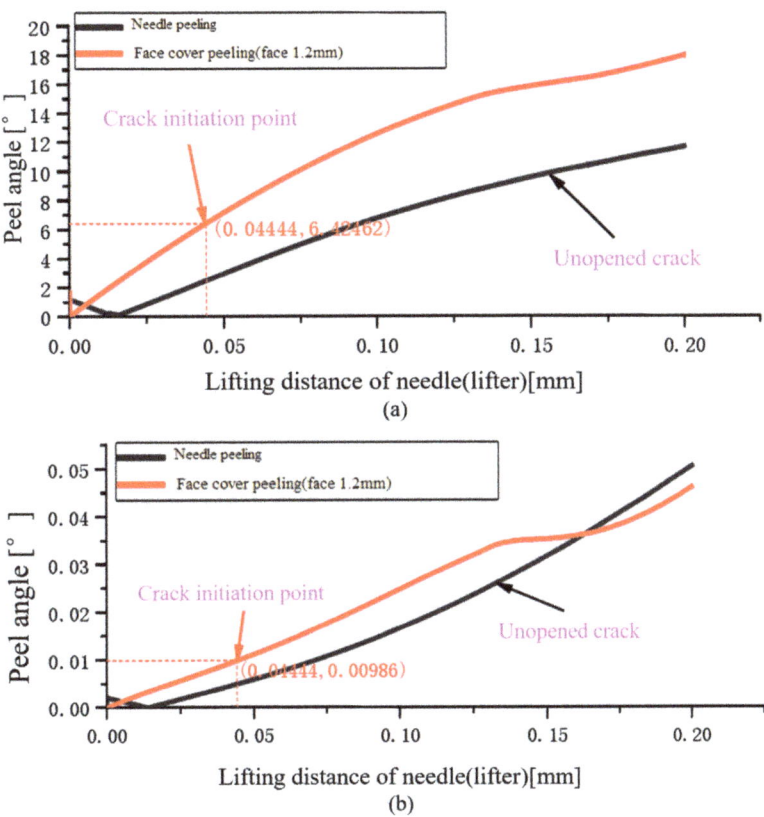

Figure 22. The coupling strategy is compared with the traditional method, when a = 2 mm, a_1 = 2.8 mm, a_0 = 1.2 mm, δ = 0.03 mm (v_{Face} = 0.2 mm/5 ms) (**a**) Peeling angle change curve; (**b**) peel force change curve.

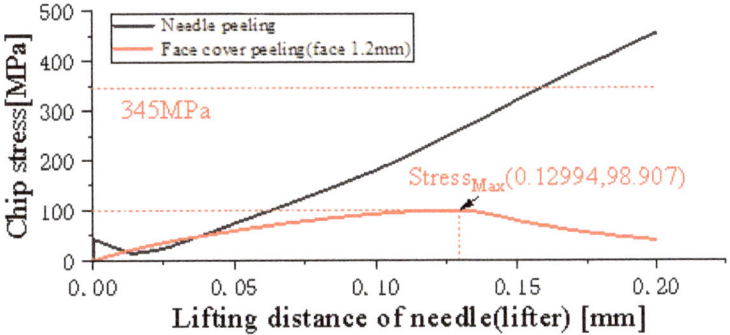

Figure 23. When v_{Face} = 0.2 mm/5 ms (a_0 = 1.2 mm), the difference in chip stress between coupling strategy and traditional method.

This shows that the coupling strategy can obviously reduce the chip stress during the crack initiation of large-size ultra-thin chip peeling. The safety and reliability have been greatly improved.

5.2. Influence of Lifting Velocity Size in Coupling Mode

We set a_0 = 1.2 mm, the lifting velocity as v_{Face1} = 0.2 mm/500 ms or v_{Face2} = 0.2 mm/5 ms, and analyzed influence of the lifting velocity.

The results are shown in Figure 24. When the lift time t = 5 ms, the crack was initiated when it was lifted to 0.04444 mm; the peeling angle and peeling force were 6.27859° and 0.0095 N, respectively. When t = 500 ms, the crack was initiated when it was lifted to 0.04344 mm; the peeling angle and peeling force were 6.42462° and 0.00986 N, respectively. With an increase in lifting distance by 0.001 mm, the corresponding peeling angle and peeling force increased by 2.33% and 3.79%. The results indicate that under the coupling strategy, secure crack initiation can be achieved without reducing the lifting velocity. This finding is consistent with the conclusion of the aforementioned analysis.

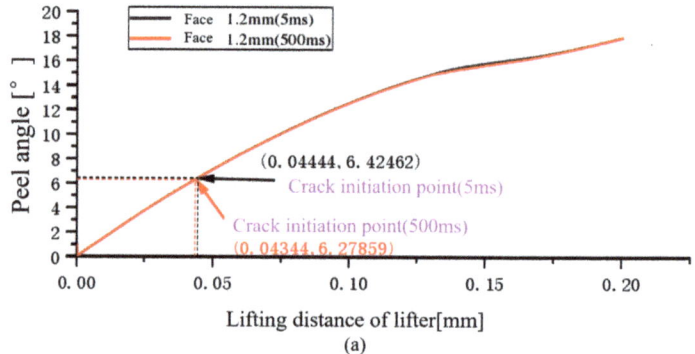

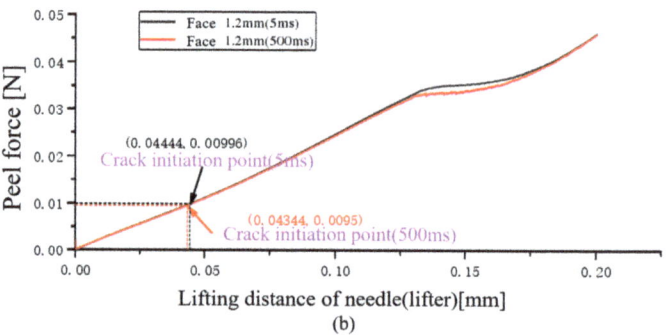

Figure 24. Under the coupling strategy, the comparison of changing the jacking rate when a = 0.4 mm, a_1 = 1.2 mm, δ = 0.03 mm. (**a**) Peeling angle change curve; (**b**) peel force change curve.

5.3. Influence of Coupling Surface Size in Coupling Mode

We set v_{Face} = 0.2 mm/5 ms, and the size of the coupling surface is set to a_0 = 0.6 mm or a_0 = 1.2 mm for comparative analysis.

The results are presented in Figure 25, where it can be observed that the 0.6 mm chip initiated a crack when lifted to 0.05169 mm, while in Figure 26, the 1.2 mm chip initiated a crack when lifted to 0.04444 mm. The peeling angle of the 1.2 mm chip was larger than that of the 0.6 mm chip, with the peeling angle and peeling force of the 0.6 mm chip being 5.631° and 0.01026 N, respectively, whereas those of the 1.2 mm chip were 6.42462° and 0.00986 N, respectively.

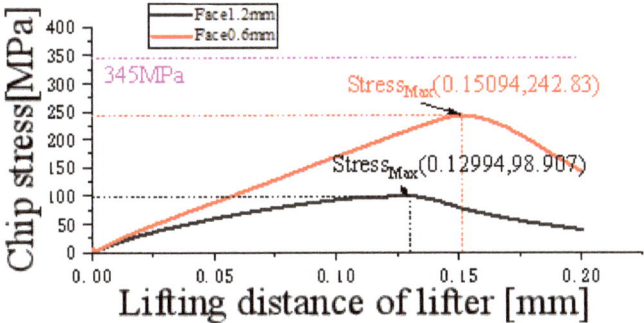

Figure 25. Under the coupling strategy, when $v_{Face} = 0.2$ mm/5 ms, the change in the chip stress with changing size of the coupling surface.

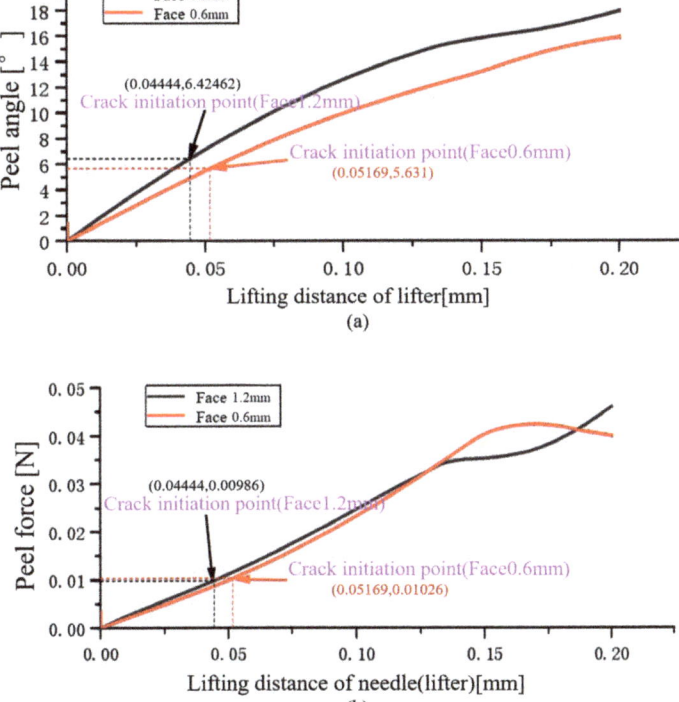

Figure 26. Under the coupling strategy, the comparison of changing the coupling surface size when $a = 2$ mm, $a_1 = 2.8$ mm, $\delta = 0.03$ mm ($v_{needle} = 0.2$ mm/5 ms). (**a**) Peeling angle change curve; (**b**) peel force change curve.

In addition, when $a_0 = 0.6$ mm, the chip stress during the lifting process was greatly increased compared with that when $a_0 = 1.2$ mm. As shown in Figure 26, the maximum stress was increased to 242.83 MPa, an increase of 145.51%, indicating that the chip stress has an important relationship with the coupling structure; thus, the design of the coupling size is very important for the safe stripping of the initial interface.

The coupling strategy improves the stiffness of the chip in the coupling area, and the chip transforms from the overall large deflection deformation to the local small deflection deformation, which essentially linearizes the nonlinear deformation problem of the large-scale ultra-thin chip and realizes controllable crack initiation. This strategy improves the peeling angle conditions in the crack initiation process, enabling effective conversion of blue film tensile force into peeling force and reducing the local stress of the chip. Consequently, even at higher peeling velocities, the crack can be safely initiated.

6. Conclusions

This paper presents an adhesive peeling model based on the cohesive zone model to analyze the fast and non-destructive peeling characteristics of large-size ultra-thin chips. The results show that large deflection deformation leads to a decrease in the peeling angle, resulting in difficulties in improving the peeling force and energy release rate. To address this, a coupling peeling strategy was proposed to increase the peeling angle during crack germination. This strategy employs structural coupling to improve the local flexibility of the chip and utilizes force surface coupling to improve the overall stiffness of the chip, especially a part of the coupling area, thereby improving the control conditions of the peeling angle. Mechanical analysis and 3D simulation based on a CZM unit showed that, compared with the traditional peeling process, the coupling strategy increased the strain mismatch effect between the chip and the soft membrane promoting rapid peeling angle excitation and increasing the peeling energy release rate. Additionally, the surge of chip stress can be effectively restrained, thereby improving the peeling safety and reliability.

The coupling strategy improved the stiffness of the chip in the coupling area, so that the chip peeling transforms from the overall large deflection deformation to a local small deflection deformation. This strategy improves the peel angle control conditions in the crack initiation process, promotes the effective conversion of the blue film tensile force into the peel force, reduces the local stress of the chip, can be safely initiated even at a higher peel rate, and improves the peeling performance. However, after the crack successfully initiates, it will rapidly expand. If handled improperly, the coupling structure will form obstacles. The following research will mainly focus on the coupled acceleration control of crack propagation in large-sized and ultra-thin chips, studying the mechanical characteristics of crack propagation acceleration, the coupled acceleration control strategy of crack propagation, selecting appropriate peeling methods, establishing a three-dimensional model, and conducting corresponding finite element analysis to determine the effect of coupled acceleration control on the variation in peeling angle.

Author Contributions: Conceptualization, T.W. and X.C.; methodology, X.C.; software, X.C.; validation, X.C., S.W. and F.L.; formal analysis, T.W.; investigation, X.C.; resources, T.W.; data curation, S.W.; writing—original draft preparation, X.C.; writing—review and editing, T.W.; visualization, F.L.; supervision, T.W. and S.L.; project administration, T.W. and X.C.; funding acquisition, T.W. and S.L. All authors have read and agreed to the published version of the manuscript.

Funding: This research was supported by the Guangdong Natural Science Foundation Project, China (No. 2021A1515010661), and the Guangdong province Science and Technology Plan Project (grant No. STKJ2021027).

Institutional Review Board Statement: Not applicable.

Informed Consent Statement: Not applicable.

Data Availability Statement: Not applicable.

Conflicts of Interest: The authors declare no conflict of interest.

References

1. Liu, X.; Li, M.; Zheng, J.; Zhang, X.; Zeng, J.; Liao, Y.; Chen, J.; Yang, J.; Zheng, X.; Hu, N. Electrochemical Detection of Ascorbic Acid in Finger-Actuated Microfluidic Chip. *Micromachines* **2022**, *13*, 1479. [CrossRef]
2. Craig, P.; Ng, R.; Tefsen, B.; Linsen, S.; Liu, Y.; Hendel, J. Information Visualisation for Antibiotic Detection Biochip Design and Testing. *Processes* **2022**, *10*, 2680. [CrossRef]

3. Ho, C.-H.; Chen, S.-M.; Wu, Y.-R. Study of the Factors Limiting the Efficiency of Vertical-Type Nitride- and AlInGaP-Based Quantum-Well Micro-LEDs. *Processes* **2022**, *10*, 489. [CrossRef]
4. Yeom, D.; Kim, J.; Kim, S.; Ahn, S.; Choi, J.; Kim, Y.; Koo, C. A Thermocycler Using a Chip Resistor Heater and a Glass Microchip for a Portable and Rapid Microchip-Based PCR Device. *Micromachines* **2022**, *13*, 339. [CrossRef]
5. Balamurugan, K.; Umamaheswaran, S.; Mamo, T.; Nagarajan, S.; Namamula, L.R. Roadmap for Machine Learning based Network-on-Chip (M/L NoC) technology and its analysis for researchers. *J. Phys. Commun.* **2022**, *6*, 022001. [CrossRef]
6. Christou, A.; Ma, S.H.; Zumeit, A.; Dahiya, A.S.; Dahiya, R. Printing of Nano- to Chip-Scale Structures for Flexible Hybrid Electronics. *Adv. Electron. Mater.* **2023**, *1*, 2201116. [CrossRef]
7. Behler, S.; Teng, W.; Podpod, A. Key Properties for Successful Ultra Thin Die Pickup. In Proceedings of the 2017 IEEE 67th Electronic Components and Technology Conference (ECTC), Orlando, FL, USA, 30 May–2 June 2017; pp. 95–101.
8. Yang, Y.; Wu, Y. Analysis of the Characteristics of UV Films and Blue Films Used in the Process of Chip Industrialization. *Mod. Electron. Technol.* **2013**, *36*, 3.
9. Canale, G.; Andrews, S.; Rubino, F.; Maligno, A.; Citarella, R.; Weaver, P.M. Realistic Stacking Sequence Optimisation of an Aero-Engine Fan Blade-Like Structure Subjected to Frequency, Deformation and Manufacturing Constraints. *Open Mech. Eng. J.* **2018**, *12*, 151–163. [CrossRef]
10. Zou, J.; Zhang, Y.; Zhang, L.; Jing, J.; Fu, Y.; Wang, Y.; Zhang, G.; Zhou, F. Numerical Simulation Research on the Effect of Artificial Barrier Properties on Fracture Height. *Processes* **2023**, *11*, 310. [CrossRef]
11. Neves, L.F.R.; Campilho, R.D.S.G.; Sánchez-Arce, I.J.; Madani, K.; Prakash, C. Numerical Modelling and Validation of Mixed-Mode Fracture Tests to Adhesive Joints Using J-Integral Concepts. *Processes* **2022**, *10*, 2730. [CrossRef]
12. Khan, S.A.; Rahimian Koloor, S.S.; King Jye, W.; Yidris, N.; Mohd Yusof, A.A.; Mohd Szali Januddi, M.A.F.; Tamin, M.N.; Johar, M. Strain Rate Effect on Mode I Debonding Characterization of Adhesively Bonded Aluminum Joints. *Processes* **2023**, *11*, 81. [CrossRef]
13. Hong, J.H.; Cheng, P.; Chen, W.; Guo, J.H.; Li, Y.L.; Liu, J.Z. Theoretical Modeling and Experimental Studies of Ultra-Thin Chip Transfer in Laser-Induced Forward Transfer. *IEEE Trans. Compon. Manuf. Technol.* **2022**, *12*, 570–577. [CrossRef]
14. Hong, J.H.; Chen, W.; Guo, J.H.; Cheng, P.; Li, Y.L.; Dong, W.T. Theoretical and experimental studies of spring-buffer chip peeling technology for electronics packaging. *Int. J. Fract.* **2022**, *236*, 109–124. [CrossRef]
15. Liu, Z.; Huang, Y.; Xiao, L.; Tang, P.; Yin, Z. Nonlinear characteristics in fracture strength test of ultrathin silicon die. *Semicond. Sci. Technol.* **2015**, *30*, 045005. [CrossRef]
16. Peng, B.; Huang, Y.; Yin, Z. Competing Fracture Modeling of Thin Chip Pick-Up Process. *IEEE Trans. Compon. Packag. Manuf. Technol.* **2012**, *2*, 1217–1225. [CrossRef]
17. Cheng, H.; Wu, J.; Yu, Q.; Kim-Lee, H.-J.; Carlson, A.; Turner, K.T.; Hwang, K.-C.; Huang, Y.; Rogers, J.A. An analytical model for shear-enhanced adhesiveless transfer printing. *Mech. Res. Commun.* **2012**, *43*, 46–49. [CrossRef]
18. Jeon, E.B.; Park, S.H.; Yoo, Y.S.; Kim, H.S. Analysis of Interfacial Peeling of an Ultrathin Silicon Wafer Chip in a Pick-Up Process Using an Air Blowing Method. *IEEE Trans. Compon. Packag. Manuf. Technol.* **2016**, *6*, 1696–1702. [CrossRef]
19. Williams, J.A.; Kauzlarich, J.J. The influence of peel angle on the mechanics of peeling flexible adherends with arbitrary load–Extension characteristics. *Tribol. Int.* **2005**, *38*, 951–958. [CrossRef]
20. Molinari, A.; Ravichandran, G. Stability of peeling for systems with rate independent decohesion energy. *Int. J. Solids Struct.* **2013**, *50*, 1974–1980. [CrossRef]
21. Molinari, A.; Ravichandran, G. Peeling of Elastic Tapes: Effects of Large Deformations, Pre-Straining, and of a Peel-Zone Model. *J. Adhes.* **2008**, *84*, 961–995. [CrossRef]
22. Kovalchick, C.; Molinari, A.; Ravichandran, G. An experimental investigation of the stability of peeling for adhesive tapes. *Mech. Mater.* **2013**, *66*, 69–78. [CrossRef]
23. Yin, H.B.; Liang, L.H.; Wei, Y.G.; Peng, Z.L.; Chen, S.H. Determination of the interface properties in an elastic film/substrate system. *Int. J. Solids Struct.* **2020**, *191–192*, 473–485. [CrossRef]
24. Zhang, L.; Wang, J. A generalized cohesive zone model of the peel test for pressure-sensitive adhesives. *Int. J. Adhes. Adhes.* **2009**, *29*, 217–224. [CrossRef]
25. Liu, Z. *Mechanism Study and Process Optimization of Non Destructive Peeling of Ultrathin Chips*; Huazhong University of Science and Technology: Wuhan, China, 2015.
26. Liu, D. *Research on Several Mechanical Problems of Multi field Coupled Laminated Structures*; Zhejiang University: Hangzhou, China, 2013.
27. Dai, W. *Analysis of Chip Stripping Process and Its Mechanism*; Huazhong University of Science and Technology: Wuhan, China, 2011.
28. Ni, J.; Wang, J. Research on the mechanism of thin chip fragmentation failure in IC cards. *Semicond. Technol.* **2004**, *04*, 40–44.
29. Shu, Z.; Peng, X.; Li, F.F.; Xu, Q. Cohesive zone model for prepreg tack based on probe test. *Acta Aeronaut. Astronaut. Sin.* **2018**, *39*, 280–292.

30. Kendall, K. Thin-film peeling-the elastic term. *J. Phys. D Appl. Phys.* **1975**, *8*, 1449–1452. [CrossRef]
31. Kendall, K. The adhesion and surface energy of elastic solids. *J. Phys. D Appl. Phys.* **1971**, *4*, 1186–1195. [CrossRef]

Disclaimer/Publisher's Note: The statements, opinions and data contained in all publications are solely those of the individual author(s) and contributor(s) and not of MDPI and/or the editor(s). MDPI and/or the editor(s) disclaim responsibility for any injury to people or property resulting from any ideas, methods, instructions or products referred to in the content.

Article

Numerical Modelling and Validation of Mixed-Mode Fracture Tests to Adhesive Joints Using *J*-Integral Concepts

Luís F. R. Neves [1], Raul D. S. G. Campilho [1,2,*], Isidro J. Sánchez-Arce [2], Kouder Madani [3] and Chander Prakash [4]

[1] Department of Mechanical Engineering, ISEP—School of Engineering, Polytechnic Institute of Porto, R. Dr. António Bernardino de Almeida, 431, 4200-072 Porto, Portugal
[2] INEGI—Institute of Science and Innovation in Mechanical and Industrial Engineering—Pólo FEUP, Rua Dr. Roberto Frias, 400, 4200-465 Porto, Portugal
[3] Department of Mechanical Engineering, University of Sidi Bel Abbes, BP 89 Cité Ben M'hidi, Sidi Bel Abbes 22000, Algeria
[4] School of Mechanical Engineering, Lovely Professional University, Phagwara 144411, India
* Correspondence: raulcampilho@gmail.com

Abstract: The interest in the design and numerical modelling of adhesively-bonded components and structures for industrial application is increasing as a research topic. Although research on joint failure under pure mode is widespread, applied bonded joints are often subjected to a mixed mode loading at the crack tip, which is more complex than the pure mode and affects joint strength. Failure of these joints under loading is the objective of predictions through mathematical and numerical models, the latter based on the Finite Element Method (FEM), using Cohesive Zone Modelling (CZM). The Single leg bending (bending) testing is among those employed to study mixed mode loading. This work aims to validate the application of FEM-CZM to SLB joints. Thus, the geometries used for experimental testing were reproduced numerically and experimentally obtained properties were employed in these models. Upon the validation of the numerical technique, a parametric study involving the cohesive laws' parameters is performed, identifying the parameters with the most influence on the joint behaviour. As a result, it was possible to numerically model SLB tests of adhesive joints and estimate the mixed-mode behaviour of different adhesives, which enables mixed-mode modelling and design of adhesive structures.

Keywords: adhesive joints; structural adhesive; fracture toughness; mixed-mode; cohesive zone modelling

1. Introduction

Automotive, construction, aeronautical, and maritime industries extensively employ adhesive bonding for structural and cosmetic purposes. The design of such adhesive joints requires an a priori characterization of the materials involved. In addition, the joints themselves are also characterised, ensuring they fulfil the requirements they were designed for. In this regard, Budzik et al. [1] reviewed standard and non-standard tests for joints employed in several technological fields while Tserpes et al. [2] reviewed failure theories employed in the design of bonded structures. The mechanical properties of the adhesive are determined through experimental testing following the applicable standards, as described by da Silva et al. [3]. Moreover, in the adherends' case, extensive testing is necessary for enhanced adherends, e.g., those studied in reference [4]. However, the behaviour of the adhesive within a joint also depends on geometric factors such as the adhesive layer's thickness (t_A) [5], material properties [6], and temperature [7]. In consequence, adhesive joints are characterised according to the expected loading conditions. Regarding the fracture behaviour of adhesive joints, there are three pure loading modes: traction (mode I), shear (mode II), and out-of-plane shear (mode III), as described by Dillard [8]. However, applications of adhesive joints often present a degree of mixing, i.e., more than one mode

is present due to load solicitation. In this case, the failure occurs in mixed mode [8]. The critical energy release rate (G_C) is among the Linear Elastic Fracture Mechanics (LEFM) methods employed to determine crack propagation, and it is necessary for computational simulations. Furthermore, G_C has to be determined for each loading mode, i.e., mode I and mode II, through experimental tests. In the adhesive joints' case, there are different experimental tests for this purpose, most of which are described by Pearson et al. [9] and Chaves et al. [10]. The tensile critical energy release rate (G_{IC}) is often determined using the double-cantilever beam (DCB) test, while the shear critical energy release rate (G_{IIC}) is determined using the end-notched flexure (ENF) test. On the other hand, Ji et al. [11] developed a mathematical model, based on the J-integral theory, to determine the tensile and shear energy release rates (G_I and G_{II}, respectively) under mixed-mode loading. Furthermore, the effect that t_A has on the cohesive laws was evaluated. The proposed methodology consisted of experimental tests using the single-leg bending (SLB) test and followed by the mathematical approach to obtain the cohesive laws. It was found that t_A has a proportional effect on G_I, G_{II}, and joint strength. However, it does not affect the normalised tension used for the cohesive laws.

Cases of experimental characterisation of adhesives using the DCB and ENF tests are often found in the literature. For example, Faneco et al. [12] employed both DCB and ENF tests to characterise a polyurethane structural adhesive, the SikaForce® 7752, for industrial use. The specimens tested for both cases were composed of aluminium adherends, and t_A = 1 mm. Six specimens of each case were tested. Upon completing the experimental campaign, good repeatability of the results was observed, indicating good control in the specimen preparation and testing. Subsequently, G_{IC} and G_{IIC} were obtained using three different methods. This approach was also followed by Cardoso et al. [13] to characterise another polyurethane structural adhesive, the SikaPower® 1277, also for industrial use. The specimen dimensions and t_A were similar to those used by Faneco et al. [12]. In addition, the results also showed good repeatability, confirming that good control was had on specimen preparation and experimental procedure. Regarding the experimental procedures, both DCB and ENF were described by da Silva et al. [3] and Pearson et al. [9]. Similarly, there are experimental tests aimed at mixed-mode loading such as the cracked-lap shear (CLS) [14], mixed-mode bending (MMB) [15], and the SLB [16]. Testing between the SLB and ENF configurations is similar, hence no extra laboratory equipment is necessary, making this test convenient [9]. Furthermore, the results obtained from the mixed-mode tests together with those from pure mode allow for determining the fracture envelopes, which show how the joint behaves under different loading conditions [9] and are useful for design purposes. The SLB test has been used to determine fracture envelopes of different adhesives. For example, Santos and Campilho [17] studied the fracture behaviour of three different adhesives, from brittle to ductile, using this test. The joints had composite adherends and t_A = 1 mm. The results from the experimental testing were repeatable and consistent. Subsequently, the values of G_I and G_{II} were obtained using six different reduction methods and, again, good repeatability was observed regardless of the method. Then, these results together with the results from G_{IC} and G_{IIC} lead to obtaining the fracture envelopes and the exponent values for the power laws. More recently, Loureiro et al. [18] performed a similar experimental campaign testing seven specimens per adhesive type, for a total of three adhesive types. In this case, the J-integral method was used to calculate G_I and G_{II}. The results showed low variability regardless of the adhesive type. Then, the fracture envelopes were obtained, and the exponents of the power laws were calculated. These results agreed with previous research, indicating their validity. Furthermore, the parameters obtained in these works are necessary for the numerical modelling of bonded joints [17,18].

Numerical modelling using the Finite Element Method (FEM) has been employed to study adhesive joints for a long time, and Adams and Peppiatt [19] are among the pioneers in this regard. More recently, cohesive zone modelling (CZM) was included in FEM, allowing one to predict joint strength, and even debonding, with good accuracy [2]. However, the cohesive laws must be properly chosen, from which the bi-linear or triangular law is a good compromise between accuracy and computational cost [20]. The parameters necessary to model the cohesive behaviour of the adhesive layer are obtained from the experimental tests listed above. Reis et al. [21] experimentally tested SLB specimens made of solid composite material (carbon fibre and polyamide), and then, numerically reproduced the experimental setup with the aim of assessing the suitability of this composite as an alternative to thermoset ones. The joints were modelled as two-dimensional (2D) plane-strain cases using FEM and CZM, trapezoidal cohesive laws were employed, and the numerical results were similar to the experimental data. Then, it was found that the chosen composite was a suitable alternative to conventional thermoset composites. Similarly, SLB for adhesive joints were modelled by Santos and Campilho [17] and Loureiro et al. [18]. In both cases, the numerical models reproduced the experimental setup performed by the authors. The numerical models were also 2D assuming plane-strain conditions and triangular cohesive laws were used. In these two cases, three different adhesive types were evaluated. Regardless of the adhesive type, the numerical results agreed with the experimental data gathered a priori, validating the numerical methodologies. Although the described research reached a good agreement between numerical and experimental data, no parametric studies of the cohesive parameters were reported. In this regard, Alfano [20] suggested that these sensitivity analyses are worth exploring. Furthermore, contrary to other joint configurations, the SLB has little presence in the literature, even though it provides data for mixed-mode fracture.

This work aims to validate the application of FEM-CZM to the analysis of SLB adhesive joints. Thus, the geometries used for experimental testing were reproduced numerically, and experimentally obtained properties were employed in these models. Upon the validation of the numerical technique, a parametric study involving the cohesive laws' parameters is performed, aiming to evaluate their influence on the overall behaviour of this type of adhesive joint.

2. Materials and Methods

2.1. Geometry

The SLB specimen has two adherends bonded together, one of them shorter and placed below, to induce mixed-mode loading during bending, i.e., three-point bending. The initial crack (a_0) should obey a relationship of 70% with respect to the half-span between supports (L), i.e., $a_0 = 0.7L$. A schematic of this specimen's geometry is shown in Figure 1 (P is the load and δ is the displacement). The SLB geometry is based on the work of Yoon and Hong [22], later expanded by Chaves et al. [10].

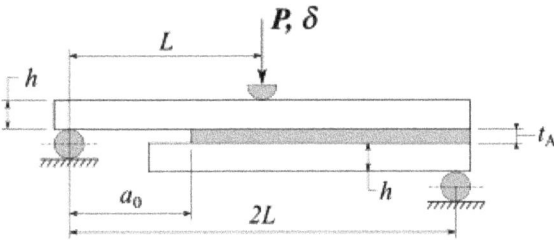

B = Out of plane width

Figure 1. Geometry and dimensions of the SLB specimens, adapted from [10].

In this work, $2L$ = 250 mm, the adherend thickness (h) = 3 mm, t_A = 1 mm, out-of-plane-width (B) = 15 mm, and $a_0 \approx$ 87.5 mm. The overall lengths of the upper and lower adherends were 280 mm and 200 mm, respectively.

2.2. Materials

For this work, three adhesive types were considered, namely the Araldite® AV138, Araldite® 2015, and SikaForce® 7752, varying from brittle to ductile. The Araldite® adhesives are epoxy-based while the SikaForce® is polyurethane-based. These adhesives were experimentally characterised in previous works [12,23,24], and their mechanical properties are listed in Table 1. In addition, the mechanical properties that were obtained employed the appropriate standards, while the fracture properties were obtained from bonded CFRP specimens. The experimental procedures for these tests were described in detail by da Silva et al. [25].

Table 1. Mechanical properties of the adhesives studied. Adapted from [12,23,24].

	AV138		2015		7752	
	Nom	Std	Nom	Std	Nom	Std
Young's modulus, E [GPa]	4890	0.81	1850	0.21	493.81	89.60
Poisson's ratio, ν	0.35	-	0.35	-	0.32	-
Tensile yield stress, σ_y [MPa]	36.49	2.47	12.63	0.61	3.24	0.48
Tensile strength, σ_f [MPa]	39.45	3.18	21.63	1.61	11.49	0.25
Tensile failure strain, ε_f [%]	1.21	0.10	4.77	0.15	19.18	1.40
Shear modulus, G [GPa]	1560	0.01	560	0.21	187.75	16.35
Shear yield stress, τ_y [MPa]	25.1	0.33	14.6	1.30	5.16	1.14
Shear strength, τ_f [MPa]	30.2	0.40	17.9	1.80	10.17	0.64
Shear failure strain, γ_f [%]	7.8	0.70	43.9	3.40	54.82	6.39
G_{IC} [N/mm]	0.2	-	0.43	0.02	2.36	0.17
G_{IIC} [N/mm]	0.38	-	4.7	0.34	5.41	0.47

The adherends were cut from carbon-fibre reinforced plastic (CFRP) plates with a thickness of 3 mm. These plates were manufactured in-house using 20 layers of carbon-epoxy pre-preg (SEAL Texipreg HS 160 RM, Legnano, Italy) with an individual thickness of 0.15 mm. The layers were manually laid-up unidirectionally, i.e., [0]$_{20}$. Then, the plates were pressed at 2 bar and 130 °C for one hour using a dedicated press with hot plates (200 kN press by Gislotica Lda; Perafita, Porto, Portugal). The manufacturing procedure for the composite plates is described in better detail by Santos and Campilho [17]. Regarding the mechanical properties of the prepreg used, these are listed in Table 2.

Table 2. Mechanical properties of the SEAL Texipreg HS 160 RM. Adapted from [26,27].

E		G		ν	
Direction	Value (MPa)	Direction	Value (MPa)	Direction	Value
1	109,000	12	4315	12	0.342
2	8819	13	4315	13	0.342
3	8819	23	3200	23	0.38

2.3. Experimental Details

In this work, three different adhesives were evaluated, and seven specimens per adhesive type were prepared. Therefore, the adherends were cut from the composite plates, mentioned in the previous section, to the appropriate sizes (Figure 1). The cutting of the specimens was done using an abrasive saw with a diamond wheel suitable for composite materials. In addition, several shims were cut and prepared to ensure the desired t_A. Once cut, the adherends and shims were prepared for the bonding process by following the procedure described by Faneco et al. [12]. Furthermore, a razor blade was placed at the end of the bond line, leading to the initial crack notch. Then, the adherends were laid on a flat surface, the spacers were placed, the respective adhesive was applied, and the second adherend was placed on top and aligned. The adherends were kept aligned during the curing time using spring-loaded clamps located in the areas where the shims are, hence ensuring the desired t_A. All the specimens were left to cure at room temperature for three days in the case of both Araldite® adhesives, and five days for the SikaForce®. After the curing process, the shims were removed, and all the excess of adhesive was carefully trimmed using mechanical means. Subsequently, each specimen was marked by adhesive type and specimen number, and the actual dimensions of each one were documented. In order to ease the visualisation and measurement of the crack propagation, one of the side faces of the specimen, including the adhesive layer, was painted in white and a scale was attached to the adherend, as shown in Figure 2. Then, the individual values of a_0 were registered. These processes are described in more detail by da Silva et al. [3].

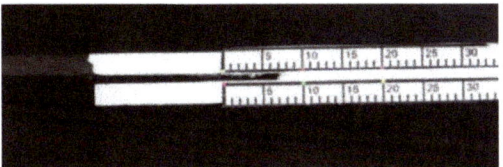

Figure 2. Painting of the specimen face and scale location to aid measuring the crack propagation.

Once all the specimens were prepared and measured, each one was tested using a universal testing machine or UTM (Shimadzu AG-X-100) with a 100 kN load cell. The bending loading was imposed through a fixture compatible with the UTM, as shown in Figure 3.

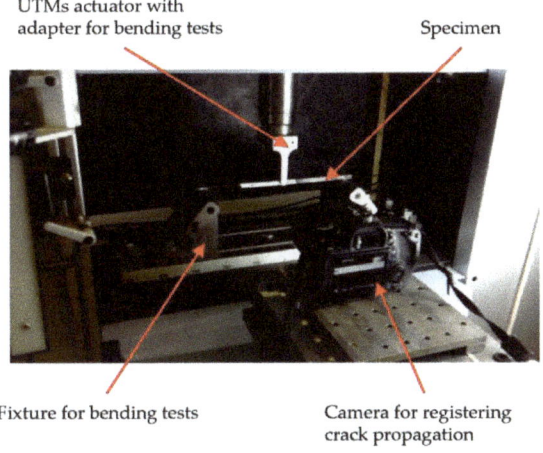

Figure 3. Experimental setup employed.

The testing speeds employed were 0.35 mm/min, 0.8 mm/min, and 3 mm/min for the Araldite® AV138, Araldite® 2015, and SikaForce® 7752, respectively. Furthermore, the crack length (a) was measured using high-resolution pictures focused on the scale attached to the specimen (Figure 2). The pictures were taken every 5 s, with the first photograph taken at the beginning of the test. Therefore, the pictures are related to the UTM data using the time stamps. The experimental setup is depicted in Figure 3. Finally, the tests were run until a reached the loading point (Figure 1).

2.4. J-Integral Formulation

The J-integral formulation was used to estimate the fracture energies from the SLB tests. This contour integral was proposed by Rice [28] in the 1960's to calculate the strain concentration near cracks and notches. Currently, this technique has been extended to several fracture tests, such as the DCB (mode I), ENF (mode II), and SLB (mixed mode). The formulae following in this work were proposed by Ji et al. [11] within the scope of adhesive layer characterisation, ultimately leading to closed-form expressions of G_I and G_{II}, enabling one to obtain the energies and mode-partitioned CZM laws by a differentiation procedure. To make this procedure possible, three relevant geometric variables, apart from the typical P and δ, should be measured during the test (Figure 4): the relative rotation between the two adherends at the loading line (θ_P), the normal separation at the crack tip (δ_n), and the shear separation at the crack tip (δ_s). G_I is given by:

$$G_I(\delta_n) = \int_0^{\delta_n} t_n(\delta_n) d\delta_n = \frac{P}{4}\theta_P \quad (1)$$

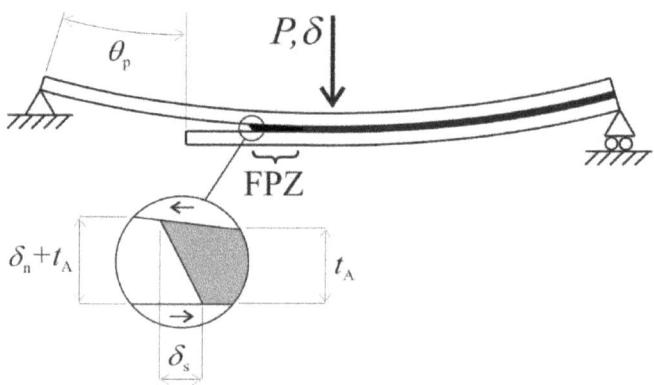

Figure 4. Schematic representation of δ_n, δ_s and θ_p.

In this expression, t_n is the current tensile stress. On the other hand, G_{II} can be calculated as:

$$G_{II}(\delta_s) = \int_0^{\delta_s} t_s(\delta_s) d\delta_s = \frac{\frac{1}{2}\left(\frac{h}{D}\right)^2 Q_T^2 a^2 + \frac{hQ_T}{2D}\delta_0}{\frac{2}{A} + \frac{h^2}{2D}} \quad (2)$$

where t_s represents the current shear stress, D is the beam bending stiffness (assuming identical adherends), A is the beam axial stiffness, and Q_T is the resultant of shear forces acting on the bonded SLB specimen. After having G_I and G_{II} as a function of δ_n and δ_s, respectively, the direct CZM law estimation method gives the tensile (t_n-δ_n) and shear cohesive laws (t_s-δ_s), by differentiating the G_I-δ_n and G_{II}-δ_s curves, respectively, resulting from the former expressions:

$$t_n(\delta_n) = \frac{dG_I(\delta_n)}{d\delta_n} = \frac{d\left\{\frac{P}{4}\theta_P\right\}}{d\delta_n} \quad (3)$$

and

$$t_s(\delta_s) = \frac{dG_{II}(\delta_s)}{d\delta_s} = \frac{d\left\{\frac{\frac{1}{2}\left(\frac{h}{D}\right)^2 Q_T^2 a^2 + \frac{hQ_T}{2D}\delta_s}{\frac{2}{A} + \frac{h^2}{2D}}\right\}}{d\delta_s} \quad (4)$$

As previously mentioned, θ_p, δ_n, and δ_s require continuous measurement during the SLB tests. Data acquisition can rely on mechanical sensors such as linear variable differential transformers (LVDT) or optical methods, including digital image correlation (DIC). The procedure in this work involved using an optical method founded on taking high-resolution pictures during the tests (Figure 3), and then performing a vector and geometric analysis of the images captured during the tests by imaging software to produce a value of θ_p, δ_n, and δ_s for each picture, which can be correlated with the testing machine data. More details about this procedure and geometrical extraction of the parameters from the pictures can be found in previous work [18].

The methodology just described allows one to obtain the current values of G_I and G_{II}, so they can be correlated within a plot, known as the fracture envelope [9]. Then, the mode-mixity is defined through a power law [29], as follows:

$$\left(\frac{G_I}{G_{IC}}\right)^\alpha + \left(\frac{G_{II}}{G_{IIC}}\right)^\beta = 1, \tag{5}$$

where the critical values of G_{IC} and G_{IIC} are known from the characterisation of the material, i.e., as reported in Table 1. The exponents α and β define the shape of the envelope, being commonly considered equal [30], so $\alpha = \beta$, with common values of 0.5, 1, 1.5, and 2. Then, the power law (Equation (5)) is plotted for each value of α. Finally, comparing the points where the current G_I and G_{II} lay in relation to the envelopes provides the exponent α for the analysed test.

2.5. Numerical Modelling

CZM modelling of the SLB specimens in Abaqus® is employed in this work to validate the CZM laws and fracture envelopes defined in the experimental part. The simulation is geometrically non-linear, which is mandatory for the magnitude of involved deformations. The mesh refinement was optimised, with higher refinement at the crack growth region and contact with the loading cylinders (as shown in Figure 5, together with the boundary conditions). Since the models are 2D, the adherends were discretised by plane-strain four-node solid finite elements (CPE4 from Abaqus®), and the adhesive by four-node cohesive elements (COH2D4 from Abaqus®). Bias effects were used to reduce the computational effort while concentrating elements where needed: six elements were considered through-thickness in the adherends with a minimum size of 0.1 mm and a maximum size of 0.2 mm, showing higher refinement at the free faces [31]. The element size in the bond line was 0.5 mm × 1 mm from the crack notch until the centre support (L from Figure 1) while the remaining size was 1 mm × 1 mm. The mesh size in the vicinities of the rollers was finer to reduce element distortion. In this case, the element size was 0.05 mm. The element size on regions of low interest was 1 mm. The models were composed of 6144 CPE4 elements, 400 COH2D4 elements, and a total of 8676 nodes. It is worth noting that the mesh sizes were chosen from the authors' previous experience with similar finite element models. Furthermore, the chosen mesh size is also in agreement with those reported in the literature for similar cases. Thus, mesh sensitivity analyses were not required.

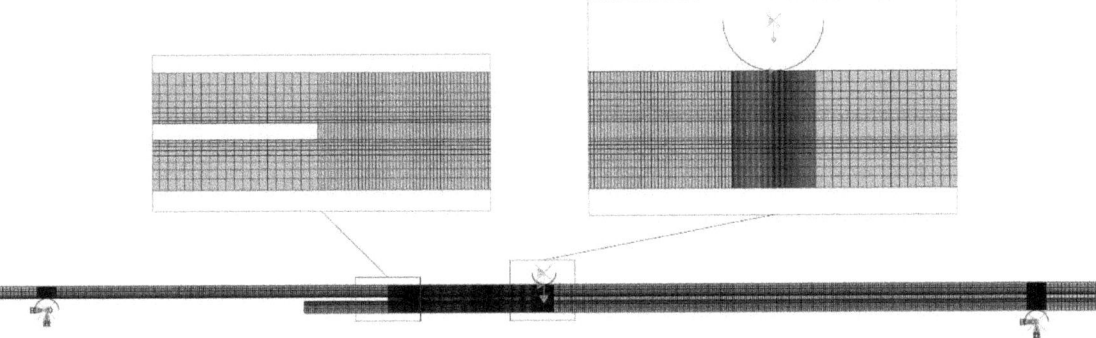

Figure 5. Mesh details and boundary conditions for the SLB model. The upper right close-up shows the horizontal constraint at the centre span.

Following the geometry shown in Figure 1, the substrates are supported and loaded through rollers. Therefore, the centres of the supporting rollers were fixed in both directions ($U_X = U_Y = 0$), which reproduces the experimental setup. In addition, the upper roller applying the displacement was constrained in the horizontal direction ($U_X = 0$) while its vertical displacement corresponds to the displacement imposed by the UTM, i.e., $U_Y = \delta$. Furthermore, the point of contact of the upper roller was also constrained in the horizontal direction, as shown in Figure 5, reducing the degrees of freedom of the system. Nevertheless, no horizontal displacement was observed during the experimental testing. The interactions between rollers and substrates were defined through surface-to-surface frictionless contact conditions with hard behaviour in the normal direction.

The modelling procedure consisted of setting one individual model for each experimental test, including the measured dimensions and a_0, for maximum accuracy. The adhesive layer was modelled by one row of four-node cohesive elements whose definition is based on the pure tensile and shear CZM laws; in this case, triangular cohesive laws were employed, of which, the relevant properties (E, G, and tensile cohesive strength or $t_n{}^0$, shear cohesive strength or $t_s{}^0$, G_{IC}, and G_{IIC}) were taken from Table 1. To numerically establish the mixed-mode behaviour, it is necessary to know the power-law exponent, which is calculated from the experimental data, namely when building the fracture envelopes for each adhesive. Thus, this exponent may differ between tested adhesives. The comparison between the experimental data and numerical predictions in the results section will be able to validate the CZM law and respective mixed-mode criteria for strength prediction of bonded joints.

2.6. Triangular CZM

CZM modelling relies on the establishment of stress-relative displacement laws or CZM laws that link paired nodes of the cohesive elements. The CZM laws reproduce the materials' elastic behaviour up to reaching the cohesive strength in the respective loading mode and the damage or softening process that follows, to simulate the material degradation until failure and respective crack growth. G_{IC} and G_{IIC} correspond to the area beneath the tensile and shear CZM laws, respectively. When considering pure mode, damage grows at a set of paired nodes when stresses are cancelled at the end of softening. On the other hand, under mixed mode damage growth is ruled by energetic criteria that combine the individual loading modes [32]. Triangular CZM laws were considered in this work, i.e., with linear softening, for pure and mixed-mode analysis. A schematic representation of this law is shown in Figure 6. In the pure mode laws, the linear part of the curve up to the cohesive strength is defined by a matrix that relates stresses with strains, and with E and ν as main parameters. Although damage initiation under mixed mode can be assessed by different criteria, this work uses the quadratic nominal stress criterion. Upon reaching the mixed-mode cohesive strength ($t_m{}^0$), the material stiffness is degraded. Damage growth, i.e., separation of the paired nodes, is predicted using a power law expression based on the current G_I and G_{II} (Equation (5)), initially proposed by Wu and Reuter [29]. In this work, it was considered that $\alpha = \beta$, whose numerical value was estimated using experimental data (Section 2.5) and subsequently validated numerically. Further details of this model are given in reference [23].

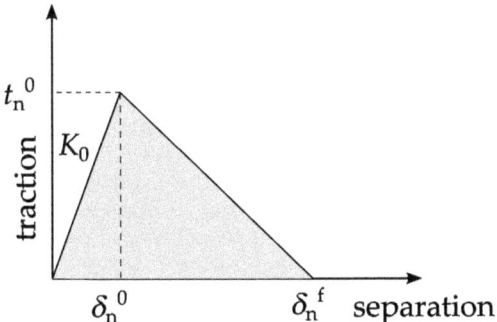

Figure 6. Schematic of a triangular cohesive law, adapted from [20].

3. Results and Discussions

3.1. P-δ Curves

The P-δ curves were the initially collected data for the tests, leading to the subsequent fracture analyses. Figure 7a gives an example of the correlation between specimens of the same adhesive (Araldite® 2015), and Figure 7b shows sample P-δ curves for each of the three adhesives, to visually reinforce the differences between adhesives. Figure 7a emphasises the repeatability of the test data, showing that the specimens were fabricated and tested under identical conditions. This agreement is also valid for the other two adhesives tested in this work. Minor elastic stiffness variations take place because of differences in a_0 between specimens. Figure 7b shows a markedly different efficiency of the adhesives, which relates to mixed-mode fracture, made visible by the different maximum load (P_m) and maximum load displacement (δP_m). In the Araldite® AV138, the evolution of P with δ is predominantly linear until the crack begins to propagate. After crack onset, few specimens showed unstable crack propagation, which is considered to be related to the presence of small defects in a brittle adhesive, triggering catastrophic failure [33]. For this adhesive, $P_m = 81.1 \pm 4.5$ N and $\delta P_m = 2.11 \pm 0.23$ mm. The Araldite® 2015 shows an improved fracture behaviour, due to much higher P_m and δP_m ($P_m = 204.2 \pm 12.8$ N and $\delta P_m = 5.6 \pm 5.58$ mm). Although this adhesive possesses lower stiffness and tensile strength than the Araldite® AV138, it also has higher ductility, hence performs better within the scope of fracture tests. Moreover, the sample P-δ curve reveals non-negligible softening up to P_m, associated to the creation of a bigger FPZ that develops at the crack tip before crack onset. Finally, the SikaForce® 7752 presents the best toughness results, with $P_m = 630.3 \pm 26.0$ N and $\delta P_m = 28.4 \pm 1.22$ mm. Compared to the previous adhesives, there is a marked softening before P_m, denoting the large dimensions' FPZ taking place before crack growth, accompanied by a softer transition to failure. These differences should reflect in the fracture measurements that follow.

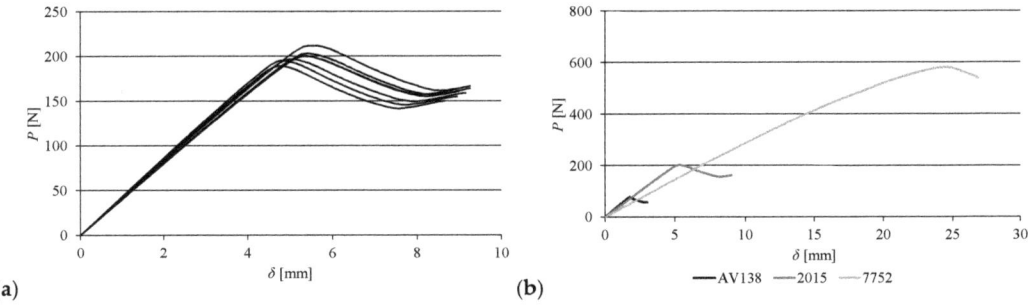

Figure 7. P-δ curves for the Araldite® 2015 (**a**) and sample P-δ curves for each adhesive (**b**).

3.2. Toughness Estimation

Estimation of G_I and G_{II} for all specimens was performed as specified in Section 2.4, beginning with plotting θ_p, δ_n, and δ_s vs. δ curves up to crack initiation, for application of the J-integral formulation. To make the curves smoother, all curves were subjected to polynomial fitting, which was successful in the sense that it was possible to replicate the experimental evolution with accuracy. It was found that the evolution of δ_n and δ_s with δ is exponential [34], while the θ_P–δ curves are nearly linear. After applying the formulae of Section 2.4, namely expressions (1) and (2), it was possible to derive the G_I–δ_n and G_{II}–δ_s plots up to crack initiation, which are on the basis of the CZM law calculation by expressions (3) and (4). Figure 8a shows sample curves for an SLB specimen bonded with the Araldite® 2015. Normally, these curves are divided into three portions: the first part with a slow increase of G_I or G_{II}, followed by a marked increase, whose maximum slope gives t_n^0 or t_s^0, and finally, the attainment of a steady-state value of G_I or G_{II}, corresponding to crack initiation. This behaviour was generally observed in the tested specimens, although with a few inconsistencies in some specimens due to experimental issues and fitting difficulties. The main problem was the curve initiation with a non-nil slope, which then reflected on non-nil stress at the initiation of the respective CZM laws. The correlation of this data with a, measured from the experimental tests, gives the R-curves, of which an example is presented in Figure 8b for the SLB bonded with the Araldite® 2015. For all adhesives, it was found that the tensile and shear plots are identical, although with $G_I > G_{II}$. All R-curves begin at the a value of a_0, corresponding to the steep increase of G_I or G_{II} triggering crack initiation, followed by a theoretically horizontal evolution of G_I or G_{II}, in which the critical values are measured by averaging. The average and standard deviation data for each adhesive (including G_I and G_{II}) are given in Table 3. The maximum coefficient of variation occurred for G_{II} of the Araldite® 2015, of 6.1%. On the other hand, the difference was high between adhesives, reflecting their known brittleness or ductility.

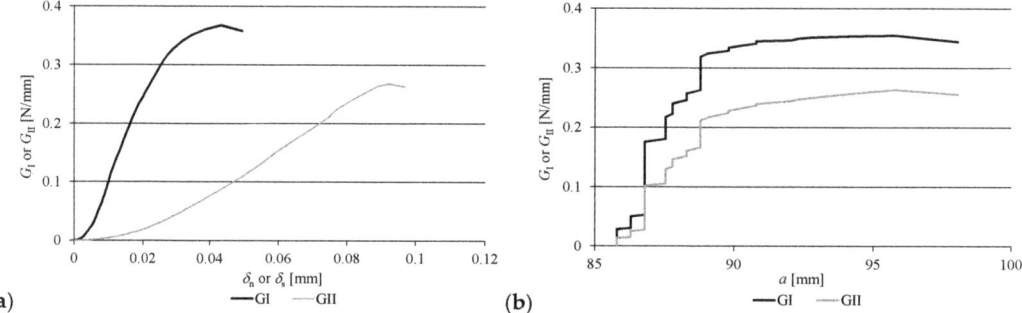

Figure 8. Sample G_I-δ_n and G_{II}-δ_s curves (**a**) and R-curves (**b**) for the Araldite® 2015.

Table 3. G_I and G_{II} for the three adhesives in the SLB test.

	Araldite® AV138		Araldite® 2015		SikaForce® 7752	
Specimen No.	G_I [N/mm]	G_{II} [N/mm]	G_I [N/mm]	G_{II} [N/mm]	G_I [N/mm]	G_{II} [N/mm]
Average	0.0657	0.0404	0.3663	0.263	3.383	2.567
Deviation	0.0024	0.0017	0.0073	0.016	0.050	0.042

3.3. Fracture Envelope

The fracture envelopes enable framing the mixed-mode behaviour of the adhesives by plotting the G_I/G_{II} data points against idealised power law criteria having as limits the G_{IC} and G_{IIC} of pure tensile (DCB) and shear (ENF) results [17]. The power law expressions are obtained from Equation (5), considering $\alpha = \beta$. Thus, from this point on, the exponent in the power law expression is cited as α. Different power laws ($\alpha = 1/2, 1, 3/2,$ and 2) are evaluated to reproduce the experimental mixed-mode behaviour of each tested adhesive. Figure 9 presents the experimental fracture envelopes for the three adhesives separately.

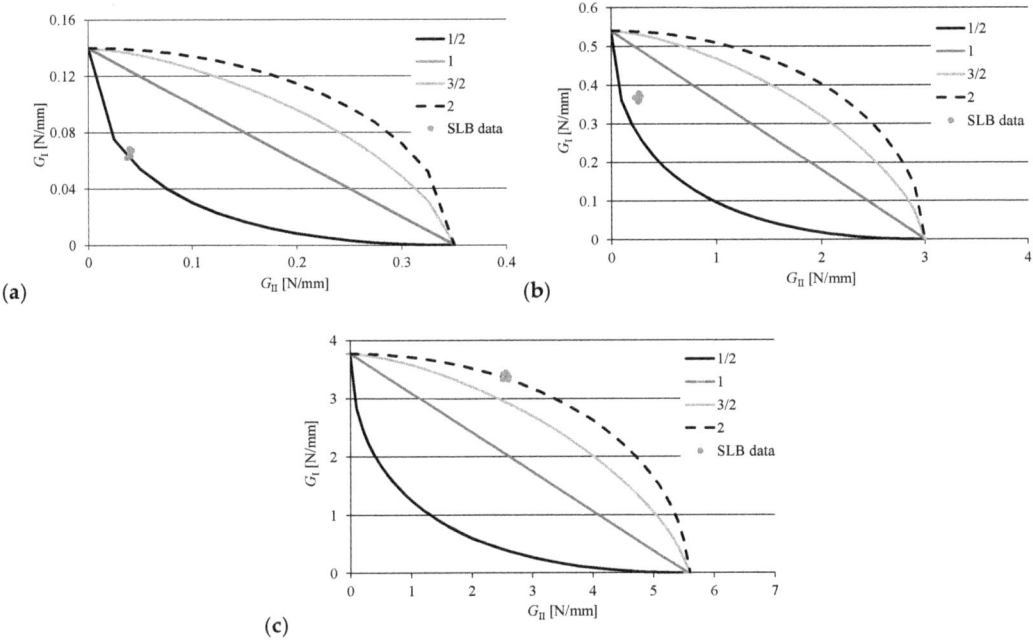

Figure 9. Experimental fracture envelopes for the adhesives Araldite® AV138 (**a**), Araldite® 2015 (**b**), and SikaForce® 7752 (**c**).

Figure 9a, relating to the Araldite® AV138, reveals proximal data points, leading to coefficients of variation of approximately 4% for both G_I and G_{II}. For this adhesive, $\alpha = 1/2$ reveals to be an accurate representation since all data points are close to this criterion. Figure 9b presents the fracture envelope for the Araldite® 2015, and highlights the good agreement between specimens, materialised by coefficients of variation of approximately 2% (G_I) and 6% (G_{II}). Although in this case $\alpha = 1/2$ is identically the best option for mixed-mode failure prediction, the data points are clearly above the criterion. Figure 9c, representing the SikaForce® 7752 fracture envelope, depicts coefficients of variation below 2% for both loading modes but reveals a markedly different behaviour to the other two adhesives. For this adhesive, the data points are situated near the $\alpha = 2$ criterion, which may be related to the polyurethane base and associated ductility.

3.4. CZM Laws

The CZM laws in both modes of loading were estimated by the direct method, as described in Section 2.4. To apply expressions (3) and (4), applicable to the mode I and II laws, respectively, it was previously necessary to approximate the data points of the G_I-δ_n and G_{II}-δ_s functions by polynomial functions, individually for each specimen, for further differentiation. Figure 10 represents, as an example, the full set of tensile (a) and shear (b) CZM laws for the Araldite® 2015, which also represents the degree of correspondence for the other two adhesives. The agreement was generally very good regarding the sets of tensile or shear CZM laws of a given adhesive, including the elastic portion up to t_n^0 or t_s^0, the values of t_n^0 and t_s^0, and also the tensile and shear failure displacements (δ_n^f

and δ_s^f, respectively). Typically, the t_n-δ_n and t_s-δ_s laws do not initiate with nil stresses, as expected, due to using polynomial approximations. The Araldite® AV138 CZM laws revealed a triangular-like form under tensile and shear assumptions. The collected data for this adhesive was as follows: $t_n^0 \approx 35$ MPa, $t_s^0 \approx 18$ MPa, $\delta_n^f \approx 0.01$ mm, and $\delta_s^f \approx 0.02$ mm. The values of δ_n^f and δ_s^f are much reduced, which can be associated with brittleness and stiff behaviour. The CZM laws of the Araldite® 2015, corresponding to the sample curves shown in Figure 10, equally depict a triangular-like shape, but ductility signs were visible near failure. The collected information for this adhesive was as follows: $t_n^0 \approx 17$ MPa, $t_s^0 \approx 7$ MPa, $\delta_n^f \approx 0.05$ mm, and $\delta_s^f \approx 0.1$ mm. Comparison of these values with those of the Araldite® AV138 gives an increase of δ_n^f of 421%, and δ_s^f of 358%. The SikaForce® 7752 CZM laws showed a significantly different shape compared to the former two adhesives, namely in the shear CZM law, which revealed a large steady-state region with significant stresses, i.e., resembling a trapezoidal shape CZM. This result arises from the large ductility of the SikaForce® 7752, and it is considered that this adhesive could be better modelled by a trapezoidal law [20]. The average data for the SikaForce® 7752 led to the smallest t_n^0 and t_s^0, and the biggest δ_n^f and δ_s^f ($t_n^0 \approx 6$ MPa, $t_s^0 \approx 5$ MPa, $\delta_n^f \approx 1.6$ mm, and $\delta_s^f \approx 1$ mm). The δ_n^f and δ_s^f values are much higher than for the other adhesives, with more significance for δ_n^f. Considering all adhesives and both loading modes, the coefficients of variations were typically under 10% for t_n^0 and t_s^0, while δ_n^f and δ_s^f could not comply with this standard and showed higher variations.

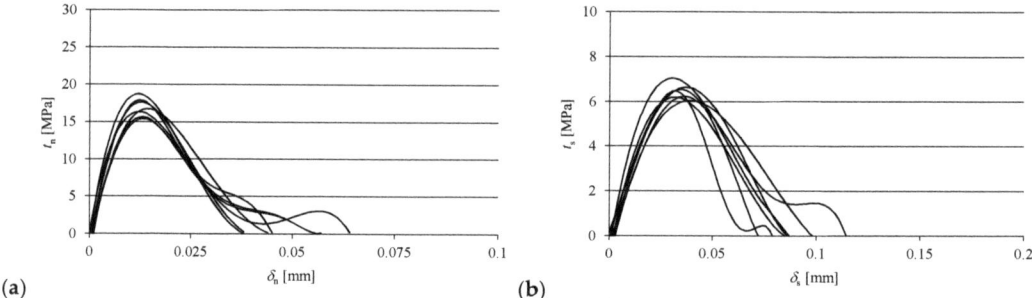

Figure 10. Experimental CZM laws for the Araldite® 2015 data: tensile (**a**) and shear (**b**).

3.5. CZM Law Validation

Validation of the cohesive laws was done through the comparison between experimental and numerical values of P_m and δP_m. Regarding the values of G_{IC} and G_{IIC}, i.e., pure mode, data from the literature were used due to the mode-mixity found in the SLB. Then, the experimentally defined α for each adhesive was assigned to each case tested. Starting with the data of the Araldite® AV138, the numerical P_m was close to the experimental one, being on average 2.6% lower (range 0.3% to −6.0%). Similarly, the numerical δ was 2.2% lower than the experimental values (range 1.4% to −7.9%). The highest difference in both values was observed on Specimen #1. Despite this fact, a good agreement between numerical and experimental data was attained. As an example, the comparison for Specimen #3 is shown in Figure 11a. Following, in the Araldite® 2015 a similar trend was observed, and the numerical P_m was on average 3.3% lower than the experimental value (range 3.8% to −7.2%). The numerical δ was on average 5.9% lower than its experimental counterpart (range 0% to −11.0%). The overall shapes of the numerical and experimental curves also matched. For example, the comparison for Specimen #3 is shown in Figure 11b. For the SikaForce® 7752, the numerical P_m was on average 7.0% lower than the experimental values (range −1.7% to −12.5%). In this case, the numerical models underpredicted δ on average by −14.0% (range −9.1% to −19.26%). Regardless of these differences, the P-δ curves, both numerical and experimental, showed similar behaviours. For example, the comparison corresponding to Specimen #3 is shown in Figure 11c.

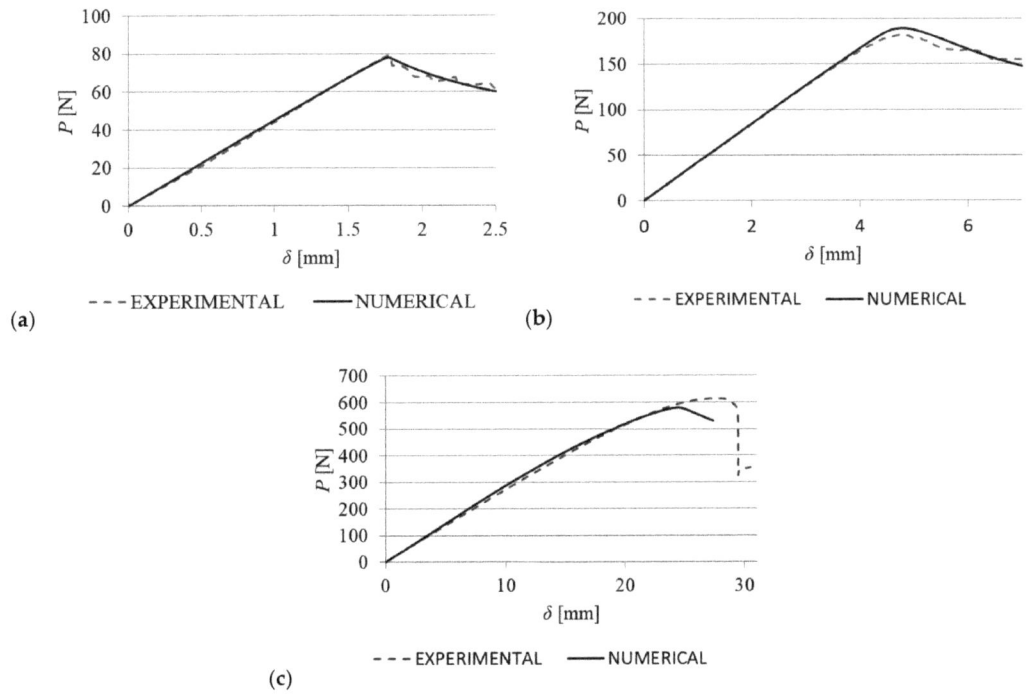

Figure 11. Comparison between numerical and experimental P-δ curves for the three adhesives studied: Araldite® AV138 (**a**), Araldite® 2015 (**b**), and SikaForce® 7752 (**c**).

Subsequently, the values of G_{IC} were determined from all the experimental and numerical cases and then compared between them. Starting with the Araldite® AV138, the value obtained from the numerical data was on average 1.0% higher than the experimental one (range 7.2% to −3.05%). For the Araldite® 2015, the numerical value was on average 0.2% lower than the experimental one (range 5.0% to −5.5%). This trend continued for the SikaForce® 7752, since the numerical value was on average −0.3% lower than the experimental one (range −0.2% to −0.8%). Overall, the variability observed in both numerical and experimental data was small, regardless of the adhesive type, as shown in Figure 12a. A similar approach was followed to determine G_{IIC}. However, this parameter presented higher variability. For the Araldite® AV138, the numerical value was on average 1.0% higher than the experimental one (range 0.0% to 5.0%). In the case of the two ductile adhesives, the average numerical value was lower than the experimental, by 11.3% (range 4.5% to −17.0%) and 2.4% (range −0.3% to −4.0%) for the Araldite® 2015 and the SikaForce® 7752, respectively. Despite these differences, a good agreement between numerical and experimental values was obtained, as shown in Figure 12b. Finally, the good agreement between numerical and experimental data regarding P_m, δP_m, G_{IC}, and G_{IIC}, observed in the described results, indicates that the chosen cohesive law is suitable for this application.

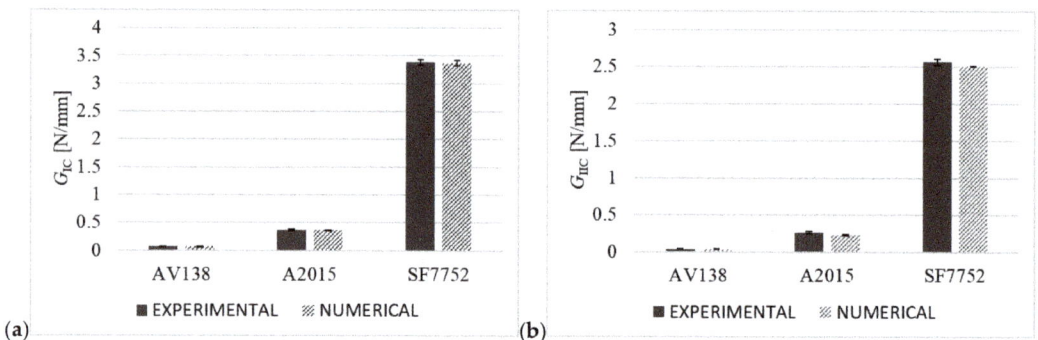

Figure 12. Comparison between experimental and numerical values of G_{IC} (**a**) and G_{IIC} (**b**), by adhesive type.

3.6. Fracture Envelope Validation

The previously obtained values of G_{IC} and G_{IIC} were related to obtain the fracture envelopes, as shown in Figure 13. In addition, small dispersion can be observed for the three adhesives studied (Figure 13), indicating the repeatability of the tests.

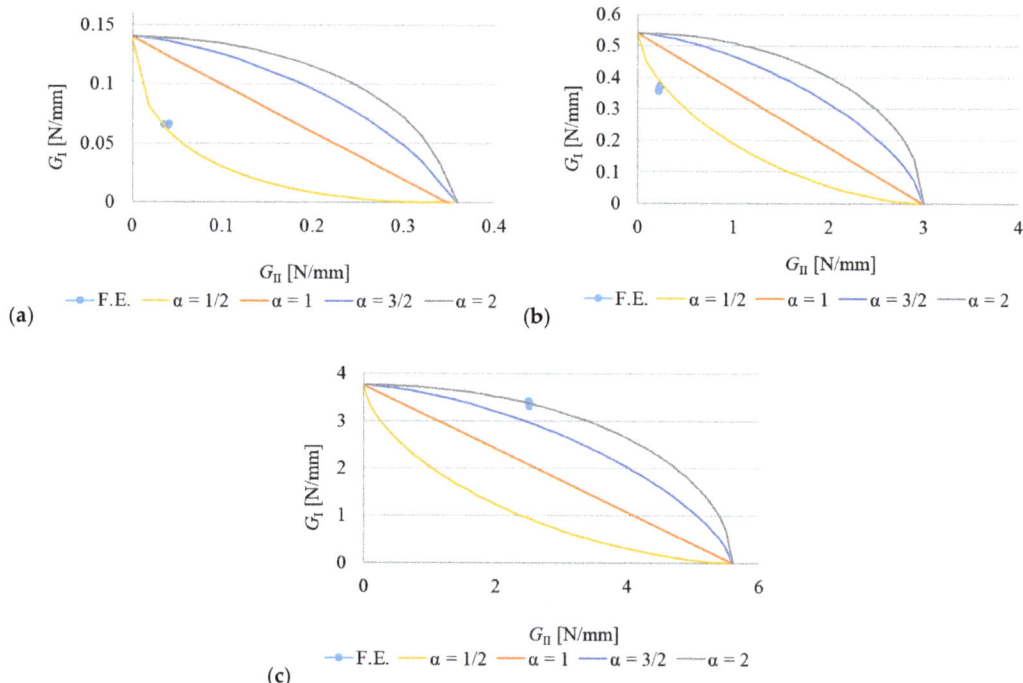

Figure 13. Fracture envelopes for the three adhesives studied: Araldite® AV138 (**a**), Araldite® 2015 (**b**), and SikaForce® 7752 (**c**).

Results for the Araldite® AV138 show that α is equal to 0.5 (Figure 13a). Furthermore, the position of the points within the fracture envelope was found in agreement with previous work [17], validating this work's fracture envelopes. Regarding the Araldite® 2015, the results present minimal scatter, as shown in Figure 13b. The position of the points within the fracture envelope indicates $\alpha = 0.5$, which is also in agreement with previous research [17], although with larger differences than those found for the Araldite® AV138. For the SikaForce® 7752, the position of the points on the fracture envelope indicates $\alpha = 2$, as shown in Figure 13c. The scatter observed in these data is also minimal, indicating good repeatability of the method. In addition, the value of α for this adhesive is also in agreement with previous work [17]. Finally, the similarities between the values in this work and those found in the literature, i.e., [17], validate the employed methodology.

3.7. CZM Parameter Analysis

The influence of G_{IC}, G_{IIC}, t_n^0, and t_s^0 on the P-δ curve was evaluated through a parametric study. In this case, four values of each parameter were tested, i.e., -50%, -25%, 25%, and 50% related to the previously described base values. The effect of these changes was evaluated per variable and with multiple variables. The variation of G_{IC} had a proportional effect on the P-δ curves, regardless of the adhesive type, while the stiffness of the joint remained constant, as shown in Figure 14. On the other hand, the variation of G_{IIC} has minimal influence on the P-δ curves and P_m, in particular, being the relative difference 6.2% for the Araldite® AV138, 6.0% for the Araldite® 2015, and 7.5% for the SikaForce® 7752. The effect was found larger as G_{IIC} was reduced. Subsequently, the combined effect of increasing or decreasing G_{IC} and G_{IIC} was studied. The increase in both parameters had a proportional effect on the P-δ curves, something similar to that observed with G_{IC} alone (Figure 14). However, the increase in P_m is higher due to the small contribution of G_{IIC}.

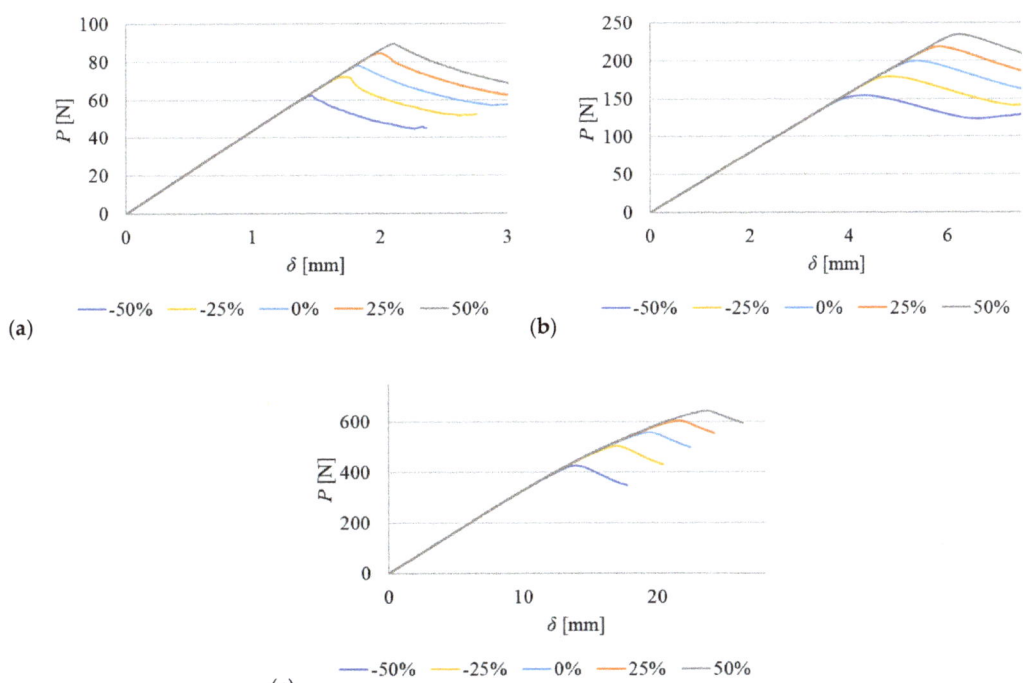

Figure 14. Effect of G_{IC} over the P-δ curve for the three adhesives: Araldite® AV138 (**a**), Araldite® 2015 (**b**), and SikaForce® 7752 (**c**). 0% corresponds to the reference case.

Considering the effect of t_n^0 on the P-δ curves, for the Araldite® AV138, the increase of this parameter has a negligible effect on joint strength. Nevertheless, the reduction in t_n^0 has a positive effect on P_m, increasing the strength of the joint, with the relative differences equal to 3.0% and 7.8% for the -25% and -50% cases, respectively. The effect of varying t_n^0 for this adhesive is shown in Figure 15a. Then, for the Araldite® 2015, a similar trend was observed (Figure 15b), but the increase was smaller. In this case, the increases in P_m were 1.3% and 3.0% for the -25% and 50% cases, respectively. In addition, the joint stiffness gradually reduced before the onset of crack propagation, as shown in Figure 15b. On the contrary, the effect of t_n^0 on P_m for the joints bonded with the SikaForce® 7752 was proportional, although negligible with a maximum increase of 0.3% for the 50% case, as shown in Figure 15c. For this adhesive, the stiffness reduction is more visible than for the Araldite® 2015, as shown in Figure 15b. Regarding the effect of t_s^0, the variation of this parameter has little effect on joint strength. However, its effect is similar to that observed with G_{IIC}, being more influential for t_s^0 reductions. A similar effect was observed in the three adhesives studied. Additionally, the combined effect of t_n^0 and t_s^0 had little influence on the overall behaviour of the joint, regardless of the adhesive type, although it should be noted that the stiffness reduced in the joints bonded with the Araldite® 2015 and SikaForce® 7752, and were more visible in the latter.

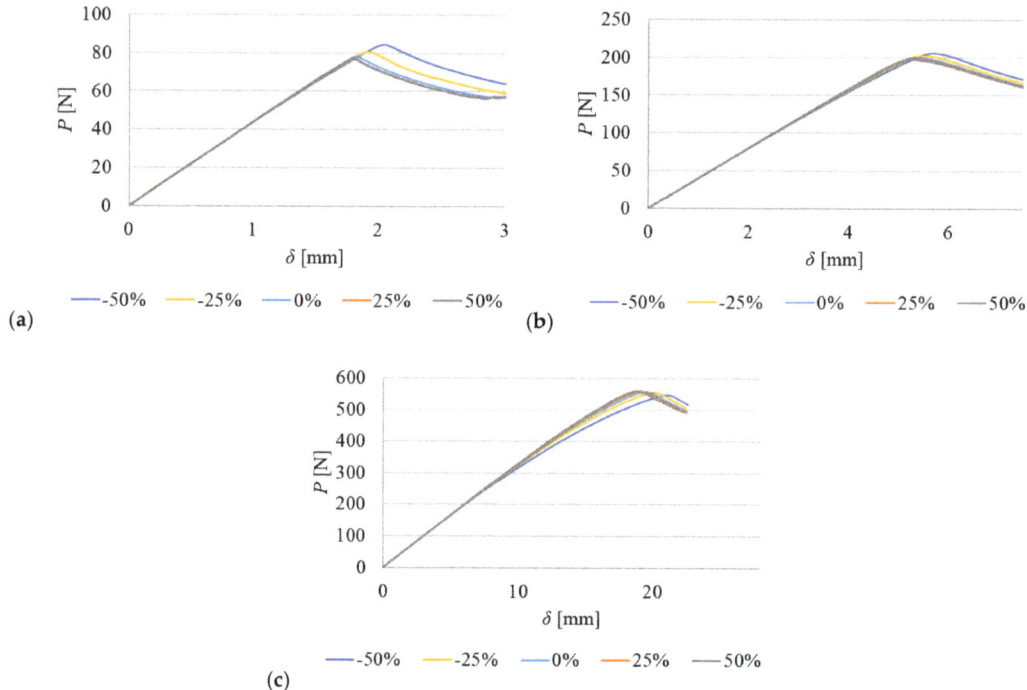

Figure 15. Effect of t_n^0 on the P-δ curves for the three studied adhesives: Araldite® AV138 (a), Araldite® 2015 (b), and SikaForce® 7752 (c). 0% corresponds to the reference case.

The combined effect of the four parameters on the P-δ curves was also evaluated. For the Araldite® AV138, P_m was proportional to the increase of the four parameters, as shown in Figure 16a. However, it can also be observed that G_{IC} influenced joint strength the most. Next, the Araldite® 2015 shows a similar pattern; however, the effect of t_n^0 and t_s^0 is observed in the gradual reduction of the stiffness prior to the crack propagation region, as shown in Figure 16b, although G_{IC} continued to be the most dominant parameter. Finally, for the SikaForce® 7752, the combined effect shows a similar trend to that observed in the Araldite® 2015, as shown in Figure 16c. From the comparison between Figures 14 and 16, it can be observed that, for the three adhesives studied, the variation of G_{IC} has the largest influence on P_m, while the variation of t_n^0 and t_s^0 affects the joint stiffness prior to crack propagation, mostly in the joints bonded with ductile adhesives. It is important to note that, in all cases, the displacements at failure (δ_n^f and δ_s^f) of the cohesive laws were automatically adjusted by the software to maintain the set value of energy (G_{IC} and G_{IIC}), hence maintaining the area beneath the triangular law [2].

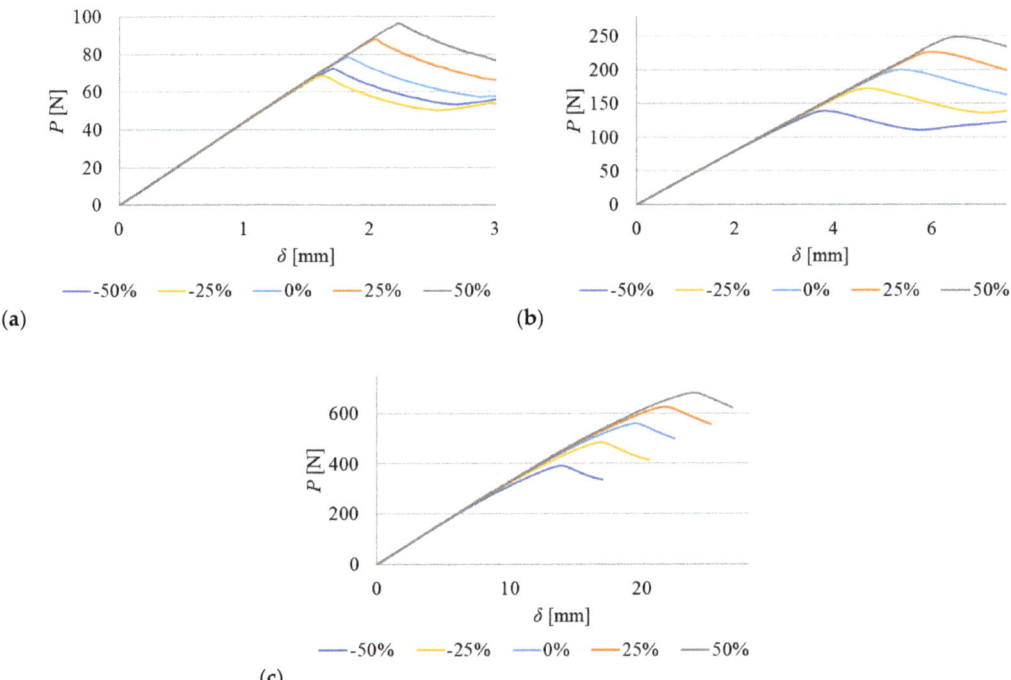

Figure 16. Combined effect of the variation of G_{IC}, G_{IIC}, t_n^0, and t_s^0 for the three adhesives studied: Araldite® AV138 (**a**), Araldite® 2015 (**b**), and SikaForce® 7752 (**c**). 0% corresponds to the reference case.

4. Conclusions

This work aimed to study the mechanical behaviour of the SLB joint through numerical analyses and to estimate the values of G_{IC} and G_{IIC} through both experimental and numerical tests. These parameters are valuable for the further design of bonded structures. Good repeatability was observed in the experimental work performed. Similarly, a good agreement between numerical and experimental results was found, indicating the suitability of the employed methodology to estimate G_{IC} and G_{IIC}. These values were significantly different, and the Araldite® AV138 presented the lowest and the SikaForce® 7752 the highest. The numerical fracture envelopes made it possible to estimate α, giving $\alpha = 1/2$ for the Araldite® AV138 and 2015, and $\alpha = 2$ for the SikaForce® 7752. These values were found similar in the tested experimental and numerical cases, further validating the methodology. The P-δ curves, and P_m in particular, were found to be sensitive to variations of G_{IC}, its effect being proportional regardless of the adhesive type, as was found through a sensitivity analysis. Similarly, variations on t_n^0 have an inversely proportional effect only in the joints bonded with the Araldite® AV138. This effect was attributed to a delay in the crack propagation, thus positively influencing P_m. Furthermore, variations of t_n^0 had no influence on P_m in the joints bonded with the ductile adhesives (Araldite® 2015 and SikaForce® 7752). Instead, these variations changed the stiffness at the initiation of the softening phases. As a result of this work, it was possible to numerically model SLB tests of adhesive joints and estimate α, which enables mixed-mode modelling and design of adhesive structures with the tested adhesives.

Author Contributions: Conceptualisation: L.F.R.N. and R.D.S.G.C.; data curation: L.F.R.N., R.D.S.G.C. and I.J.S.-A.; formal analysis: L.F.R.N. and I.J.S.-A.; methodology: R.D.S.G.C.; resources: C.P. and K.M.; software: L.F.R.N. and I.J.S.-A.; supervision: R.D.S.G.C., C.P. and K.M.; writing—original draft: L.F.R.N., R.D.S.G.C. and I.J.S.-A.; writing—Review and editing: C.P. and K.M. All authors have read and agreed to the published version of the manuscript.

Funding: This research received no external funding.

Informed Consent Statement: Not applicable.

Data Availability Statement: Not applicable.

Conflicts of Interest: The authors declare no conflict of interest.

References

1. Budzik, M.K.; Wolfahrt, M.; Reis, P.; Kozłowski, M.; Sena-Cruz, J.; Papadakis, L.; Nasr Saleh, M.; Machalicka, K.V.; Teixeira de Freitas, S.; Vassilopoulos, A.P. Testing mechanical performance of adhesively bonded composite joints in engineering applications: An overview. *J. Adhes.* **2022**, *98*, 2133–2209. [CrossRef]
2. Tserpes, K.; Barroso-Caro, A.; Carraro, P.A.; Beber, V.C.; Floros, I.; Gamon, W.; Kozłowski, M.; Santandrea, F.; Shahverdi, M.; Skejić, D.; et al. A review on failure theories and simulation models for adhesive joints. *J. Adhes.* **2022**, *98*, 1855–1915. [CrossRef]
3. Da Silva, L.F.M.; Giannis, S.; Adams, R.D.; Nicoli, E.; Cognard, J.-Y.; Créac'hcadec, R.; Blackman, B.R.K.; Singh, H.K.; Frazier, C.E.; Sohier, L.; et al. Manufacture of Quality Specimens. In *Testing Adhesive Joints: Best Practices*; da Silva, L.F.M., Dillard, D.A., Blackman, B., Adams, R.D., Eds.; Wiley-VCH Verlag & Co. KGaA: Weinheim, Germany, 2012; pp. 1–78.
4. Shahapurkar, K.; Alblalaihid, K.; Chenrayan, V.; Alghtani, A.H.; Tirth, V.; Algahtani, A.; Alarifi, I.M.; Kiran, M.C. Quasi-Static Flexural Behavior of Epoxy-Matrix-Reinforced Crump Rubber Composites. *Processes* **2022**, *10*, 956. [CrossRef]
5. Banea, M.D.; da Silva, L.F.M.; Campilho, R.D.S.G. The Effect of Adhesive Thickness on the Mechanical Behavior of a Structural Polyurethane Adhesive. *J. Adhes.* **2015**, *91*, 331–346. [CrossRef]
6. Kafkalidis, M.S.; Thouless, M.D. The effects of geometry and material properties on the fracture of single lap-shear joints. *Int. J. Solids Struct.* **2002**, *39*, 4367–4383. [CrossRef]
7. Astrouski, I.; Kudelova, T.; Kalivoda, J.; Raudensky, M. Shear Strength of Adhesive Bonding of Plastics Intended for High Temperature Plastic Radiators. *Processes* **2022**, *10*, 806. [CrossRef]
8. Dillard, D.A. *Fracture Mechanics of Adhesive Bonds*; Adams, R.D., Ed.; Woodhead Publishing Limited: Sawston, UK, 2005; pp. 189–208.
9. Pearson, R.A.; Blackman, B.R.K.; Campilho, R.D.S.G.; de Moura, M.F.S.F.; Dourado, N.M.M.; Adams, R.D.; Dillard, D.A.; Pang, J.H.L.; Davies, P.; Ameli, A.; et al. Quasi-Static Fracture Tests. In *Testing Adhesive Joints: Best Practices*; da Silva, L.F.M., Dillard, D.A., Blackman, B.R.K., Adams, R.D., Eds.; Wiley-VCH Verlag & Co. KGaA: Weinheim, Germany, 2012; pp. 163–272.
10. Chaves, F.J.P.; da Silva, L.F.M.; de Moura, M.F.S.F.; Dillard, D.A.; Esteves, V.H.C. Fracture mechanics tests in adhesively bonded joints: A literature review. *J. Adhes.* **2014**, *90*, 955–992. [CrossRef]
11. Ji, G.; Ouyang, Z.; Li, G. On the interfacial constitutive laws of mixed mode fracture with various adhesive thicknesses. *Mech. Mater.* **2012**, *47*, 24–32. [CrossRef]
12. Faneco, T.M.S.; Campilho, R.; Silva, F.J.G.; Lopes, R. Strength and fracture characterization of a novel polyurethane adhesive for the automotive industry. *J. Test. Eval.* **2017**, *45*, 398–407. [CrossRef]
13. Cardoso, M.G.; Pinto, J.E.C.; Campilho, R.D.S.G.; Nóvoa, P.J.R.O.; Silva, F.J.G.; Ramalho, L.D.C. A new structural two-component epoxy adhesive: Strength and fracture characterization. *Procedia Manuf.* **2020**, *51*, 771–778. [CrossRef]
14. Shiino, M.Y.; Alderliesten, R.C.; Donadon, M.V.; Cioffi, M.O.H. The relationship between pure delamination modes I and II on the crack growth rate process in cracked lap shear specimen (CLS) of 5 harness satin composites. *Compos. Part A Appl. Sci. Manuf.* **2015**, *78*, 350–357. [CrossRef]
15. Bennati, S.; Fisicaro, P.; Valvo, P.S. An enhanced beam-theory model of the mixed-mode bending (MMB) test—Part I: Literature review and mechanical model. *Meccanica* **2013**, *48*, 443–462. [CrossRef]
16. Oliveira, J.J.G.; Campilho, R.D.S.G.; Silva, F.J.G.; Marques, E.A.S.; Machado, J.J.M.; da Silva, L.F.M. Adhesive thickness effects on the mixed-mode fracture toughness of bonded joints. *J. Adhes.* **2020**, *96*, 300–320. [CrossRef]
17. Santos, M.A.S.; Campilho, R.D.S.G. Mixed-mode fracture analysis of composite bonded joints considering adhesives of different ductility. *Int. J. Fract.* **2017**, *207*, 55–71. [CrossRef]
18. Loureiro, F.J.C.F.B.; Campilho, R.D.S.G.; Rocha, R.J.B. J-integral analysis of the mixed-mode fracture behaviour of composite bonded joints. *J. Adhes.* **2020**, *96*, 321–344. [CrossRef]
19. Adams, R.D.; Peppiatt, N.A. Stress analysis of adhesive-bonded lap joints. *J. Strain Anal.* **1974**, *9*, 185–196. [CrossRef]
20. Alfano, G. On the influence of the shape of the interface law on the application of cohesive-zone models. *Compos. Sci. Technol.* **2006**, *66*, 723–730. [CrossRef]
21. Reis, J.P.; de Moura, M.F.S.F.; Moreira, R.D.F.; Silva, F.G.A. Mixed mode I + II interlaminar fracture characterization of carbon-fibre reinforced polyamide composite using the Single-Leg Bending test. *Mater. Today Commun.* **2019**, *19*, 476–481. [CrossRef]

22. Yoon, S.H.; Hong, C.S. Modified end notched flexure specimen for mixed mode interlaminar fracture in laminated composites. *Int. J. Fract.* **1990**, *43*, R3–R9. [CrossRef]
23. Campilho, R.D.S.G.; Banea, M.D.; Pinto, A.M.G.; da Silva, L.F.M.; de Jesus, A.M.P. Strength prediction of single- and double-lap joints by standard and extended finite element modelling. *Int. J. Adhes. Adhes.* **2011**, *31*, 363–372. [CrossRef]
24. Campilho, R.D.S.G.; Moura, D.C.; Gonçalves, D.J.S.; da Silva, J.F.M.G.; Banea, M.D.; da Silva, L.F.M. Fracture toughness determination of adhesive and co-cured joints in natural fibre composites. *Compos. Part B Eng.* **2013**, *50*, 120–126. [CrossRef]
25. Da Silva, L.F.M.; Dillard, D.A.; Blackman, B.; Adams, R.D. *Testing Adhesive Joints*; Wiley: Hoboken, NJ, USA, 2012.
26. Ribeiro, T.E.A.; Campilho, R.D.S.G.; da Silva, L.F.M.; Goglio, L. Damage analysis of composite–aluminium adhesively-bonded single-lap joints. *Compos. Struct.* **2016**, *136*, 25–33. [CrossRef]
27. Campilho, R.D.S.G.; de Moura, M.F.S.F.; Domingues, J.J.M.S. Modelling single and double-lap repairs on composite materials. *Compos. Sci. Technol.* **2005**, *65*, 1948–1958. [CrossRef]
28. Rice, J.R. A path independent integral and the approximate analysis of strain concentration by notches and cracks. *J. Appl. Mech.* **1968**, *35*, 379–386. [CrossRef]
29. Wu, E.M.; Reuter, R.C.J. *Crack Extension in Fiberglass Reinforced Plastics*; T&AM Report No. 275; Department of Theoretical and Applied Mechanics, University of Illinois: Urbana, IL, USA, 1965.
30. Alfano, G.; Crisfield, M.A. Finite element interface models for the delamination analysis of laminated composites: Mechanical and computational issues. *Int. J. Numer. Methods Eng.* **2001**, *50*, 1701–1736. [CrossRef]
31. Leitão, A.C.C.; Campilho, R.D.S.G.; Moura, D.C. Shear Characterization of Adhesive Layers by Advanced Optical Techniques. *Exp. Mech.* **2016**, *56*, 493–506. [CrossRef]
32. Kim, K. Softening behaviour modelling of aluminium alloy 6082 using a non-linear cohesive zone law. *Proc. Inst. Mech. Eng. Part L J. Mater. Des. Appl.* **2015**, *229*, 431–435. [CrossRef]
33. Constante, C.J.; Campilho, R.D.S.G.; Moura, D.C. Tensile fracture characterization of adhesive joints by standard and optical techniques. *Eng. Fract. Mech.* **2015**, *136*, 292–304. [CrossRef]
34. Leffler, K.; Alfredsson, K.S.; Stigh, U. Shear behaviour of adhesive layers. *Int. J. Solids Struct.* **2007**, *44*, 530–545. [CrossRef]

Influence of Loading Rate on the Cohesive Traction for Soft, Rubber-Like Adhesive Layers Loaded in Modes I and III

Peer Schrader [†], Dennis Domladovac [†] and Stephan Marzi *

Institute of Mechanics and Materials, Technische Hochschule Mittelhessen, University of Applied Sciences, Wiesenstraße 14, 35390 Gießen, Germany
* Correspondence: stephan.marzi@me.thm.de; Tel.: +49-641-309-2124
† These authors contributed equally to this work.

Abstract: To date, the fracture behaviour of soft, polyurethane-based adhesive joints has rarely been investigated. This work contributes to the experimental investigation of such joints in modes I and III by performing double cantilever beam (mode I) and out-of-plane loaded double cantilever beam (mode III) tests at various loading rates. The tests were evaluated using a *J*-integral method, which is well established for testing stiff adhesive layers and is conventionally used to determine the cohesive traction at the crack tip. Additionally, fibre-optics measurements were conducted to provide crack extension, process zone length, and cohesive traction from the measured backface strain of the adherends. It was found that the energy release rate seems to be largely independent of the loading mode. However, differences were observed regarding process zone length and resistance curve behaviour. Furthermore, the backface strain measurement allows the determination of the cohesive traction along with the complete adhesive layer as well as separation and separation rate, yielding rate-dependent cohesive laws. A comparison indicated that the cohesive traction obtained from the *J*-integral method does not match the measured benchmark from the backface strain measurements because the underlying theoretical assumptions of the *J*-integral method are likely violated for soft, rubber-like adhesive joints.

Keywords: adhesive joints; polyurethane; fracture mechanics; backface strain measurement; rate-dependency; cohesive parameters; experimental testing of adhesives

1. Introduction

The literature contains a large number of studies investigating the fracture behaviour of epoxy-based adhesives but comparatively few works investigating soft, rubber-like polyurethane-based adhesives. However, many authors agree that polyurethane adhesives have various advantages in terms of the more even load distribution of peel loads, higher elongation at break, good damping properties and fatigue resistance, and energy consumption during impact [1–3]. The latter is of particular importance in passenger protection, as increased fracture energy leads to a greater amount of energy being absorbed by the adhesive layer in the event of a crash accompanied by finite deformations in the adhesive layer, which could potentially help to minimise personal injuries. Despite these important factors, only a few studies have investigated the fracture behaviour of polyurethane-based adhesive joints, whereas numerous studies have been conducted on polyurethane adhesives in their bulk form. It is assumed that this lack of research may be due to issues such as large process zones and energy dissipation through viscoelastic or viscoplastic effects, as well as creep processes complicating both the experimental investigation of the fracture behaviour and the extraction of fracture mechanical parameters.

The determination of cohesive laws is of particular importance for the design of adhesively bonded joints, because from these, by use of cohesive zone modelling, the behaviour of the joint can be predicted efficiently in finite element analyses. The aim of cohesive zone modelling is to reproduce the macroscopic fracture behaviour of the adhesive layer by the use of traction separation relations, which, in the best-case scenario, can be evaluated directly from mechanical fracture experiments such as, e.g., the double cantilever beam (DCB) test, the end-notched flexure test, or the out-of-plane loaded double cantilever beam (ODCB) test in modes I, II, and III, respectively. Commonly, an evaluation method based on the J-integral according to Rice [4], in which the cohesive laws are obtained by taking the derivative of the externally measured J-integral with respect to the crack opening displacement (COD), is used for this purpose in the single mode testing of both stiff, epoxy-based, e.g., [5–11], and soft, rubber-like adhesives, e.g., [12–14]. The approach assumes a purely non-linear elastic material behaviour, with the crack tip being the only inhomogeneity in the body, which, however, could be a problematic assumption for testing soft, rubber-like adhesive systems because the effects of the loading rate and energy dissipation outside of the crack tip, i.e., in the process zone, may not be taken into account accordingly. For pure mode I loading, Rosendahl et al. [14] showed that the approach can, indeed, approximately be used for thick, hyperelastic adhesive layers under quasi-static conditions using finite element analyses. However, this finding remains to be verified experimentally. Furthermore, in the mode III testing of rubber-like adhesives, in which the process zones are significantly larger than in mode I [15], the approach has not yet been used. To experimentally investigate the applicability of the J-integral method, the aim of our study is to propose an alternative methodology for determining cohesive laws based on the deflection curve of the adherends in DCB and ODCB tests to circumvent the underlying assumptions of the J-integral approach, e.g., rate-independent material behaviour and negligible effects of the process zone. As we will show, this novel method also has some additional advantages in accounting for rate-dependent fracture behaviour, as it can also be used to directly measure rate-dependent cohesive laws.

The dependency on loading rate and mode on the energy release rate (ERR) of rubber-like adhesives has also been investigated in some recent studies: In pure mode I testing, Schmandt and Marzi [12,13] investigated the effect of loading rate and adhesive thickness on the fracture energy, cohesive strength, and joint stiffness of polyurethane-based adhesives with DCB tests using the above-mentioned method of evaluation and found that fracture energy and cohesive strength show dependencies on both the loading rate and layer thickness. Boutar et al. [16] investigated the quasistatic single mode I and mode II fracture of a polyurethane-based adhesive system and found a significant dependency of the obtained fracture energy on the loading mode, with the mode II fracture energy being over three times larger than the mode I fracture energy at a layer thickness of 1 mm. In contrast, Loh and Marzi [15] investigated the mixed-mode I+III behaviour at a layer thickness of 3 mm and found that there could be an indication that the critical fracture energy of thick polyurethane-based joints does not depend on the mode-mix ratio. However, they also stated that the experimental scatter in their results did not allow a definitive statement about this issue. Furthermore, because of a pronounced resistance curve behaviour, they were unable to determine the cohesive traction in the adhesive layer with the J-integral approach, which also indicates that finding another methodology that allows the determination of the cohesive traction for such soft, rubber-like adhesive layers is an important advance in the state of research.

As hinted at earlier, the determination of process zone length and crack tip position is also of interest for the investigation of the fracture behaviour of adhesive joints: considering the determination of crack length, Schrader et al. [17] found that the crack extension measurement for rubber-like adhesive joints proved to be a difficult task with both optical methods of crack length measurement and the enhanced simple beam theory approach according to Škec et al. [18], leading to the conclusion that other methods for determining an equivalent crack tip position could be advantageous. Hence, as an alternative, we rely on an approach based on measurements on the adherends' backface strain (BFS) within this study, as the measurement of the BFS also allows the determination of the deflection curve of the adherends, which is crucial for our aim of determining the cohesive traction without underlying J-integral assumptions. Similar approaches have already been established in some other studies with a focus on the pure mode I testing of stiff adhesive systems: Ben Salem et al. [19] used several strain gauges along the top surface of a DCB specimen bonded by a structural adhesive joint for crack tip detection and identified the crack tip position from the position of the maximum bending strain. Similarly, Bernasconi et al. [20] and Lima et al. [21] used optical backscatter reflectometry to obtain the adherends' BFS. Truong et al. [22] also calculated the resistance curve for a composite specimen from BFS measurements. To obtain a deflection curve during DCB experiments, Reiner et al. [23] and Sun and Blackman [24] used digital image correlation (DIC) to obtain the displacement profiles, enabling the calculation of the ERR from the obtained displacement data. Additionally, especially for the investigation of soft adhesive systems, a measurement of strain along the adherends allows the investigation of the process zone shape, as performed, e.g., by Jumel et al. [25]. Schrader and Marzi [11] recently investigated a stiff, epoxy-based adhesive system in mode III loading and also calculated both crack length and process zone length from the measured BFS of the adherends. They also noted that the investigation of the process zone using BFS measurements could be of particular interest for investigating soft, rubber-like adhesive layers. Hence, the state of research indicates that BFS measurements seem to offer valuable data for determining cohesive laws from the experimental results.

Building on the mentioned studies, the present work aims to, for the first time, holistically investigate the effects of crack opening velocity and loading mode on a soft, rubber-like polyurethane-based adhesive joint, especially considering the determination of cohesive laws. Differences between the different evaluation methods, i.e., the J-integral method and BFS measurements, shall be investigated, highlighted, and discussed in order to gain insight into the applicability of the J-integral approach for soft, rubber-like adhesive layers, because, as hinted at earlier, some of its underlying assumptions may be violated for such adhesive systems. Furthermore, measuring the BFS along with the adhesive layer offers the hitherto unprecedented opportunity to investigate whether the cohesive law measured at the crack tip is at least similar to the cohesive traction separation relations along with the complete adhesive layer.

For this reason, we performed DCB and ODCB experiments on a soft, polyurethane-based adhesive system (Wiko Ultimate Elongation GLUETEC Industrieklebstoffe GmbH & Co. KG, Greußenheim, Germany) in both DCB and ODCB tests at different loading rates, i.e., 0.05 mm/s, 0.5 mm/s, and 5 mm/s in mode I and 0.05 deg/s, 0.5 deg/s, and 5 deg/s in mode III. In each of the test series, one experiment with a fibre-optics-based BFS measurement was performed to investigate the deformation behaviour of the adherends and to compare the results with the conventionally used evaluation methods for the determination of cohesive laws based on the J-integral.

We shall begin by briefly presenting the necessary theoretical background on the evaluation methods based on the J-integral and BFS measurements of the DCB and ODCB experiments. After stating the materials and methods, we shall present and thoroughly discuss the most important experimental findings. This includes the observed fracture patterns, the bending strain measured by the optical fibres, the rate-dependency of the ERR in modes I and III, the obtained resistance curves, the measured process zone lengths, and the cohesive laws. Furthermore, the BFS measurements are compared to the globally measured data to verify the used evaluation approaches. As we will show, the determination of cohesive laws from the deflection curve of the adherends is a valuable addition to fracture mechanical testing, as the conventional J-integral method of determining the cohesive traction may be prone to error because the underlying assumptions could be violated for soft, rubber-like adhesive layers. Additionally, the presented method based on the BFS measurement allows the determination of a rate-dependent cohesive law, which, to the authors' knowledge, has not been achieved elsewhere.

2. Theory

2.1. J-Integral and Cohesive Traction

The J-integral of an arbitrarily shaped, non-linear elastic body—following the notation of Rice [4]—is defined as

$$J = \int_S \left(W\, dy - t_i \frac{\partial \Delta_i}{\partial x} ds \right), \qquad (1)$$

where S describes an arbitrary path circumscribing the crack tip in a counter-clockwise direction, t_i are components of the (nominal) traction vector, Δ_i are components of the displacement vector, and W is the strain energy density; see Figure 1. The integration is performed in the reference configuration and, per the definition, provides the sum of all inhomogeneities in the body. As the above equation is written in index notation, it shall be summed over $i = 1, \ldots, 3$ to compute the total value of the J-integral.

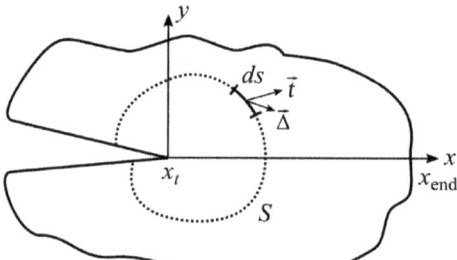

Figure 1. Schematic representation of the line J-integral around a notch for a plane problem.

Considering the testing of adhesive layers, determining the traction vector is of special interest for modelling the fracture behaviour of adhesive joints using cohesive zone models. Briefly, if the integration path is chosen around the boundary between the adherend and adhesive layer parallel to the x-axis ($dy = 0$) and exploiting the symmetry of a specimen (i.e., identical adherends), the above equation can be expressed as

$$J^{(\text{loc})} = 2 \int_{x_{\text{end}}}^{x_t} t_i(x) \frac{\partial \Delta_i}{\partial x} dx = \int_{x_{\text{end}}}^{x_t} t_i(x) \frac{\partial \delta_i(x)}{\partial x} dx \qquad (2)$$

where $\delta_i = 2\Delta_i$ are the components of the separation vector, i.e., the relative displacement of the upper and lower boundary, x_{end} is the (unloaded) end of the adhesive layer, and x_t is the crack tip position. The assumption of the elastic behaviour of the adhesive layer implies that, given single mode loading, the traction depends solely on the deformation state, $t_I(\delta_I(x))$ and $t_{III}(\delta_{III}(x))$, respectively. Inserting this into the above equation and substituting $\frac{\partial \delta_i(x)}{\partial x} dx = d\delta_i(x)$ then yields

$$J_I^{(\text{loc})} = \int_0^{\delta_I(x_t)} t_I(\delta_I(x))\, d\delta_I(x) \quad \text{and} \quad J_{III}^{(\text{loc})} = \int_0^{\delta_{III}(x_t)} t_{III}(\delta_{III}(x))\, d\delta_{III}(x) \qquad (3)$$

under the assumption that the end of the adhesive layer x_{end} is unloaded. It should be noted that a transition between Equations (2) and (3) is only possible under the condition of integrability, $\nabla \times \vec{t} = \vec{0}$. This integrability condition is automatically fulfilled if the cohesive traction depends only on the separation in the respective loading mode (decoupled behaviour), i.e., $t_I(\delta_I(x))$ and $t_{III}(\delta_{III}(x))$; a dependence on, e.g., the separation rate would violate the integrability condition and a conversion from Equation (2) to (3) would not be feasible. Using the mode I and mode III COD, $\delta_{I,t} = \delta_I(x_t)$ and $\delta_{III,t} = \delta_{III}(x_t)$, the equation can be rewritten in differential form and rearranged for the cohesive traction,

$$t_I(\delta_{I,t}) = \frac{dJ_I^{(\text{loc})}}{d\delta_{I,t}} \quad \text{and} \quad t_{III}(\delta_{III,t}) = \frac{dJ_{III}^{(\text{loc})}}{d\delta_{III,t}}, \qquad (4)$$

thus yielding the so-called cohesive laws in the individual loading modes I and III.

It shall be noted that the Equations (2)–(4) apply locally in the vicinity of the crack tip. For the experimental evaluation, it is demanded that $J^{(loc)}$ is in equilibrium with the sum of contributions from external loads, which should apply as long as no energy is dissipated outside of the adhesive layer. This method is straightforward, as by measuring the J-integral over external loads (cf. Section 2.2) and the COD, cohesive laws can be determined directly by the derivation of the measured quantities.

It should be highlighted, however, that it may be difficult to justify the validity of the assumptions behind Equation (4) for a soft, polyurethane-based, rubber-like adhesive. For such adhesives, the assumption of purely elastic behaviour behind the presented derivations is deemed problematic: firstly, the implication that the cohesive traction solely depends on the separation may neglect the effects of loading rate on the material behaviour, wherefore the integrability condition for transitioning between Equations (2) and (3) would be violated. Secondly, the assumption of a non-linear elastic body implies that the crack tip is the only material inhomogeneity in the body. This could also be deemed problematic, as soft adhesive layers may develop process zones of finite length before ultimate failure. As the J-integral provides the sum of all inhomogeneities in the elastic body, inhomogeneities in the process zone, e.g., plastic effects, viscoelasticity, and damage, could also contribute to the value of the externally measured J-integral and could, hence, falsely be ascribed to the crack tip when calculating the cohesive traction from Equation (4).

Because the assumptions behind Equation (4) may be violated during the testing of soft, rubber-like adhesives, it can already be assumed that the approach of taking the derivative of the externally measured value of J for the COD could be error-prone. However, as this approach to the determination of cohesive laws is deemed very pragmatic and was already used successfully in studies investigating the mode I fracture of polyurethane-based adhesive joints [12–14], it is worthwhile to check this approach as it could at least provide a good approximation for the traction at the crack tip. This work aims to assess the quality of the approximation by using additional methods of measurement, i.e., BFS measurements, which allow a determination of the nominal traction along with the adhesive layer.

2.2. Determination of the ERR in DCB and ODCB Experiments

Consider the DCB and ODCB specimens displayed schematically in Figure 2. Briefly, if the specimen of width b is loaded in pure mode I during a DCB test, as found by Paris and Paris [26], the J-integral according to Equation (2) reduces to

$$J_I = \frac{F_y(\theta_1 + \theta_2)}{b}. \tag{5}$$

For pure mode III loading during ODCB tests, Loh and Marzi [9] derived that the J-integral yields

$$J_{III} = \frac{M_y^2}{b}\frac{1}{EI_y} \tag{6}$$

with the applied moment M_y and the bending stiffness EI_y of the adherends. Loh and Marzi [27] found in a later study that unintended contributions to J can occur during testing in mode III, which result from a mode I contribution due to the specimen twisting under an out-of-plane deformation, J_{I^*}, and a contribution in modes I and II due to the finite width of the adhesive layer, J_{I+II}:

$$J_{I^*} = \frac{1}{2b}\frac{{}^1M_x^2 + {}^2M_x^2}{\mu I_{yz}} \quad \text{and} \quad J_{I+II} = \frac{{}^2M_z^2}{2b}\frac{1}{EI_z} \tag{7}$$

Here, μ denotes the shear modulus of the adherends, and I_{yz} and I_z denote the torsional second moment of area and the second moment of area of the adherend around the bending axis z, respectively. From this, the total value of the J-integral is obtained from the sum of mode III and unintended contributions:

$$J = J_{III} + J_{I^*} + J_{I+II} \tag{8}$$

It should be noted that in the subsequent studies by Loh and Marzi [9,27,28] and Schrader and Marzi [10], the contributions J_{I^*} and J_{I+II} were found to be negligible at the point of fracture during pure mode III investigations of both epoxy-based and polyurethane-based adhesive systems, i.e., $J \equiv J_{III}$. Hence, the cohesive law can then be determined from Equation (4) as the externally measured value of J from the outer loads is in equilibrium with the value of J in the adhesive layer given that the adherends do not deform plastically.

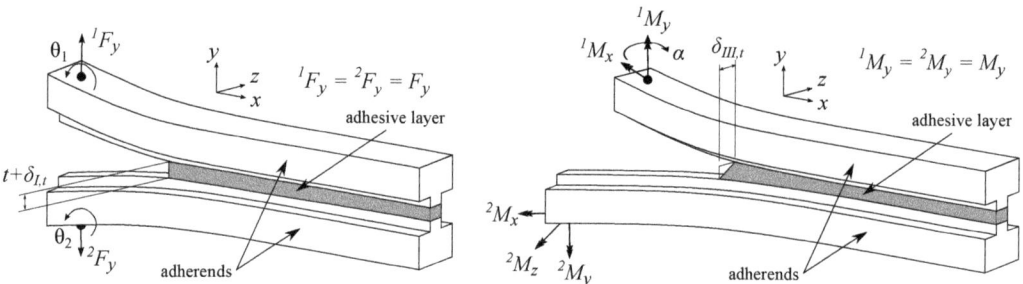

Figure 2. Schematical representation of the used specimens with applied loads: (**left**) DCB specimen, (**right**) ODCB specimen.

2.3. Determination of ERR and Cohesive Traction from BFS Measurements

To gain better insight into the deformation behaviour of the specimen in each loading configuration, a measurement of the adherends' BFS ε at discrete measuring points along with the specimen is used to determine the deflection curve at different times during the experiment. For each measurement in time, from the distance c between the position of strain measurement and the neutral axis of the adherend, which is assumed to be a Euler–Bernoulli beam, the beam curvature κ is obtained via $\kappa(x) = \varepsilon(x)/c$, ultimately yielding the bending moment

$$M_b(x) = -\kappa(x)EI \tag{9}$$

from the bending stiffness EI around the bending axis of interest (y-axis in the ODCB and z-axis in the DCB tests, respectively). From this, transverse force $Q(x)$ and line load $q(x)$ are obtained by the differentiation of the bending moment for the x-position along the adherends, giving

$$Q(x) = \frac{dM_b(x)}{dx} \quad \text{and} \quad q(x) = -\frac{d^2 M_b(x)}{dx^2}. \tag{10}$$

Furthermore, integrating the curvature along the beam provides the slope $\varphi(x)$ of one lever arm and the separation $\delta(x)$ between the two adherends via

$$\varphi(x) = -\int_{x_{end}}^{x_t} \kappa(x)\,dx \quad \text{and} \quad \delta(x) = 2\int_{x_{end}}^{x_t} \varphi(x)\,dx. \tag{11}$$

It should be mentioned that, for a specimen with an unloaded end, it can be reasonably assumed that the integration constants for slope and deflection become nought, allowing the calculation of both quantities without further restrictions. As a result, a measurement of the beam curvature provides an additional possibility of obtaining cohesive traction at discrete measuring points along with the length of the beam via

$$t(x) = \frac{q(x)}{b} \tag{12}$$

under the assumption that the load is distributed equally on the width of the adhesive layer for both peel and shear loads. Thus, a comparison can be made to check the applicability of Equation (4) with this measurement. Furthermore, considering Equation (2), from the stress in the cohesive zone according to Equation (12) and the relationship $\partial \Delta_i / \partial x = \varphi$, the J-integral is obtained via

$$J = 2 \int_{x_{\text{end}}}^{x_t} t(x) \varphi(x) \, dx. \tag{13}$$

It is therefore evident that the measurement of the elongation at the marginal fibres of the adherends can be used to gain better insight into the fracture behaviour of the adhesive layer. By investigating the deformation behaviour at different times during the experiment and points along with the specimen, the traction, separation, and separation rate can be obtained at each discrete measuring point along with the specimen.

3. Materials and Methods

3.1. Specimen Manufacturing

Within this study, both DCB and ODCB tests were performed on the polyurethane-based adhesive system Wiko Ultimate Elongation (GLUETEC Industrieklebstoffe GmbH & Co. KG, Greußenheim, Germany) at various loading velocities. The tested adhesive system is a one-component, moisture-curing adhesive that exhibits a high elongation at a break of about 800%, according to the manufacturer's data. The substrates of the used specimens were made of the high-strength aluminium alloy AlZn5,5MgCu (material grade number 3.4365, E = 70 GPa). The used specimens are displayed in Figure 3 with the corresponding dimensions. The adherends had a T-shaped cross-section to achieve a smaller adhesive layer width compared to the width of the adherends, avoiding plastification in the aluminum during the experimental investigation. Furthermore, the length of the specimens was chosen to be shortly below a meter, ensuring that the process zone did not reach the end of the specimen during the crack initiation phase, even in the case of finite deformations at the crack tip, ensuring an unloaded end of the specimen.

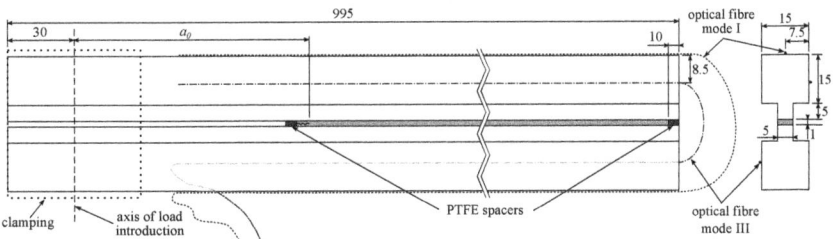

Figure 3. Dimensions of the tested specimens; $EI_y = 2.98 \times 10^8$ Nmm², $EI_z = 4.56 \times 10^8$ Nmm², $\mu I_{yz} = 2.75 \times 10^8$ Nmm².

Before applying the adhesive, the bonding surfaces of the substrates were sandblasted with corundum (grain size of 100–150 μm) and degreased with isopropyl alcohol. The adhesive was then applied with an electric caulking gun. To define the layer thickness, PTFE spacers with a nominal thickness of 1 mm were placed at the beginning and the end of the adhesive layer and removed after curing. Screw clamps were used to hold the substrates in place during the curing procedure. The specimens were cured in a lab for 1–2 weeks under laboratory conditions, i.e., at a room temperature of (23 ± 3) °C and relative humidity of about (50 ± 5)%, in line with the manufacturer's data of the moisture-curing adhesive system. Before testing, a sharp pre-crack was introduced at the beginning of the adhesive layer by inserting a thin razor blade in the middle of the adhesive layer parallel to the bonding surfaces. This was done to achieve a fracture mechanical specimen, provoke cohesive failure, and define a sharp initial pre-crack for the evaluation of the COD. With the described procedure, an initial crack length of (135.7 ± 1.2) mm, i.e., the distance between the initial crack tip and the axis of load introduction, and an adhesive layer thickness of (0.88 ± 0.08) mm were achieved.

3.2. Experimental Setups and Test Matrix

The DCB and ODCB tests were performed in a biaxial tension-torsional servo-hydraulic test machine (MTS Landmark Bionix, MTS Systems, Eden Prairie, USA). The experimental setups are displayed in Figure 4. To measure the rotations θ_1 and θ_2 of the specimens at the load introduction points in the DCB tests, incremental rotary encoders (BDH 1P.05A320000-L0-5, Baumer AG, Frauenfeld, Switzerland) with a resolution of 320,000 steps per full turn were used. The applied force was measured below the lower clamping device with a six-axis load cell (K6D110 4 kN/250 Nm, ME-Messsysteme GmbH, Hennigsdorf, Germany). To examine the rate-dependency of the adhesive, the DCB tests were performed at external loading rates of 0.05 mm/s, 0.5 mm/s, and 5 mm/s.

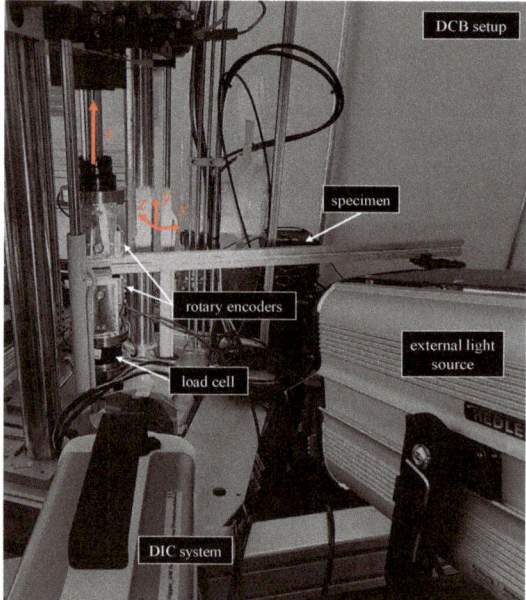 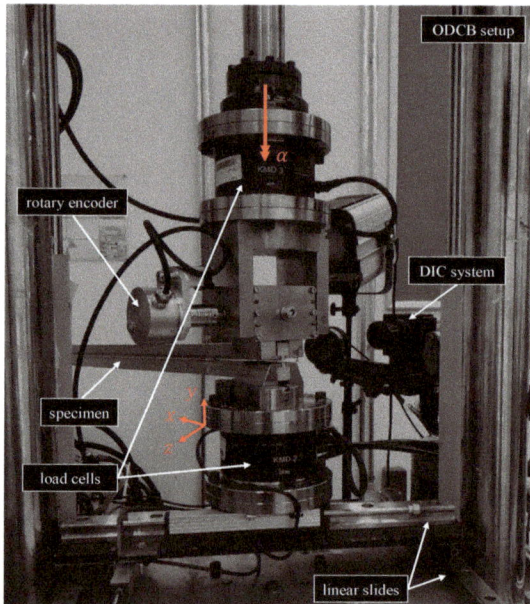

Figure 4. Experimental setups: (**left**) mode I DCB setup, (**right**) mode III ODCB setup.

During the ODCB tests, the applied moments were measured using two of the above-mentioned six-axis load cells, one at each load introduction point of the specimen. To avoid lateral forces on the specimen, the bottom clamping of the specimen was mounted on two orthogonally placed linear slides. Throughout the ODCB tests, the axial force was controlled to be nought by the used testing machine. At the time of carrying out the experiments, it was assumed that the floating support would ensure that the transverse forces would not influence the experimental results akin to the results of Schrader and Marzi [10], wherefore the measurement of the transverse forces was omitted. As we will show later, however, it was found during the post-processing of the BFS measurement that this assumption is problematic for the tested soft, rubber-like adhesive layer. The ODCB tests were performed at external loading rates of 0.05 deg/s, 0.5 deg/s, and 5 deg/s, respectively.

To investigate the deflection curve, the BFS along with the specimen was measured using a fibre-optics system (ODiSI-B 5500, Luna Innovations Inc., Roanoke, VA, USA, positional resolution of 2.5 mm). The fibre was bonded to the adherends along the upper and lower surface of the adherends for the DCB tests and on the tensile-loaded outer surface of the adherends for the ODCB tests. As the experimental effort largely increases with the additional use of this measuring system, we refrained from increasing sample sizes with BFS measurements for this pilot study. The results were evaluated following the procedure described in Section 2.3. It must be stated that numerically taking the derivative of the measured curvature for the x-position along the beam produces numerical noise. To counteract this, the measurements were filtered with a Savitzky–Golay filter before each derivation step.

The COD was measured with stereo camera systems in all cases. To evaluate the COD, the relative distance between two measuring points at the position of the initial pre-crack was determined through DIC measurements, with one point being on the lower and one on the upper substrate. In the mode III experiments, the measurement of the COD was adjusted for the rigid body rotation of the specimens. Two DIC systems were used based on the desired rate of image acquisition: for the experiments at lower image acquisition rates between 1 and 20 fps, a 12 MP ARAMIS 3D Motion and Deformation Sensor with the corresponding evaluation software (GOM Aramis, GOM GmbH, Braunschweig, Germany) was used. For the tests with image acquisition rates between 30 and 125 fps, two 1 MP Photron FASTCAM Nova S6 (Photron USA, San Diego, CA, USA) and the evaluation software VIC-3D 8 (Correlated Solutions, Irmo, SC, USA) were used. Within the course of this study, to reduce numerical errors during differentiation, the experimental results of $t_i(\delta_t)$ were obtained with the procedure proposed by Biel [29], in which the experimental results of J vs. δ_t were fitted with a Prony series before taking the derivative.

It shall be stated that the external loading rates were selected so that, starting with a quasi-static loading rate of 0.05 mm/s in mode I and 0.05 deg/s in mode III, the rates increased by powers of ten with each test series. Although higher loading rates could have been achieved with the given test setups and the used servo-hydraulic test machine, testing at larger rates was refrained from because the fibre-optics system could not provide a sufficient temporal resolution.

For a better overview, the number of the performed experiments is summarized in Table 1 with the external loading rate, sample size, used DIC systems, and image acquisition rates. As stated earlier, in each of the conducted test series, one BFS measurement was conducted using the fibre-optics system.

Table 1. Test matrix and used DIC systems.

External Loading Rate		Sample Size	DIC Sensors	Image Acquisition Rate
Mode I (mm/s)	0.05	5	Aramis 3D Sensor [1]	1 fps
	0.5	4	Photron FASTCAM [2]	30 fps
	5	5	Photron FASTCAM [2]	125 fps
Mode III (deg/s)	0.05	4	Aramis 3D Sensor [1]	1 fps
	0.5	4	Aramis 3D Sensor [1]	20 fps
	5	4	Photron FASTCAM [2]	125 fps

Referrals for DIC setups: [1] ARAMIS 3D Motion and Deformation Sensor, GOM Correlate (GOM GmbH, Braunschweig, Germany). [2] Photron FASTCAM Nova S6 (Photron USA, San Diego, USA), VIC-3D 8 (Correlated Solutions, Irmo, USA).

3.3. Determination of Crack Extension and Process Zone Length

From the BFS measurement, the crack extension and the length of the loaded region within the adhesive layer can also be determined using the Euler–Bernoulli beam theory (cf. Figure 5). It can reasonably be assumed that the transverse force in the lever arms of the adherends is constant during the DCB experiments, yielding a linear increase in the measured strain along with the optical fibre. Hence, to measure the crack extension, linear regression can be performed in the linear region of the measured strain, where the point of 0.5% deviation from linearity is defined as the crack tip position x_t.

To determine the length of the loaded region within the adhesive layer, similarly to the method of Schrader and Marzi [11], the maximum fibre strain in the pressure zone was used. For the sake of brevity, we will refer to this loaded region within the adhesive layer with the term "process zone" in the context of this study. It shall be highlighted that the wording should not be confused with the term "fracture process zone", i.e., the region in the adhesive layer in which the material exhibits plastic deformations, damage, etc. The end position of the process zone is defined as the fibre position at which the threshold of 10% below the maximum fibre strain in the pressure zone is undercut. The process zone length l_p is then computed from the difference between the current crack tip position and the end position of the process zone. As stated by Schrader and Marzi [11], the definition of the process zone length will likely overestimate the length of the fracture process zone due to, e.g., bondline elastic deformations and early non-linear shear stress-strain behaviour, but give a reasonably accurate measurement of the length of the loaded region within the adhesive layer.

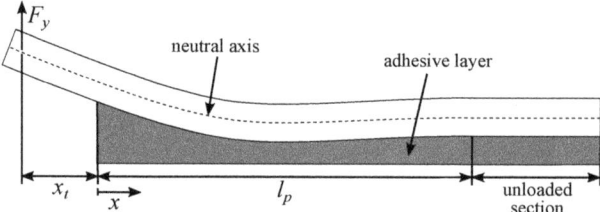

Figure 5. Determination of crack tip position and process zone length.

The procedure in the mode III experiments is analogous, with the difference that the beam curvature (and, hence, the bending strain in the optical fibre) in the lever arms before the crack tip is assumed to be constant under pure mode III loading from an external bending moment. Hence, linear regression is performed in the region of constant beam curvature. In this case, the crack tip position is defined as the point of 1% deviation from linearity.

Additionally, the crack length is also calculated analytically under the assumption of simple beam theory, i.e., the assumption of the adherends being Euler–Bernoulli perfectly clamped at a point-like crack tip. A comparison is deemed worthwhile as the ERR for stiffer adhesive layers is often calculated from the crack length (e.g., akin to the methods standardised in ISO 25217 [30]), and analytically determining the crack length for soft, rubber-like adhesive layers instead of measuring it with great experimental effort could be beneficial in practice. For the DCB experiments, the crack length was calculated from the load-point separation s and the rotational angle θ at the load introduction points via

$$a_I = \frac{3s}{4\theta}. \tag{14}$$

In the mode III ODCB experiments, the crack length was computed analytically via

$$a_I = \frac{\alpha E I_y}{2 M_y} \tag{15}$$

with α being the rotational angle of the biaxial testing machine.

4. Results and Discussion

4.1. General Observations and Fracture Surfaces

In all cases, large displacements at the initial crack tip are observed before the crack starts to propagate. During quasi-static mode I loading, the crack travels directly to the nearby interface, followed by adhesive failure, which is commonly observed regarding the quasi-static peeling of adhesive joints [12]. With increasing loading rate, the mode I failure becomes more cohesive (cf. Figure 6). In the ODCB experiments, due to finite deformations at the crack tip, the mode III shear transitions into a peel load accompanied by partly adhesive failure at highly stretched parts of the joint (cf. Figure 7). Interestingly, the large displacement aspect during mode III loading indicates crack propagation perpendicular to the actual bonding surface, accompanied by partly adhesive failure at the outer edges of the adhesive layer. This is probably related to the general tendency of the adhesive to fail adhesively at particularly slow rates. It is assumed that during loading, highly stretched parts of the joint at the outer boundary fail adhesively, hence, reducing its effective width before an ultimate cohesive failure occurs. However, this behaviour ceases at an increased rate of 5 deg/s, as the fracture surfaces show a tilted fracture surface with purely cohesive failure. In Figure 7, it can also be observed that the outer edges of the adhesive layer opposing the side of partly adhesive failure are tilted and plastically deformed.

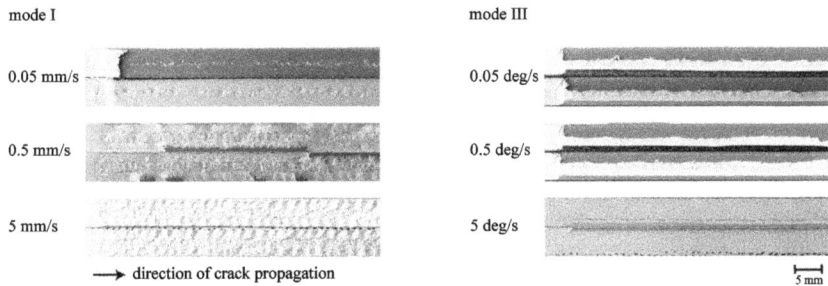

Figure 6. Representative fracture surfaces observed in the experiments.

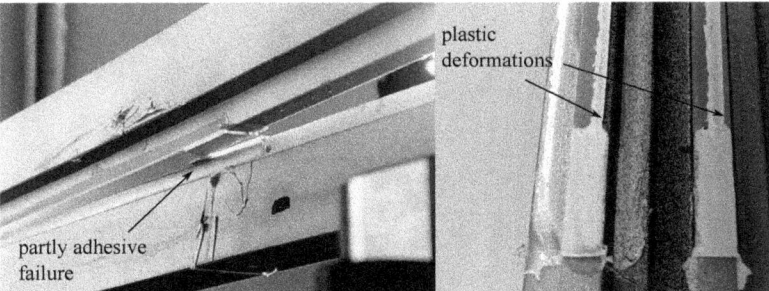

Figure 7. Partly adhesive failure and plastically deformed, tilted side surfaces observed during mode III loading at the loading rates of 0.05 deg/s and 0.5 deg/s.

4.2. BFS Obtained from the Fibre-Optics Measurements

Figure 8 shows the development of the bending strain in the optical fibre over the runtime of a DCB and an ODCB test at different selected times during the measurement. For better visualization, the zero value of the abscissa is set at the initial crack tip position. Independently from the loading mode, the process zone is already quite large at the beginning of the crack propagation, strongly indicating that the assumption of an infinitesimally small process zone is violated.

In the mode I experiments, the maximum strain first increases with the applied load and then begins to shift along with the specimen as the crack progresses. Furthermore, the bending strain behaves linearly in front of the crack tip, indicating that a constant transverse force is applied in the lever arm. Deviations from linearity can hence be ascribed to the adhesive layer, indicating that the selected criterion for the detection of the crack tip position delivers satisfactory results.

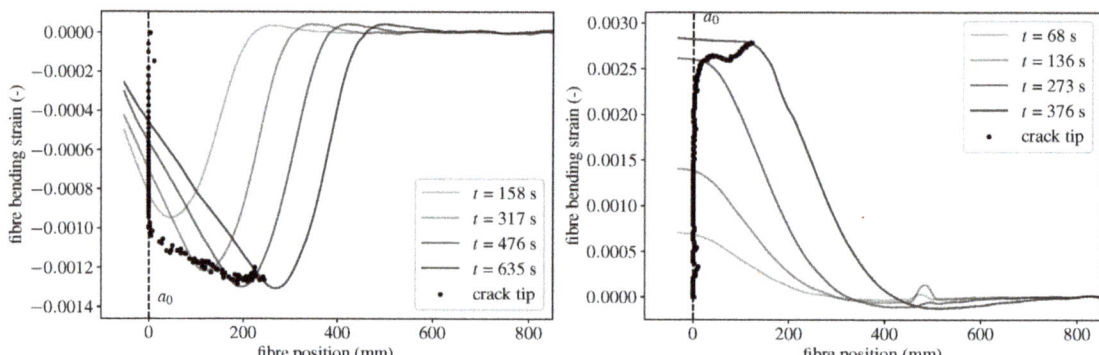

Figure 8. Development of the bending strain measured by the optical fibre: (**left**) quasi-static DCB test, (**right**) quasi-static ODCB test.

In the mode III experiments, although one would expect a constant bending strain in the region of the lever arms because of the applied bending moment M_y, a linear growth of the measured strain can be observed, indicating that an additional transverse force, probably due to friction in the lateral slides below the lower clamping device, acts on the specimen. The transverse force obtained from the BFS measurement, i.e., the slope of the measured strain in the region of the lever arms, is displayed in Figure 9 for the different loading rates. As the slope is determined through the numerical differentiation of the strain data, the measurement noise is amplified, yielding the observed fluctuations in the displayed transverse force. The assumption of friction being the main reason for the transverse forces is supported by the fact that the resisting force is relatively constant after a certain break-away force of the linear slides is reached. Because this resisting force is counter-directed to the applied moment component M_y, it will inevitably reduce the traction and the value of J in the adhesive layer. This result, which unfortunately only became apparent during post-processing, was rather unexpected. While this will not influence the BFS evaluation, it must be assumed that the influence has a significant impact on the evaluation of J from the external measurements, as it cannot be considered with the used method of evaluation for the ODCB tests.

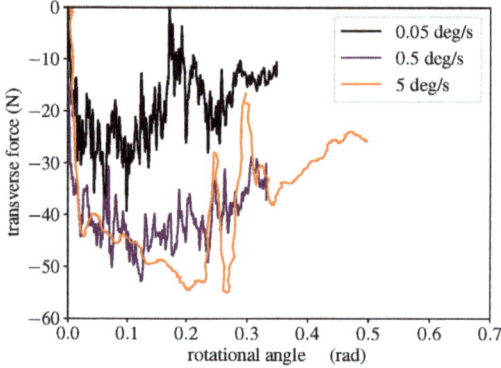

Figure 9. Transverse forces obtained from the bending strain of the optical fibre during the ODCB tests.

4.3. Comparison between BFS, Load, and DIC Measurements

Before further investigating the fracture behaviour of the tested adhesive joints with the BFS measurements, it shall be investigated whether the results can be verified with the globally measured data of the COD and applied load. In Figure 10, the results for two representative specimens (nominal adhesive layer thickness of 1 mm at the lowest loading rate) are shown for both the mode I and the mode III experiments. As can be observed, the separation at the initial crack tip obtained from both the DIC measurement as well as the values from the BFS measurement show a good agreement, indicating that the separation of the adherends can be determined from the BFS measurement with good accuracy. As the measurement data of the BFS measurements are integrated along with the complete specimen to obtain the COD at the position of the crack tip x_t, cf. Equation (11), this means that the separation at each measurement point along the adhesive layer can be determined reliably. As the shear force in mode I is constant in the lever arms in front of the crack tip, the values obtained by the BFS measurement may be compared with the values measured on the external load cells as well, also showing a very good agreement. To compare the moments in mode III, the observed slope in the fibre bending strain in front of the crack tip is extrapolated to the point of load introduction. Here, the external moment measurement also agrees well with the moment obtained from the BFS measurement. Overall, the good agreement of the external measurement of COD and applied load with the BFS measurements indicate that the methodology proposed in Section 2.3 delivers valid results.

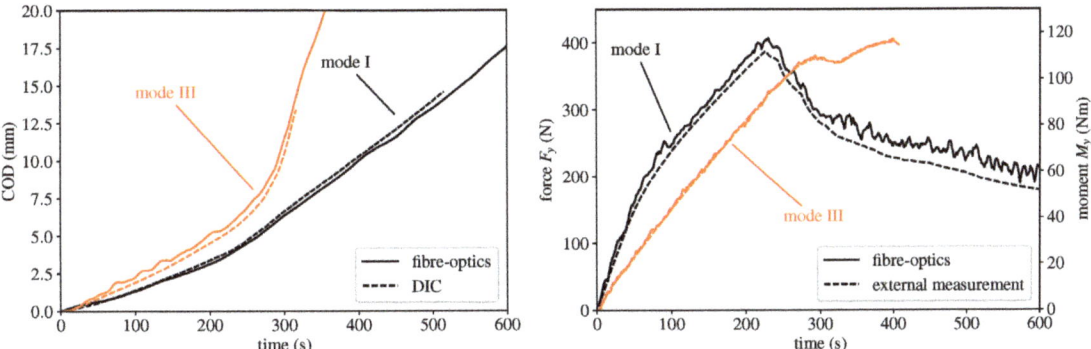

Figure 10. Comparison between externally measured values and BFS measurement: (**left**) separation of the adherends at the crack tip, (**right**) applied external force/bending moment.

4.4. Influence of Loading Mode and Loading Velocity on the ERR

The mode I ERR obtained from Equation (5) is shown in Figure 11 over the measured rotational angle θ; for a better overview, the tests conducted with BFS measurements are highlighted. As expected, the measured values for J at fracture initiation increase with the loading rate. The large discrepancy between the obtained ERR at 0.05 mm/s and 0.5 mm/s can be related to the adhesive failure observed during the quasi-static experiments.

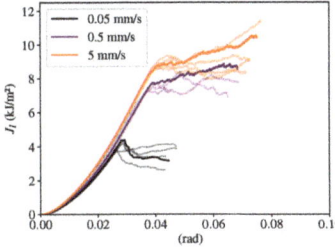

Figure 11. Measured ERR during the mode I experiments; experiments with additional BFS measurements are highlighted.

In Figure 12, the mode III ERR according to Equation (6) over the rotational angle α and the relative influence of the unintended contributions to J at the onset of fracture according to Equation (7) are displayed for each loading rate. Here, it can be observed that, during the experiments at 0.05 deg/s and 0.5 deg/s, the ERR does not reach a steady plateau throughout the experimental investigation, already indicating that the ERR is rising with crack propagation, yielding a resistance curve (cf. Section 4.5). It can also be observed that the unintended contributions from the transverse moments are indeed negligible at the point of fracture, which is in good agreement with the results of prior investigations [10,15,27,28]. This also allows the conclusion that the BFS measurement, although affected by a transverse force, is not influenced significantly by the moment components responsible for the unintended contributions to J.

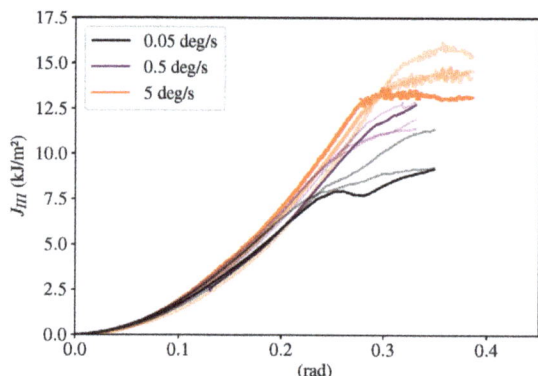
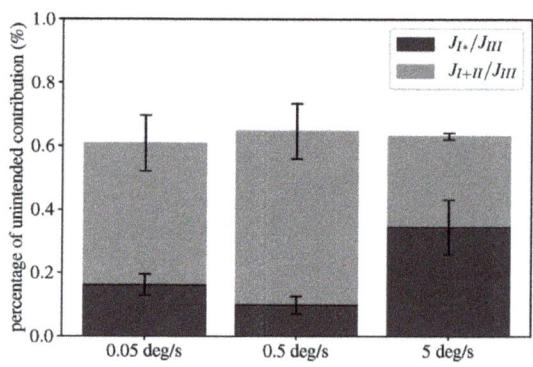

Figure 12. Results of the ODCB experiments: (**left**) measured ERR and (**right**) relative influence of the unintended contributions to J at the start of crack propagation. Experiments with additional BFS measurement are highlighted.

Figure 13 presents the values for J obtained from Equation (5) and (6) in comparison to the value obtained from the BFS measurement according to Equation (13) for the mode I and mode III experiments. Here, a good correspondence between both methods of evaluation can be seen for pure mode I loading. For the mode III experiments, however, it can be observed that the value for J according to Equation (6) and the BFS measurement differ greatly from another, with the BFS J, Equation (13), being approx. 20% lower than the externally measured value throughout the experiments. As hinted at earlier, this is likely due to the transverse force (cf. Figure 9), which was observed during the post-processing of the BFS measurements but not recorded during the experiments. This is also undermined by the fact that both the transverse force and the difference between the evaluation methods are the smallest at the loading rate of 0.05 deg/s; for the tests at 0.5 deg/s and 5 deg/s, in which the transverse force is larger, the difference also increases. As the external measurement seems to be strongly influenced by the friction within the lateral slides, the results from the BFS measurements clearly show that the evaluation of the ODCB test has to be revised for the testing of soft, rubber-like adhesive layers in future investigations.

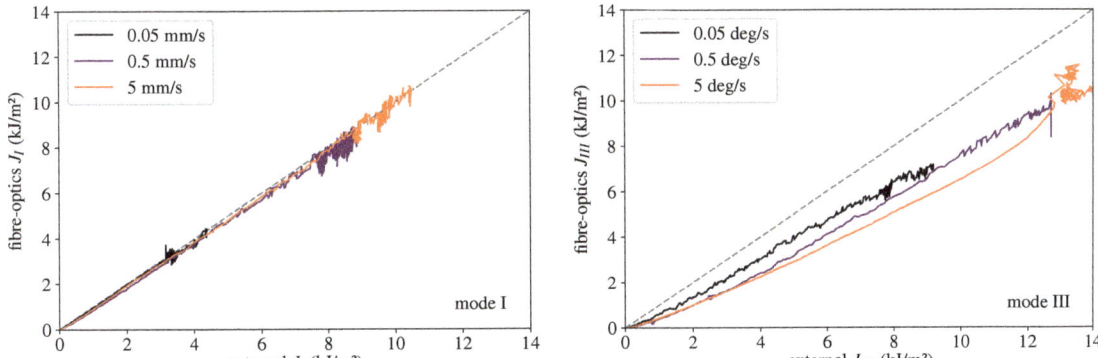

Figure 13. Comparison between externally measured values and BFS measurement of J: (**left**) mode I DCB tests, (**right**) mode III ODCB tests.

To estimate the dependency of J on the loading rate and loading mode, the values of J at crack initiation are displayed over the representative crack opening velocity in Figure 14. The representative crack opening velocity was determined from a linear regression of the COD vs. time in the initial linear region of $dJ/d\delta_{i,t}$, akin to the approaches of Schmandt and Marzi [13] and Schrader and Marzi [10], respectively. Hence, it shall be highlighted that the representative crack opening velocity is determined locally at the crack tip from the COD measurements and cannot be easily assessed from the external loading rates before testing. Generally, a large discrepancy between the mode I and mode III results is visible if the externally measured values for J_I and J_{III} are considered. However, the values obtained from the BFS measurements indicate that the differences between mode I and mode III mainly result from neglecting the transverse forces due to friction in the lateral slides. Hence, given the limitations of this study, a similar rate-dependency is obtained for both modes I and III, indicating that the ERR could be independent of loading mode, as was also hypothesised by Loh and Marzi in [15]. This also correlates with the large deformations at the crack tip observed during the mode III experiments, which ultimately yield a local peel load at the crack tip at fracture initiation.

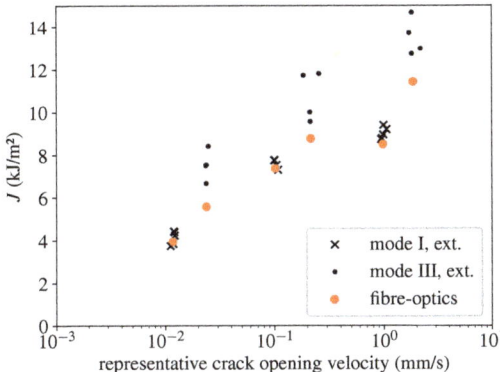

Figure 14. Influence of loading mode and representative crack opening velocity on the externally measured ERR.

4.5. Crack Propagation, Resistance Curve Behaviour and Process Zone Length

The resistance curves for all tested specimens are shown in Figure 15. Whereas a constant ERR can be observed in the mode I experiments during crack propagation, the

ERR increases with crack extension in the mode III experiments at the lower loading rates of 0.05 deg/s and 0.5 deg/s. Due to the presence of crack extension before reaching the critical value of J, the cohesive traction cannot be calculated from $dJ/d\delta_{III,t}$, Equation (4), for these experiments, as although the crack already started to propagate, the cohesive traction would be unequal to nought until the J-plateau was reached.

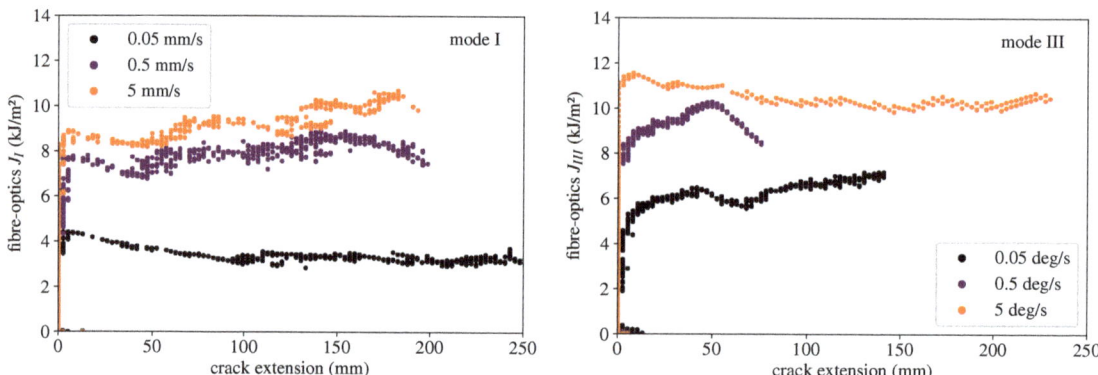

Figure 15. Resistance curves obtained from the fibre-optics measurements: (**left**) mode I DCB tests, (**right**) mode III ODCB tests.

The process zone lengths obtained from the BFS measurements are shown in Figure 16 over the measured crack extension. Generally, the length of the process zone increases until the start of crack propagation and remains constant over the experiment in good approximation in all cases, indicating stationary conditions behind the crack tip even in the case of an observed resistance curve. During mode I testing, the process zone length seems to be largely independent of the loading rate. In the mode III experiments, however, it is noticeable that the process zone length drastically decreases at the loading rate of 5 deg/s, which can likely be ascribed to the partly adhesive failure during the experiments at 0.05 deg/s and 0.5 deg/s. In these experiments, the partly adhesive failure before cohesive crack propagation causes a decrease in the stiffness of the joint and, hence, larger process zones. Additionally, it should be noted that the process zone lengths in mode III are significantly larger than in mode I at the start of crack propagation, which can generally be related to a lower stiffness of the adhesive in shear than in peel.

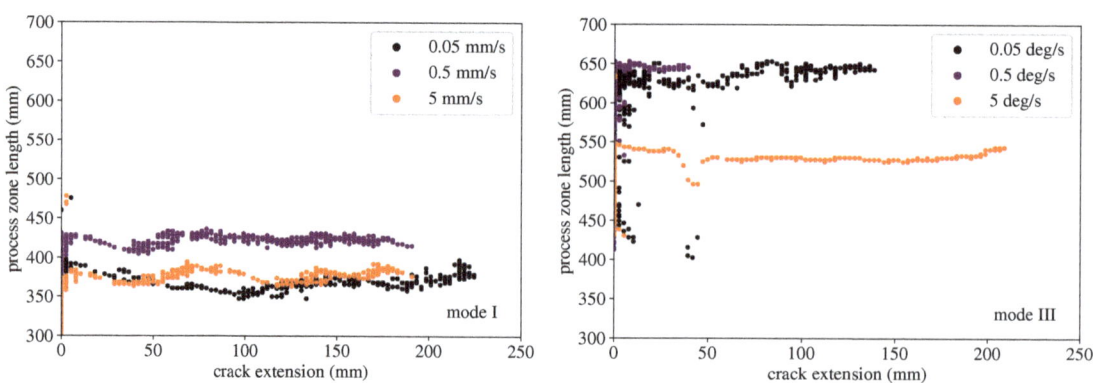

Figure 16. Development of the process zone during the experiments: (**left**) mode I DCB tests, (**right**) mode III ODCB tests.

As stated earlier, a comparison between the crack length obtained from the BFS measurement and the analytical crack length according to simple beam theory is sought. In Figure 17, the crack extension according to the BFS measurement is displayed over the analytical crack extension during crack propagation. It can be observed that the slope of the curves is relatively close to one in the range of crack propagation in both modes I and III, which correlates with the results of Schrader et al. [17], who also found that the crack extension can be approximated for soft, rubber-like adhesive systems with simple beam theory assumptions. Figure 17 also shows that the initial crack length is heavily overestimated by the analytical approach, with the error being around 160 mm in mode I and 180 mm in mode III, which, in all cases, is significantly larger than the initial crack length. As this offset seems to be constant, however, it could be argued that analytically calculating the equivalent crack length would be possible for the given soft, rubber-like adhesive system if the crack length were corrected for the determined offset. Hence, it could be argued that G-based evaluation methods relying on a corrected beam theory approach could also pose an option for the determination of the ERR for soft, rubber-like adhesive systems. However, it shall be stated that using the J-integral approach of determining the ERR is likely still favourable in this case, as it allows determining the ERR without the necessity of inferring virtual crack extensions or similar correction factors.

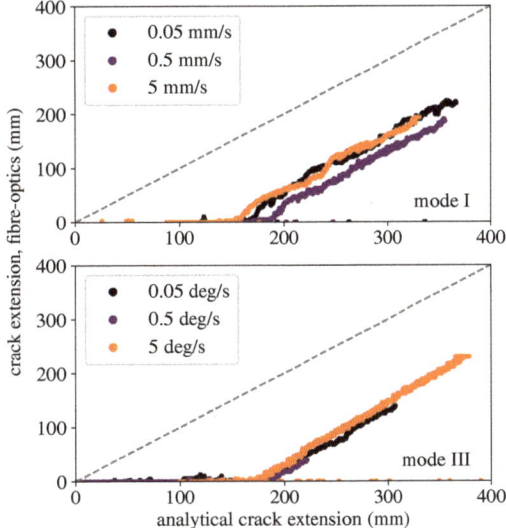

Figure 17. Comparison between BFS crack extension and analytical crack extension.

It is advised that future studies investigate the influence of the specimen geometry on the process zone length and crack propagation more closely. As stated earlier, if the process zone reaches the end of the specimen, the assumption of an unloaded end behind the J-integral evaluation of the cohesive traction is violated (cf. Equation (3)). A future experimental investigation could, hence, be valuable, especially for the practical design of joints with shorter adhesive layers.

4.6. Cohesive Traction in the Adhesive Layer

The traction at the initial crack tip obtained from the "conventional" method according to Equation (4) (bold lines) as well as the mean cohesive traction in the complete adhesive layer according to the BFS measurements, cf. Equation (12), (scatter bands) is shown in Figure 18 for both modes I and III. It can already be observed that the measured cohesive traction changes with loading mode, as the initial stiffness of the joint is significantly lower in mode III than in mode I. Furthermore, the measured cohesive traction is dependent on the loading rate in both cases, already violating the underlying assumption of Equation (4) that the cohesive traction must strictly depend only on the deformation and not on the deformation rate. As can be observed, the traction obtained from $dJ/d\delta_{I,t}$ approximately correlates with the BFS measurement in pure mode I, as both the stiffness of the adhesive layer and the plateau stress fit well with each other. For the lowest loading velocity, however, a clear discrepancy in the range of falling traction can be observed, which can probably be related to an increased influence of material inhomogeneities or creep effects in the process zone on the material behaviour. At the loading rates of 0.5 mm/s and 5 mm/s, their influence may be less pronounced in the process zone, which could explain the better agreement between the BFS measurement and $dJ/d\delta_{I,t}$. Overall, the rough correspondence between methods of traction determination correlates with the investigations of Rosendahl et al. [14], who also found that calculating $dJ/d\delta_{I,t}$ can be used to approximate the cohesive traction of soft, rubber-like adhesive layers in pure mode I.

For the mode III experiments at 0.05 deg/s and 0.5 deg/s, as hinted at earlier, the cohesive traction cannot be calculated from $dJ/d\delta_{III,t}$ due to the observed resistance curve behaviour. In contrast, the BFS measurement can still be used to calculate the cohesive traction within the adhesive layer in these experiments, which is a clear methodological advantage. Additionally, at the highest mode III loading rate of 5 deg/s, the traction obtained from $dJ/d\delta_{III,t}$ differs greatly from the BFS measurements, allowing the conclusion that the determination of the cohesive traction from $dJ/d\delta_{III,t}$ is not feasible in mode III for such soft, rubber-like adhesive layers.

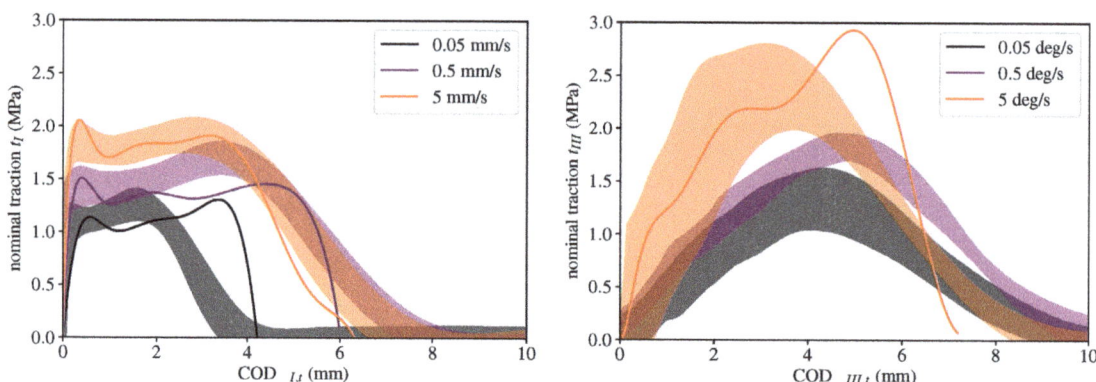

Figure 18. Comparison between traction at the crack tip obtained from $dJ/d\delta_{i,t}$, Equation (4), (bold lines) and mean and standard deviation curves from the BFS measurements (scatter bands): (**left**) mode I DCB tests, (**right**) mode III ODCB tests.

It is generally assumed that the differences between both methods of evaluation arise from violating the underlying theoretical assumptions behind the J-integral method. The BFS measurement, however, can circumvent these assumptions and, by capturing the deformation behaviour of the complete specimen, allows the determination of the traction at the crack tip and along the complete cohesive zone from beam theory without neglecting the influences of energy dissipation in the process zone or influences of the loading rate on the cohesive traction. It can therefore be assumed that the determination of cohesive stresses using Equation (4), i.e., $dJ/d\delta_{i,t}$, for such soft, rubber-like adhesive layers is prone to error and can, within the limitations of this study, only be considered an approximation in pure mode I loading. Furthermore, the goodness of the approximation cannot be estimated a priori, as neither the rate development nor the influence of dissipative effects in the process zone on the material behaviour is known if the traction is calculated from Equation (4).

Another benefit of the BFS measurement shall be noted: As stated earlier, the BFS measurement also allows determining the separation of the adherends at each point of the optical fibre from the curvature of the adherends at discrete measurements in time. Hence, by calculating the time derivative of the measured separation at each measuring point along with the specimen, the separation rate within the complete adhesive layer is obtained, also allowing the investigation of the rate-dependency of the joint's behaviour. In Figure 19, the cohesive traction at each point of measurement is displayed over the separation and separation rate. Interestingly, the differences between the measurements at each measurement point seem to be very small, indicating that the cohesive traction is, in good approximation, independent of the position along with the specimen and that separation and separation rate are relatively similar for each point along with the specimen throughout the measurement. Hence, it can be concluded that the modelling of the joint can theoretically be performed relatively straightforwardly with a rate-dependent cohesive law.

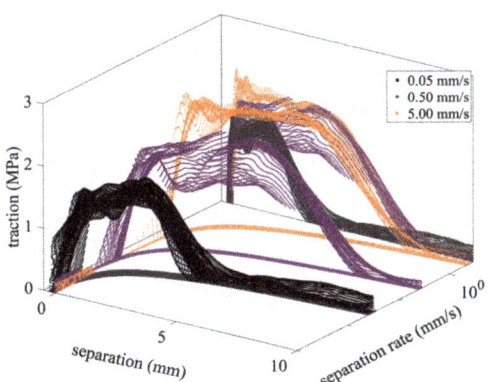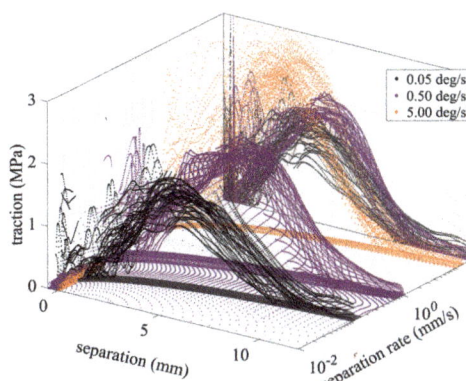

Figure 19. Traction in the cohesive zone over current separation and separation rate: (**left**) mode I DCB tests, (**right**) mode III ODCB tests.

Although we were unable to implement the traction obtained from the BFS measurement into a cohesive zone model within the scope of this study, a future implementation is deemed worthwhile. With a rate-dependent cohesive zone model formulated from the measured values, it might be possible to better reproduce the behaviour of the adhesive joint. It is also assumed that, if observed, it might even be possible to approximately reproduce stick-slip or resistance curve behaviour due to the accurate determination of cohesive traction and separation rate from the BFS measurement. However, these assumptions remain to be addressed in the context of a future simulative study.

4.7. Methodological Critique

As discussed earlier, a distinctive feature of the presented methodology using BFS measurements for the investigation of the fracture behaviour of adhesive layers is that both the nominal traction within the adhesive layer as well as the current separation and the separation rate can be obtained for a large amount of measuring points along with the complete specimen. This is particularly important for the numerical modelling of the fracture behaviour using cohesive zone models, as the conventional method of using the *J*-integral to obtain the cohesive traction cannot be used for the soft, rubber-like adhesive layer under investigation because, as shown, the fundamental assumptions of the method are violated. Hence, the results of this study heavily imply that the use of BFS measurements for the investigation of adhesive joints may serve as a window to a better understanding of their fracture behaviour.

However, it has to be stated that the use of BFS measurements, especially for the almost 1 m long adherends used in this study, requires a very high experimental effort for specimen preparation and investigation. As there are separate measured values for each measurement point on the optic fibre, a large amount of data has to be evaluated. Furthermore, the numerical derivation of the measurement data to obtain cohesive traction produces large amounts of numerical noise, which must carefully be removed using suitable filters before processing the data. As a result, the evaluation of the measurement data is very complex and time-consuming.

As just described, filtering the BFS measurement data for further processing is a major challenge in evaluation. It could therefore be appropriate to first approximate the measured beam curvature using an analytical relationship (polynomials, exponential functions, etc.) to facilitate numerical integration and differentiation. We have refrained from this in the context of this study to introduce as few assumptions as possible into the evaluation of the data a priori. In future studies, however, it is argued that the evaluation process could be simplified by carefully selecting appropriate fit functions, e.g., [24,31].

Finally, we would like to state that the determination of the beam curvature with BFS offers additional possibilities in other areas of application within fracture mechanics testing, which have not—or only to a limited extent—been addressed in this study: it is argued that besides the primary focus of this work, i.e., the determination of the cohesive traction, changes in the beam curvature due to damage evolution behind the crack tip as observed by Schrader et al. [17] could be detected by fibre-optics measurements, allowing the researcher to gain better insight in the damage processes within the adhesive layer behind the major crack tip.

Especially for G-based approaches to determining the fracture energy of an adhesive layer, the current crack length must be measured with good accuracy. As stated earlier, for stiff adhesive layers, crack length measurements using BFS measurement techniques have already been successfully applied in various studies in both modes I and III, e.g., [11,19–22]. The determination of an equivalent crack length from the BFS measurements was also shown to be possible in this study for soft, rubber-like adhesive layers in both modes I and III, which could allow the determination of crack tip position and crack propagation rate for adhesive systems or test setups in which optical methods for the evaluation of crack length fail due to a lack of space or lack of visibility of the current crack tip position. Furthermore, compared to analytical methods for the determination of the equivalent crack length from load point displacement and/or applied loads, the approach presented here does not require any assumptions to be made about the boundary conditions of the substrates' beam bending, such as cantilever beams that are perfectly clamped at the crack tip, which, considering the finite length of the process zone, was shown to be problematic in this work.

It was also shown within the course of this study that, in theory, fibre-optics measurements could even eliminate the need for other COD measurement systems, such as DIC systems or COD gauges, as the system can also provide information about these quantities. Particularly if the entire process zone is to be examined, measurement employing DIC is very difficult, as a very large measurement window is required to cover the entire length of the specimen, which will negatively affect the accuracy of the DIC measurement. Furthermore, considering the mode III investigation, the large out-of-plane deformations are difficult to capture with DIC measurements due to the limited depth of focus. A calculation of the COD from the beam deflection curve is, therefore, a worthwhile option for evaluation when investigating adhesive layers that exhibit finite deformations before ultimate failure.

Overall, we believe that implementing the use of fibre-optics for the mechanical fracture investigation of adhesive joints could be a valuable addition to current research practice, because, as was shown in this study under mode I and III loading, the BFS measurement provides detailed insight into the behaviour of the adhesive. The novel approach we presented based on the BFS measurement allows the determination of rate-dependent traction separation relations directly from DCB and ODCB experiments, which provides a valuable database for inputting into cohesive zone models. Hence, in future investigations, these experimental results could be used to develop new or improve existing cohesive zone models for predicting the fracture of soft, rubber-like adhesive joints, which is crucial for the design of adhesively bonded components.

5. Conclusions

In our study, we investigated the effects of loading rate and mode on the fracture behaviour of a soft polyurethane adhesive joint subjected to peel and shear loading. The rate-dependency was investigated at external loading rates over three orders of magnitude in peel and shear. Next to the conventional evaluation methods employing the J-integral, crack extension, process zone length, and cohesive traction were determined from BFS measurements. Within the limitations of this study, the following conclusions can be drawn:

- The results indicate that the ERR of the tested adhesive system may be largely independent of loading mode in pure modes I and III. This is probably due to the shear loads in mode III testing ultimately transitioning into a peel load at finite deformations.
- The process zone can be investigated thoroughly by the use of BFS measurements. It was observed that the process zone is fully developed at the start of crack propagation in all cases. During the mode III investigations, the process zones are significantly larger than in mode I, which is probably related to the stiffness of the adhesive being lower in shear than in peel.
- The BFS measurements allow the determination of cohesive laws along with the complete adhesive layer based on the Euler–Bernoulli beam theory. Differences between the evaluation method using the proposed BFS and the J-integral method were observed, which is likely due to a violation of the underlying theoretical assumptions of the J-integral method when investigating soft, rubber-like adhesive layers. Furthermore, from the BFS measurement, the rate development along with the complete adhesive layer can be measured, which enables determining a rate-dependent cohesive law.
- As the cohesive laws could not be determined reliably from the J-integral method in the mode III experiments, a determination of cohesive traction with BFS measurements or similar methods is deemed mandatory for soft, rubber-like adhesive layers subjected to mode III loading.
- Although the ERR remains relatively independent of loading mode, the measured cohesive laws are not. Users should bear this in mind when designing and numerically investigating soft, rubber-like adhesive layers and must not assume that the cohesive laws in modes I and III are equivalent.

We were able to show that the investigation of the fracture behaviour of soft, rubber-like adhesive joints using the J-integral method involves complications that require investigation in more detail in future studies. For the time being, the BFS measurements were used as proof of concept, from which, in future investigations, further insights can certainly be gained. Hence, we advise that further research is undertaken in the following areas:

- It became apparent from the BFS measurements during the mode III investigations that transverse forces in the lateral slides influence the external determination of the ERR for the tested soft, rubber-like adhesive system. If ODCB experiments are conducted on similar adhesive systems in the future, the transverse forces should be included in the external evaluation of the J-integral.
- Although it was not possible to implement the measured cohesive laws in finite element analyses in the scope of this study, an implementation using cohesive zone models is deemed worthwhile. A simulative study could investigate whether the experimental results (and especially the observed resistance curve behaviour) can be reproduced with the rate-dependent model.
- It should be investigated whether local effects, i.e., damage behind the crack tip in creep tests or geometric influences due to defects in the adhesive layer, can be investigated more thoroughly using the proposed methodology from BFS measurements.

Author Contributions: Conceptualization, P.S. and D.D.; methodology, P.S. and D.D.; formal analysis, P.S. and D.D.; investigation, P.S. and D.D.; writing—original draft preparation, P.S. and D.D.; writing—review and editing, P.S. and D.D.; visualization, P.S. and D.D.; supervision, S.M.; funding acquisition, S.M. All authors have read and agreed to the published version of the manuscript.

Funding: This project is supported by the Federal Ministry for Economic Affairs and Climate Action (BMWK) on the basis of a decision by the German Bundestag [grant number ZB-ZF4283703]. The financial support is gratefully acknowledged.

Data Availability Statement: The raw and processed data required to reproduce these findings are shown in the present manuscript or cited in the reference section where taken from literature and are available from the corresponding author on request.

Acknowledgments: This article is part of P. Schrader's doctoral thesis at the Doctoral Center for Engineering Sciences of the Research Campus of Central Hessen under the supervision of the Justus-Liebig-University Giessen in cooperation with the University of Applied Sciences of Central Hessen (Technische Hochschule Mittelhessen). The authors would like to thank Maike Sapotta (GLUETEC Industrieklebstoffe GmbH & Co. KG, Germany) for supplying the tested adhesive. Furthermore, we want to thank Jens Minnert and the Institute of Civil Engineering (Technische Hochschule Mittelhessen) for kindly lending us their fibre-optics system for the experimental investigation.

Conflicts of Interest: The authors declare no conflict of interest.

Abbreviations

The following abbreviations are used in this manuscript:

COD	Crack opening displacement
DCB	Double cantilever beam
ODCB	Out-of-plane loaded double cantilever beam
BFS	Backface strain
ERR	Energy release rate
DIC	Digital image correlation

References

1. Loureiro, A.L.; Da Silva, L.F.M.; Sato, C.; Figueiredo, M.A.V. Comparison of the Mechanical Behaviour Between Stiff and Flexible Adhesive Joints for the Automotive Industry. *J. Adhes.* **2010**, *86*, 765–787.
2. Banea, M.D.; Da Silva, L.F.M.; Campilho, R.D.S.G. The Effect of Adhesive Thickness on the Mechanical Behavior of a Structural Polyurethane Adhesive. *J. Adhes.* **2015**, *91*, 331–346. [CrossRef]
3. Banea, M.D.; Da Silva, L.F.M. Mechanical Characterization of Flexible Adhesives. *J. Adhes.* **2009**, *85*, 261–285. [CrossRef]
4. Rice, J.R. A Path Independent Integral and the Approximate Analysis of Strain Concentration by Notches and Cracks. *J. Appl. Mech.* **1968**, *35*, 379–386. [CrossRef]
5. Andersson, T.; Stigh, U. The stress–elongation relation for an adhesive layer loaded in peel using equilibrium of energetic forces. *Int. J. Solids Struct.* **2004**, *41*, 413–434. [CrossRef]
6. Leffler, K.; Alfredsson, K.S.; Stigh, U. Shear behaviour of adhesive layers. *Int. J. Solids Struct.* **2007**, *44*, 530–545. [CrossRef]
7. Marzi, S.; Biel, A.; Stigh, U. On experimental methods to investigate the effect of layer thickness on the fracture behavior of adhesively bonded joints. *Int. J. Adhes. Adhes.* **2011**, *31*, 840–850. [CrossRef]
8. Stigh, U.; Biel, A.; Svensson, D. Cohesive zone modelling and the fracture process of structural tape. *Procedia Struct. Integr.* **2016**, *2*, 235–244. [CrossRef]
9. Loh, L.; Marzi, S. An Out-of-plane Loaded Double Cantilever Beam (ODCB) test to measure the critical energy release rate in mode III of adhesive joints: Special issue on joint design. *Int. J. Adhes. Adhes.* **2018**, *83*, 24–30. [CrossRef]
10. Schrader, P.; Marzi, S. Mode III testing of structural adhesive joints at elevated loading rates. *Int. J. Adhes. Adhes.* **2022**, *113*, 103078. [CrossRef]
11. Schrader, P.; Marzi, S. Novel mode III DCB test setups and related evaluation methods to investigate the fracture behaviour of adhesive joints. *Theor. Appl. Fract. Mech.* **2022**, *123*, 103699. [CrossRef]
12. Schmandt, C.; Marzi, S. Effect of crack opening velocity on fracture behavior of hyperelastic semi-structural adhesive joints subjected to mode I loading. *Procedia Struct. Integr.* **2018**, *13*, 799–805. [CrossRef]
13. Schmandt, C.; Marzi, S. Effect of crack opening velocity and adhesive layer thickness on the fracture behaviour of hyperelastic adhesive joints subjected to mode I loading: Special issue on joint design. *Int. J. Adhes. Adhes.* **2018**, *83*, 9–14. [CrossRef]
14. Rosendahl, P.L.; Staudt, Y.; Odenbreit, C.; Schneider, J.; Becker, W. Measuring mode I fracture properties of thick-layered structural silicone sealants. *Int. J. Adhes. Adhes.* **2019**, *91*, 64–71. [CrossRef]

15. Loh, L.; Marzi, S. Mixed-mode I+III tests on hyperelastic adhesive joints at prescribed mode-mixity. *Int. J. Adhes. Adhes.* **2018**, *85*, 113–122. [CrossRef]
16. Boutar, Y.; Naïmi, S.; Mezlini, S.; da Silva, L.F.; Ben Sik Ali, M. Characterization of aluminium one-component polyurethane adhesive joints as a function of bond thickness for the automotive industry: Fracture analysis and behavior. *Eng. Fract. Mech.* **2017**, *177*, 45–60. [CrossRef]
17. Schrader, P.; Schmandt, C.; Marzi, S. Mode I creep fracture of rubber-like adhesive joints at constant crack driving force. *Int. J. Adhes. Adhes.* **2022**, *113*, 103079. [CrossRef]
18. Škec, L.; Alfano, G.; Jelenić, G. Enhanced simple beam theory for characterising mode-I fracture resistance via a double cantilever beam test. *Compos. Part B Eng.* **2019**, *167*, 250–262. [CrossRef]
19. Ben Salem, N.; Budzik, M.K.; Jumel, J.; Shanahan, M.; Lavelle, F. Investigation of the crack front process zone in the Double Cantilever Beam test with backface strain monitoring technique. *Eng. Fract. Mech.* **2013**, *98*, 272–283. [CrossRef]
20. Bernasconi, A.; Kharshiduzzaman, M.; Comolli, L. Strain Profile Measurement for Structural Health Monitoring of Woven Carbon-fiber Reinforced Polymer Composite Bonded joints by Fiber Optic Sensing Using an Optical Backscatter Reflectometer. *J. Adhes.* **2016**, *92*, 440–458. [CrossRef]
21. Lima, R.; Perrone, R.; Carboni, M.; Bernasconi, A. Experimental analysis of mode I crack propagation in adhesively bonded joints by optical backscatter reflectometry and comparison with digital image correlation. *Theor. Appl. Fract. Mech.* **2021**, *116*, 103117. [CrossRef]
22. Truong, H.T.; Martinez, M.J.; Ochoa, O.O.; Lagoudas, D.C. Mode I fracture toughness of hybrid co-cured Al-CFRP and NiTi-CFRP interfaces: An experimental and computational study. *Compos. Part A Appl. Sci. Manuf.* **2020**, *135*, 105925. [CrossRef]
23. Reiner, J.; Torres, J.P.; Veidt, M. A novel Top Surface Analysis method for Mode I interface characterisation using Digital Image Correlation. *Eng. Fract. Mech.* **2017**, *173*, 107–117. [CrossRef]
24. Sun, F.; Blackman, B. A DIC method to determine the Mode I energy release rate G, the J-integral and the traction-separation law simultaneously for adhesive joints. *Eng. Fract. Mech.* **2020**, *234*, 107097. [CrossRef]
25. Jumel, J.; Budzik, M.K.; Shanahan, M.E. Process zone in the Single Cantilever Beam under transverse loading. *Theor. Appl. Fract. Mech.* **2011**, *56*, 7–12. [CrossRef]
26. Paris, A.J.; Paris, P. Instantaneous evaluation of J and C. *Int. J. Fract.* **1988**, *38*, R19–R21. [CrossRef]
27. Loh, L.; Marzi, S. A Mixed-Mode Controlled DCB test on adhesive joints loaded in a combination of modes I and III. *Procedia Struct. Integr.* **2018**, *13*, 1318–1323. [CrossRef]
28. Loh, L.; Marzi, S. A novel experimental methodology to identify fracture envelopes and cohesive laws in mixed-mode I + III. *Eng. Fract. Mech.* **2019**, *214*, 304–319. [CrossRef]
29. Biel, A. Constitutive Behaviour and Fracture Toughness of an Adhesive Layer. Doctoral Dissertation, Chalmers University of Technology, Göteborg, Sweden, 2005.
30. *ISO 25217*; Adhesives—Determination of the Mode 1 Adhesive Fracture Energy of Structural Adhesive Joints Using Double Cantilever Beam and Tapered Double Cantilever Beam Specimens. International Organization for Standardization: Geneva, Switzerland, 2009.
31. Mao, J.; Nassar, S.; Yang, X. An improved model for adhesively bonded DCB joints. *J. Adhes. Sci. Technol.* **2014**, *28*, 613–629. [CrossRef]

Disclaimer/Publisher's Note: The statements, opinions and data contained in all publications are solely those of the individual author(s) and contributor(s) and not of MDPI and/or the editor(s). MDPI and/or the editor(s) disclaim responsibility for any injury to people or property resulting from any ideas, methods, instructions or products referred to in the content.

Article

Evaluation of XD 10 Polyamide Electrospun Nanofibers to Improve Mode I Fracture Toughness for Epoxy Adhesive Film Bonded Joints

Stefania Minosi *, Fabrizio Moroni and Alessandro Pirondi

Department of Engineering and Architecture, University of Parma, Parco Area delle Scienze 181/A, 43124 Parma, Italy
* Correspondence: stefania.minosi@unipr.it

Abstract: The demand for ever-lighter structures raises the interest in bonding as a joining method, especially for materials that are difficult to join with traditional welding and bolting techniques. Structural adhesives, however, are susceptible to defects, but can be toughened in several ways: by changing their chemical composition or by adding fillers, even of nanometric size. Nanomaterials have a high surface area and limited structural defects, which can enhance the mechanical properties of adhesives depending on their nature, quantity, size, and interfacial adhesion. This work analyzes the Mode I fracture toughness of joints bonded with METLBOND® 1515-4M epoxy film and Xantu-Layr electrospun XD 10 polyamide nanofibers. Two joint configurations were studied, which differed according to the position of the nanomat within the adhesive layer: one had the nanofibers at the substrate/adhesive interfaces, and the other had the nanofibers in the center of the adhesive layer. Double cantilever beam joints were manufactured to evaluate the Mode I fracture toughness of the bonding with and without nano-reinforcement. The nanofibers applied at the substrate/adhesive interface improved the Mode-I fracture toughness by 32%, reaching the value of 0.55 N/mm. SEM images confirm the positive contribution of the nanofibers, which appear stretched and pulled out from the matrix. No fracture toughness variation was detected in the joints with the nanofibers placed in the middle of the adhesive layer.

Keywords: bonded joints; bonding reinforcement; nanomaterials; fracture toughness; epoxy; electrospinning

Citation: Minosi, S.; Moroni, F.; Pirondi, A. Evaluation of XD 10 Polyamide Electrospun Nanofibers to Improve Mode I Fracture Toughness for Epoxy Adhesive Film Bonded Joints. *Processes* **2023**, *11*, 1395. https://doi.org/10.3390/pr11051395

Academic Editor: Raul D. S. G. Campilho

Received: 30 March 2023
Revised: 20 April 2023
Accepted: 25 April 2023
Published: 4 May 2023

Copyright: © 2023 by the authors. Licensee MDPI, Basel, Switzerland. This article is an open access article distributed under the terms and conditions of the Creative Commons Attribution (CC BY) license (https://creativecommons.org/licenses/by/4.0/).

1. Introduction

Advanced composite materials are commonly used for various applications due to their high strength-to-weight ratio. Glass fiber-reinforced polymers (GFRPs), carbon fiber-reinforced polymers (CFRPs), and sandwich structures are widely used in the aerospace, automotive, marine, and railway industries, as well as in the production of wind blades and sports equipment. The development of these materials has led to the advancement of structural adhesives and bonding techniques [1–5] used to join complex and multi-material structures, replacing traditional mechanical fastening [6–8]. Composite materials, metal fiber laminates, sandwich composites, and adhesive joints are subject to delamination. This refers to interlayer failure in the case of composite laminates, while for metal fiber laminates, sandwiches, and bonded joints, it refers to interface failure [9–11]. However, adhesive bonding is prone to delamination failure under high peel loads, which can be improved by developing new adhesive materials and bonding techniques [11].

Epoxy adhesive is widely used as a structural adhesive. However, in its neat formulation, it undergoes brittle fracturing, with a low fracture toughness, which represents a significant limitation for its application in the structural field [11,12]. Joints bonded with brittle epoxy adhesive are defect-sensitive and exhibit broad strength dispersion due to scatter in flaw sizes. To enhance the toughness of structural adhesives, particularly epoxy systems, various methods are commonly employed, such as adding fillers or thermally expandable particles (TEPs), or modifying the chemical resin composition [13–15].

The addition of rubber is also a common method used to enhance the fracture toughness of adhesives [16]. The rubbery phase can be introduced in the form of cross-linked [17,18] or core–shell rubbery particles [19], or liquid rubber can be mixed with resin precursors to allow the precipitation of rubbery particles during resin cross-linking [18,20]. To achieve the toughening effect for epoxy systems, a rubbery fraction between 5 and 20 wt. % is added. However, adding a high amount of rubber can lead to a reduction in the glass transition temperature (Tg), elastic modulus, and strength of the resin [21].

Adding organic and inorganic fillers, such as metallic micro- and nano-particles, nanoclays or short fibers, is another method used to improve the fracture toughness of structural adhesives [13,22–24]. Nanoparticles can also increase the fracture toughness, strength, and stiffness of bonded joints [13]. However, it is crucial to develop a strong interfacial adhesion to correctly transfer the load from the polymeric matrix to the nano-reinforcement [13]. Carbon-based nanoparticles, such as carbon nanofibers (CNFs), carbon nanotubes (CNTs) and graphene nanoplatelets (GNPs), are widely used for this purpose [25–39]. These nanoparticles enhance the fatigue life of bonded joints and can be used for damage detection, as they also improve the electrical properties of the resin they are dispersed in [29–39]. Recently, studies have shown that the application of hybrid nanoparticles is a viable approach to designing tougher, stronger and more durable bonded joints [40–44].

The integration of polymeric nanofibers has been shown to be an effective method for toughening epoxy matrices and composite materials [45]. Many studies have shown that composite laminates reinforced with electrospun polymeric nanofibers exhibit enhanced mechanical properties, including improved fracture toughness and delamination strength, with the interposition of a thermoplastic nanomat between composite layers promoting the ply-to-ply bridging effect [46–53]. Hamer et al. studied laminates of CFRP interleaved with electrospun Nylon 66 nanofibers. They performed DCB tests to evaluate the effects of the nanofibers on Mode I fracture toughness. The mat of naofibers embedded in the midplane improved the toughness by about 3 times [47]. Beckermann and Pickering studied the effects of interleaved nanofiber plies on the mode I and mode II interlaminar fracture toughness of carbon and epoxy resin laminates. Various types of electrospun nanofibers were placed in the midplane planes of the laminates. The results show that the best performance was achieved using 4.5 g/m^2 PA66 ply, with fracture toughness improvements of 156% and 69% for Mode I and Mode II, respectively [49]. Saghafi et al. studied the effect of Nylon 6,6 nanofibers interleaved in the midplane of glass/epoxy laminates on mode I and mode II fracture toughness. Nylon 6,6 nanofibers improved the initial G_{IC} and G_{IIC} energy release rates by 62% and 109%, respectively [50]. Daelemans et al. demonstrated that nanofibrous veils of PA 6.9 with different morphologies interleaved in UD carbon/epoxy laminates cause an increase in mode II interlaminar fracture toughness. Mode II interlaminar fracture toughness is doubled by randomly deposited PA 6.9 nanofibers [51]. Goodarz et al. demonstrated that the interfacial incorporation of aramid nanofibers significantly increases the absorbed impact energy, compared to laminates without nanofibers [54].

These results suggest that polymeric nanofibers could be effective in improving crack toughening for bonded joints as well. There have been limited studies on the use of electrospun nanofibers in adhesive bonding, particularly with medium–low-fracture toughness epoxy resins [55–58]. The works [55,56] analyze the effect of a core/shell structure of electrospun meta-aramid fibers integrated into the adhesive layer of the epoxy-bonded joint. Single-lap shear test results show that electrospun meta-aramid nanofibers decreased joint strength, while those with core/shell structure restored the strength of pure epoxy. Razavi et al. found that incorporating polyacrylonitrile (PAN) nanofibers into an aluminum DCB joint bonded with 2k epoxy resin resulted in a two-fold increase in fracture toughness [57]. Ekrem and Avci demonstrated that incorporating polyvinyl alcohol (PVA) nanofibrous mats into the adhesive layer of single lap joints (SLJ) and DCB joints improved shear strength by 13.5% and increased mode I fracture toughness by about two times that of the neat adhesive [58]. In a previous work, the authors demonstrated that electrospun nylon nanofibers can act as reinforcements and support for the adhesive layer, improving the fracture toughness of low-toughness resins in DCB joints made with pre-impregnated nanofibers [59–62]. In a previous work, impregnation of the nanofibers was initially performed with low- and medium-viscosity epoxy resins to facilitate the wetting of the nanomat [59]. Then, an unfilled medium viscosity two-component epoxy adhesive was used before using a high-viscosity, high-strength two-component epoxy adhesive system [60–62]. However, tests performed on bonded joints were characterized by extensive areas of adhesive failure, at the interface between substrate and adhesive. The improvement of interfacial adhesion is critical for the evaluation of the effect of nanofibrous structures.

Despite the non-marginal scientific literature on the application of nanofibers for interface toughening, the application to adhesive bonding is still limited, and in-house procedures for the embedding are used, for which the possibility of scaling up to an industrial level is either impractical or unclear. Additionally, home-made electrospun nanofibers are often used, which may have a more limited reproducibility than when manufactured on an industrial scale.

Based on the previous considerations, this study investigates the effect of commercial XD 10 polyamide electrospun nanofibers (XantuLayr™, NANOLAYR LTD, Auckland, New Zealand) on composite joints bonded with epoxy film, commonly used in the aerospace industry. The XantuLayr nanomat is known to improve the interlaminar fracture toughness of composite laminates, resulting in higher delamination resistance and damage tolerance [49,53].

In this work, a XantuLayr electrospun nanomat was used for the first time as a toughening element of composite adhesive joints made by secondary bonding. Furthermore, the joints were produced using bonding techniques employed in the automotive and aerospace industries, and were thus compatible with current industrial practices. The manufacturing technique of these joints is therefore replicable in an industrial environment and not just on a laboratory scale. In this work, two joint configurations were studied: one with nanofibers applied at the adhesive/adherend interfaces, and one with nanofibers placed in the center of the adhesive layer. Double Cantilever Beam (DCB) joints were produced to evaluate the mode-I fracture toughness of the bond with and without nano-reinforcement. A morphological analysis was also performed to understand the phenomena occurring during crack propagation.

2. Materials and Method

2.1. Adherends

In this study, CYCOM® 977-2 prepreg (Solvay Specialty Polymers SpA, Bollate (MI), Italy) was used to fabricate composite adherents. This prepreg is suitable for aerospace and aircraft applications that require impact resistance and light weight. The unidirectional tape used had a nominal thickness of 0.186 mm and a density of 1.55 g/cm^3. A quasi-isotropic laminate was produced using 32 layers of prepreg with a lamination sequence of $[45/0/-45/90]_{4s}$. The panels were prepared for bonding using a peel ply, a sacrificial layer of fabric put on the surface of the composite. The panel was vacuum-bagged and cured in an autoclave using the cycle specified by the prepreg technical datasheet. The vacuum bag was realized as reported in Figure 1. After curing, the part was debagged and cut to size. The peel ply was removed from the panel surface prior to bonding.

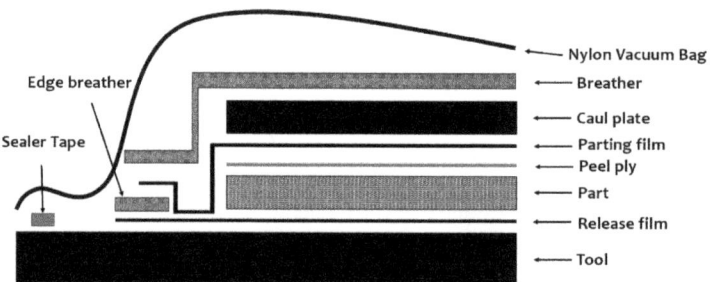

Figure 1. Vacuum bag.

The elastic modulus of the laminate was evaluated to be 58 GPa using tensile testing in accordance with the ASTM D3039 standard [63].

2.2. Adhesive

The epoxy adhesive used for the joint fabrication was the METLBOND® 1515-4M (Solvay Specialty Polymers SpA, Bollate (MI), Italy). This adhesive is mainly employed for bonding composites, although it is suitable for various metal bonding applications. The adhesive has a nominal weight of 242 g/m², with a nylon web carrier accounting for 7.5% of the total weight of the film. The elastic modulus of the adhesive film is closely related to the curing pressure and temperature. Based on recommendations from the supplier and various studies in the literature, the elastic modulus of METLBOND® 1515-4M was determined to be 3.5 GPa [64–66]. The yield strength, ultimate strength and strain at failure of the adhesive were not available in the data sheet and, since they are not essential to the purposes of the work and the manufacturing of a tensile test specimen out of a film adhesive is not straightforward, they were not evaluated.

The curing cycle for the bonded joints included an autoclave at 6 bar pressure at 180 °C for 210 min. This cycle is similar to that proposed by the adhesive manufacturer's datasheet, and is the same as that employed for the adherents' manufacturing. The chosen cure cycle is of industrial significance as it enables the consolidation and cure of prepreg and secondary bonding simultaneously, leading to time and energy savings, and it is suitable for co-bonded joints.

2.3. Nanofibers

XantuLayr® (NANOLAYR LTD, Auckland, New Zealand) is a thermoplastic nanofiber veil produced using Sonic Electrospinning Technology. It consists of XD10 polyamide nanofibers that form an ultra-thin non-woven web. For this study, a XantuLayr® nanomat with an areal density of 3 g/m² was used. The thickness of the nanomat was measured using a digital indicator (ALPA, Pontoglio (BS), Italy), with a preload of 0.65 N, and was found to be in the range of 120 to 160 µm.

2.4. Double Cantilever Beam Fabrication

To assess the impact of integrating commercial nanofibers on the fracture toughness of the adhesive system, four series of DCB joints were manufactured. Table 1 and Figure 2 provide details of the various configurations of the DCB joints.

Table 1. Composite DCB joints configurations.

Series ID	Adhesive Layer	Number of Samples
1S	METLBOND® 1515-4M	7
2S	2 METLBOND® 1515-4M	8
1S–2NF	XantuLayr® + METLBOND® 1515-4M + XantuLayr®	8
2S–1NF	METLBOND® 1515-4M + XantuLayr® + METLBOND® 1515-4M	8

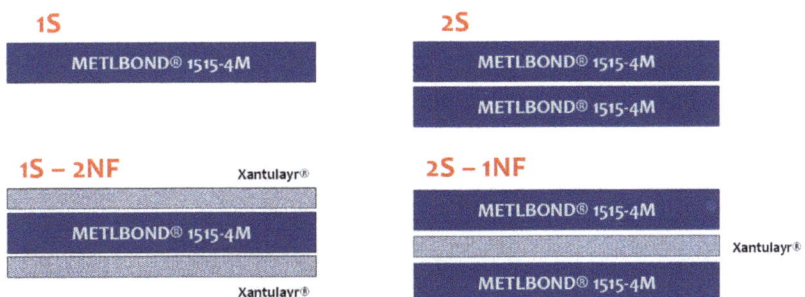

Figure 2. Configurations of the adhesive layers.

The four configurations chosen were 1S, 2S, 1S–2NF and 2S–1NF. The 1S and 2S configurations refer to the virgin specimens, bonded with one and two layers of adhesive, respectively. The 1S–2NF and 2S–1NF configurations refer to the bonded joints reinforced with nanofibers. The 1S–2NF configuration involves the positioning of nanofibers at the adhesive/support interfaces of joints bonded with one layer of adhesive. The 2S–2NF configuration involves the positioning of nanofibers at the center of the adhesive layer of joints bonded with a layer of epoxy film.

The number of samples was defined in such a way as to derive at least five useful samples for calculating the average fracture toughness value and assessing the repeatability of the failure type.

To manufacture the specimens for testing, two composite panels measuring 190 × 150 mm^2 were bonded together and placed in a vacuum bag. A peel-ply was used in the preparation of the panels, which was then removed after the composite had cured, prior to bonding. This method proved to be an effective means of ensuring a strong bond between the composite parts. The joint curing process was carried out in an autoclave at 6 bar pressure and 180 °C temperature for 210 min. During the bonding stage, a 25 mm initial defect was introduced by placing a 0.1 mm-thick Teflon patch on one end of the joint. The nanofibers were placed manually on the adhesive film. Since they were supported on a paper backing, they were easy to handle. Once the nanofibers were properly positioned, the paper backing was removed. After curing, the panels were cut to form DCB joints with a length of 150 mm and width of 25 mm. Two pairs of bonded panels were manufactured for each configuration. Holes were machined for each joint to enable the attachment of steel blocks, which were utilized to secure the specimen in the testing apparatus. Steel blocks were glued to the DCBs using Loctite Hysol 9466 adhesive. To ensure the correct positioning of the blocks, they were fastened to the adherents with screws and bolts. Once the adhesive was polymerized after 24 h at room temperature, screws and nuts were removed, and the DCB was ready to be tested. The adherent dimensions were reduced if compared with ASTM D3433 standards. These dimensions were chosen on the basis of the available material. The DCB geometry is illustrated in Figure 3. Figures 4 and 5 show an example of a tested specimen and the same one mounted on the test machine.

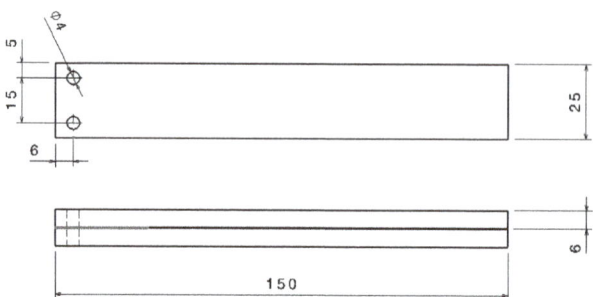

Figure 3. Composite DCB geometry.

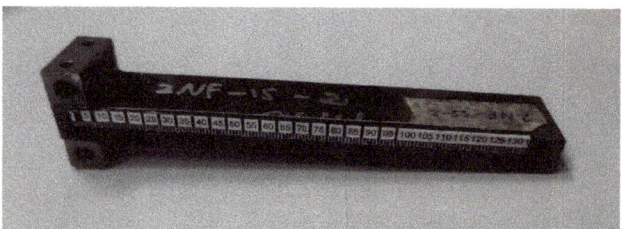

Figure 4. Example of a DCB specimen.

Figure 5. Specimen mounted on the testing machine.

The Table 2 shows the average thicknesses of the tested specimens. The calculation of the adhesive layer's thickness was carried out using the difference between the average joint thickness and the average thickness, t, of the individual adhesives. The presence of the nylon cloth inside the adhesive layer ensures joints with a constant adhesive cross-section.

Table 2. Samples thickness.

Sample ID	t (mm)
1S	0.17
2S	0.48
1S–2NF	0.18
2S–1NF	0.52

2.5. DCB Test

The DCB test was conducted on a servo-hydraulic MTS 810 machine with a 3 kN load cell, using displacement control at a constant crosshead velocity. The loading and unloading rates were 2 mm/min and 5 mm/min, respectively. The correct determination of mode I fracture toughness is crucial to assess the integration effect of the nanomaterial. There are several data reduction methods that can be used to overcome the problem of direct crack length monitoring during the DCB test. The data reduction schemes include the compliance calibration method, in which compliance is calibrated as a polynomial function of crack length, and the compliance-based beam method, which considers the influence of the fracture process zone [67–69]. In this paper, according to the same procedure used in the past by the authors, partial unloadings are performed to determine the specimen compliance and actual crack length using Krenk's model, reported in [70], which accounts for the out-of-plane deformation of the adhesive layer and rotation at the crack tip. The model is represented by Equation (1):

$$C = \frac{\delta'}{P} = 2\left[\frac{2\lambda_\sigma}{k}(1+\lambda_\sigma a) + (a+g)\frac{(2\lambda_\sigma^2)}{k}(1+2\lambda_\sigma a) + \frac{a^3}{3EJ} + g\frac{a^2}{2EJ}\right] \quad (1)$$

The joint compliance (C (mm/N)) is determined by dividing the Crack Mouth Opening Displacement (CMOD) measurement at the front of the specimen (δ' (mm)) by the load (P (N)). Other variables in the equation include the actual crack length (a (mm)), Young's modulus of the adherents (E (MPa)), and area moment of inertia of the adherent (J (mm^4)). A clip gage was used to measure the CMOD during testing. The model presented in Equation (1) has been modified from the one proposed by Krenk to account for the distance (g (mm)) between the measurement point and the load axis, as well as the effect of shear.

The joint geometry is shown in Figure 6.

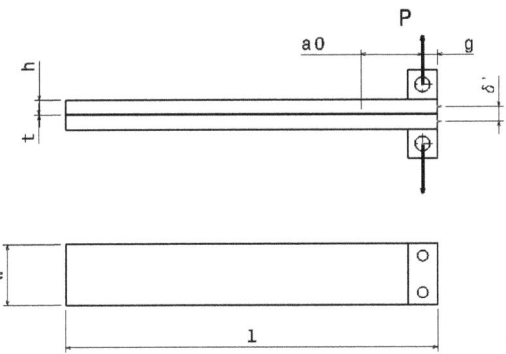

Figure 6. DCB geometry [71,72], where a_0 is 25 mm and g is 6 mm.

The dimensionless parameters λ_σ and k are reported in Equations (2) and (3):

$$\lambda_\sigma = \sqrt[4]{\frac{6}{h^3 t}\frac{E_a}{E(1-v_a^2)}} \quad (2)$$

$$k = \frac{2E_a t}{t(1-v_a^2)} \quad (3)$$

where all the sizes are expressed in mm, while E_a (MPa) and v_a (dimensionless) are the Young's modulus and the Poisson's ratio of the adhesive, respectively. The Mode I strain energy release rate G (N/mm) is:

$$G_I = \frac{(Pa)^2}{tEJ}\left(1 + \frac{1}{\lambda_\sigma a}\right)^2 \quad (4)$$

Since the fiber volume fraction is negligible, for the rule of mixtures, the Young's modulus of the nanomat prepreg is also approximately the same as that of the adhesive alone.

3. Results and Discussion

Figures 7–9 show the load against CMOD of virgin and nanomodified specimens 1S, 2S, 1S–NF and 2S–1NF, taken as representative. The load peaks of virgin samples are slightly lower than 600 N, and the employment of two layers instead of one does not significantly affect adhesive performance. The nanomodified sample has a slightly higher load peak than the neat joint during crack propagation, and the behavior of the 2S–1NF joint is comparable to that of the virgin samples.

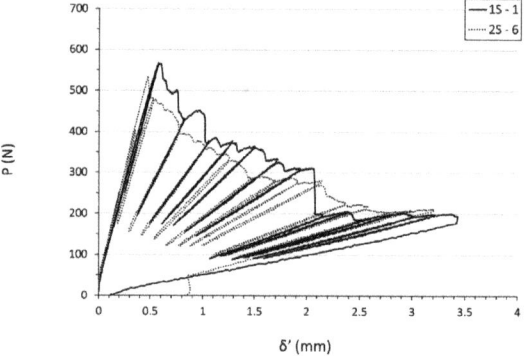

Figure 7. Load against CMOD (δ') for virgin specimens 1S–1 and 2S–6.

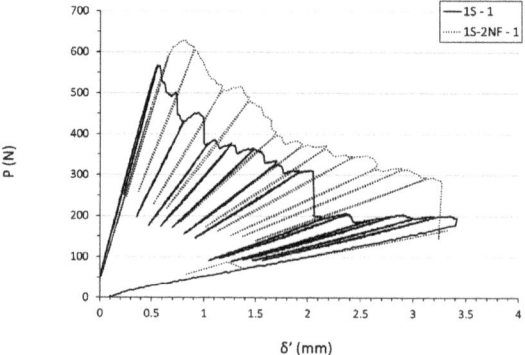

Figure 8. Load against CMOD (δ') for both virgin (1S–1) and nylon-nanomodified (1S–2NF–1) specimens.

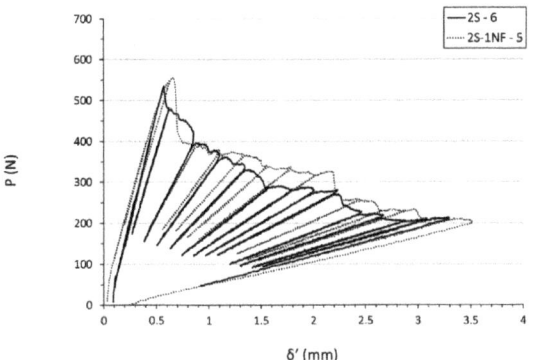

Figure 9. Load against CMOD (δ') for both virgin (2S–6) and nylon-nanomodified (2S–1NF–5) specimens.

Figure 10 displays the fracture surfaces of 1S, 2S, 1S–2NF, and 2S–1NF samples. As can be seen from the picture, the fibers of the bonded surface are at 45°. This solution was preferred as it represents a more general case of joining. The blue areas highlight the presence of the adhesive on the substrate under examination. The 1S specimens failed cohesively during the first stage of the crack propagation, but the crack deviated inside the composite support generally after 30 mm of propagation inside the adhesive. The failure mode of 2S samples was more scattered, with cohesive failure observed in joints 2S–1, 2S–4, and 2S–5 in the initial stage of crack propagation, interfacial fracture in joints 2S-2 and 2S-3, and cohesive fracture in joints 2S–6, 2S–7, and 2S–8. Figure 11 shows SEM images of the fracture surfaces of the 1S–1 and 2S–6 samples, respectively, which show micro-dimples and broken nylon fibers. The 1S–2NF specimens failed cohesively, but the crack deviated inside the composite supports of the samples 1S–2NF 3, 4 and 5 after 30 mm of propagation inside the adhesive. Samples 2S–1NF exhibited cohesive failure, but half of them were subjected to crack propagation inside the composite layer at Δa values of 10–20 mm, making the results less reproducible. Figure 12 shows SEM images of the fracture surfaces of 1S–2NF–1 and 2S–1NF–5 samples, respectively, which showed micro-dimples, the presence of nylon cloth, and areas rich in nanofibers. The nanofibers in the sample 2S–1NF appear less stretched and were broken inside the matrix without evident pull-out, resulting in fracture toughness values lower than those of 1S–2NF samples and comparable to those of virgin ones.

Figure 13 shows a comparison of the R-Curves of one representative specimen for each configuration. The black markers identify the G_{IC} values used for calculating the average fracture toughness during the steady-state crack propagation phase, while the grey markers represent the excluded ones. Considering all the specimens tested (see Table 1), the average fracture toughness of the neat adhesive is 0.42 ± 0.07 N/mm for 1S and 0.42 ± 0.10 N/mm for 2S. The average fracture toughness of the nanomodified 1S–2NF series is instead 0.55 ± 0.16 N/mm, while for 2S–1NF, it is about 0.44 ± 0.8 N/mm. The average G_{IC} values are reported in Figure 14. The samples 1S–2NF have more scattered values, but higher average GIC than 2S–1NF. The average maximum load values are reported in Figure 15. Again, the highest maximum load was achieved by the 1S–2NF samples with an average maximum load value of 580 N ± 47 N, which is 10% higher than that of the 1S specimens.

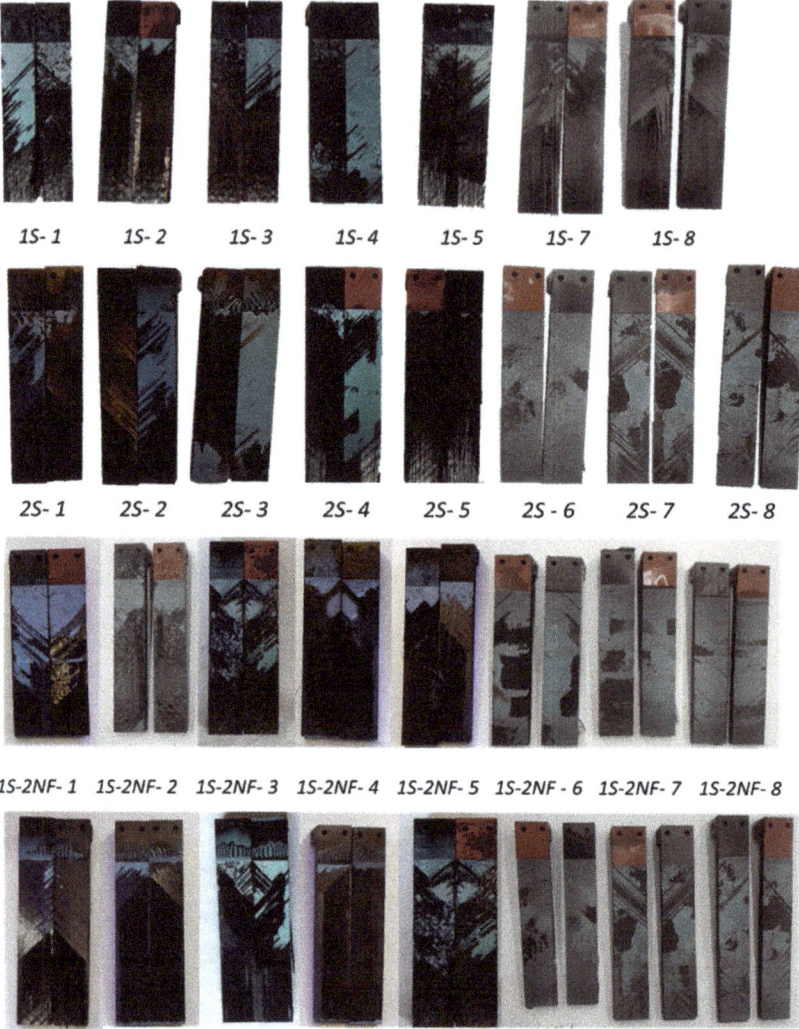

Figure 10. Fracture surface.

The results obtained confirm that the virgin samples exhibit the same fracture toughness values, regardless of the number of adhesive layers used for bonding. The highest values of fracture toughness were obtained by the 1S–2NF samples. Nanofibers placed at the adhesive/adhesive interface deformed and contributed to the joint toughness. The configuration 1S–2NF exhibited a 32% improvement compared to 1S samples. The lower deformation of the nanofibers placed between the two adhesive layers of the 2S–1NF samples resulted in a lower toughness value of the system, which, however, was still comparable with the virgin samples.

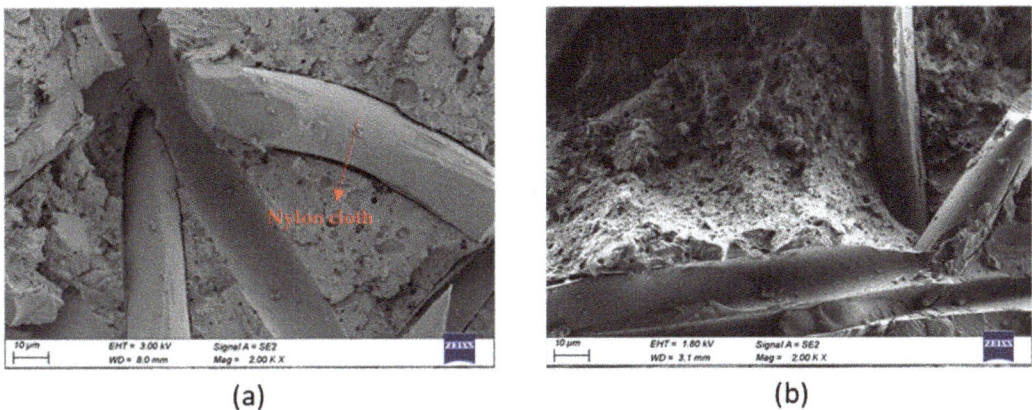

Figure 11. SEM images of fracture surfaces of 1S–1 (**a**) and 2S–6 (**b**) at 2000×.

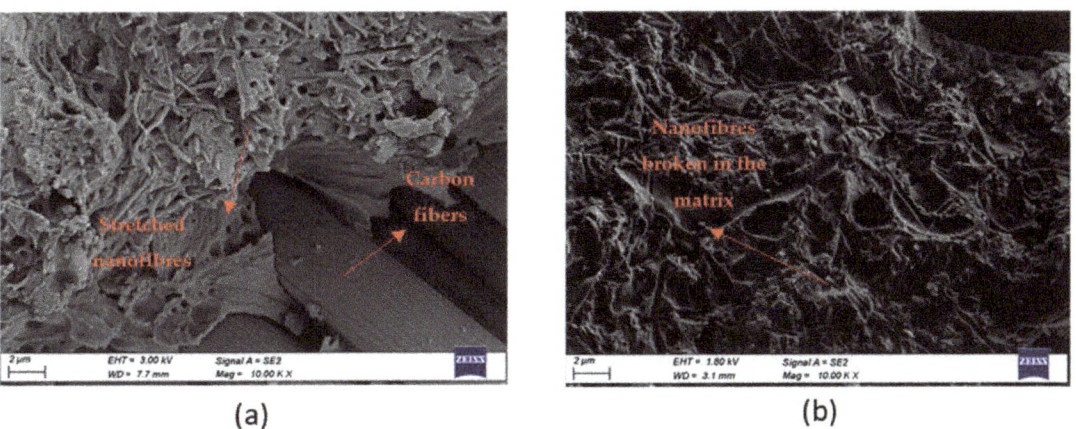

Figure 12. SEM images of fracture surfaces of 1S–NF–1 (**a**) and 2S–1NF–5 (**b**) at 10,000×.

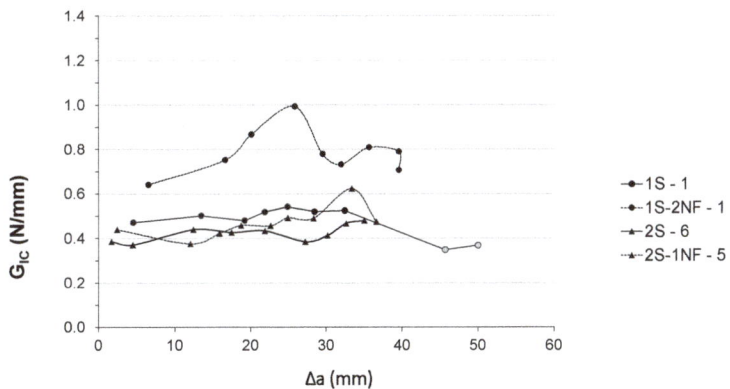

Figure 13. Comparison of R-Curves of representative virgin (1S and 2S) and nanomodified (1S–2NF and 2S–1NF) specimens.

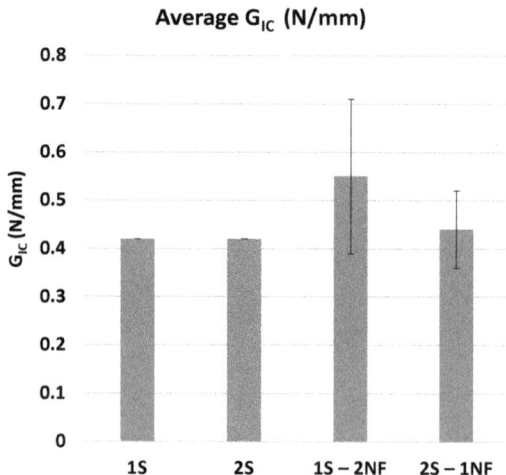

Figure 14. G_{IC} average values for 1S, 2S, 1S–2NF, and 2S–1NF samples.

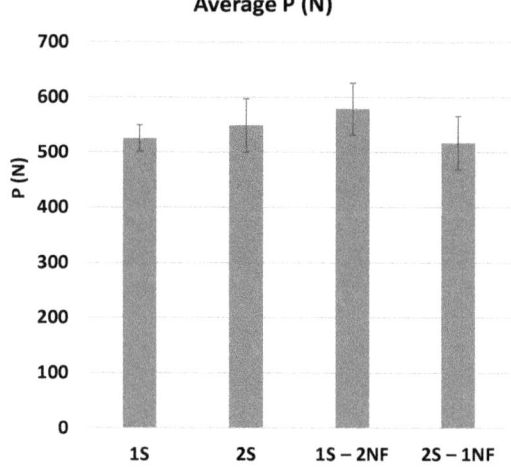

Figure 15. Maximum load average values for 1S, 2S, 1S–2NF, and 2S–1NF samples.

4. Conclusions

The effect of commercial XD10 PA (XantuLayr®) nanofibers within composite joints bonded with the epoxy film was studied. Materials and bonding techniques commonly employed in the automotive and aerospace sector were used for joint manufacturing. Since the adhesive was in film form, nanofiber integration could be approached in two ways. The first was to apply the nanofiber at the adherents/adhesive interface. The second was to interleave the nanomat between two layers of adhesive.

The main results are reported below:

1. The behaviors of virgin samples are similar, and are not influenced by the number of adhesive layers used for bonding;
2. The application of commercial XD10 PA nanofibers (XantuLayr®) at the adhesive/adherent interface improves the mechanical performance of the composite joints, which exhibited higher fracture toughness and fracture resistance than virgin samples. In

particular, the 1S-2NF specimens exhibited 10% and 32% higher maximum strength and fracture toughness values, respectively;
3. SEM images confirm the contribution of the nanofibers that appear elongated and detached from the matrix. The deformation of the nanomat contributes to the joint toughness;
4. The same nanomaterial applied to the center of the adhesive layer does not contribute to the fracture toughness of the joint, as the nanomodified joints show the same G_{IC} and standard deviation values as the virgin samples.

The application of commercial XD10 PA nanofibers (XantuLayr®) was very simple, and is definitely compatible with materials and bonding techniques used for composite materials.

Further developments of this work will involve evaluating mode II fracture toughness through End-Notched-Flexure (ENF). Further analyses could be conducted on thinner substrate bonds by performing T-Peel tests.

Author Contributions: Conceptualization, S.M. and A.P.; methodology, S.M., F.M. and A.P.; software, N/A; validation, A.P. and F.M.; formal analysis, S.M.; investigation, S.M.; resources, S.M., F.M. and A.P.; data curation, S.M.; writing—original draft preparation, S.M.; writing—review and editing, A.P. and S.M.; visualization, S.M.; supervision, A.P.; project administration, N/A; funding acquisition, N/A. All authors have read and agreed to the published version of the manuscript.

Funding: This research received no external funding.

Data Availability Statement: Data is contained within the article.

Conflicts of Interest: The authors declare no conflict of interest.

References

1. Da Silva, L.F.M.; Öchsner, A.; Adams, R.D. (Eds.) *Handbook of Adhesion Technology*; Springer: Berlin/Heidelberg, Germany, 2011; ISBN 978-3-642-01168-9.
2. Kupski, J.; Teixeira de Freitas, S. Design of Adhesively Bonded Lap Joints with Laminated CFRP Adherends: Review, Challenges and New Opportunities for Aerospace Structures. *Compos. Struct.* **2021**, *268*, 113923. [CrossRef]
3. De Oliveira, L.Á.; Vieira, M.M.; dos Santos, J.C.; Freire, R.T.S.; Tonatto, M.L.P.; Panzera, T.H.; Zamani, P.; Scarpa, F. An Investigation on the Mechanical Behaviour of Sandwich Composite Structures with Circular Honeycomb Bamboo Core. *Discov. Mech. Eng.* **2022**, *1*, 7. [CrossRef]
4. Soltannia, B.; Mertiny, P.; Taheri, F. Static and Dynamic Characteristics of Nano-Reinforced 3D-Fiber Metal Laminates Using Non-Destructive Techniques. *J. Sandw. Struct. Mater.* **2021**, *23*, 3081–3112. [CrossRef]
5. Zamani, P.; Jaamialahmadi, A.; da Silva, L.F.M. Fatigue Life Evaluation of Al-GFRP Bonded Lap Joints under Four-Point Bending Using Strain-Life Criteria. *Int. J. Adhes. Adhes.* **2023**, *122*, 103338. [CrossRef]
6. Turan, K.; Örçen, G. Failure Analysis of Adhesive-Patch-Repaired Edge-Notched Composite Plates. *J. Adhes.* **2017**, *93*, 328–341. [CrossRef]
7. Olajide, S.O.; Kandare, E.; Khatibi, A.A. Fatigue Life Uncertainty of Adhesively Bonded Composite Scarf Joints—An Airworthiness Perspective. *J. Adhes.* **2017**, *93*, 515–530. [CrossRef]
8. Rasane, A.R.; Kumar, P.; Khond, M.P. Optimizing the Size of a CFRP Patch to Repair a Crack in a Thin Sheet. *J. Adhes.* **2017**, *93*, 1064–1080. [CrossRef]
9. De Cicco, D.; Taheri, F. Effect of Functionalized Graphene Nanoplatelets on the Delamination-Buckling and Delamination Propagation Resistance of 3D Fiber-Metal Laminates Under Different Loading Rates. *Nanomaterials* **2019**, *9*, 1482. [CrossRef]
10. Gong, Y.; Chen, X.; Zou, L.; Li, X.; Zhao, L.; Zhang, J.; Hu, N. Experimental and Numerical Investigations on the Mode I Delamination Growth Behavior of Laminated Composites with Different Z-Pin Fiber Reinforcements. *Compos. Struct.* **2022**, *287*, 115370. [CrossRef]
11. Neto, J.A.B.P.; Campilho, R.D.S.G.; da Silva, L.F.M. Parametric Study of Adhesive Joints with Composites. *Int. J. Adhes. Adhes.* **2012**, *37*, 96–101. [CrossRef]
12. Dadian, A.; Rahnama, S.; Zolfaghari, A. Experimental Study of the CTBN Effect on Mechanical Properties and Mode I and II Fracture Toughness of a New Epoxy Resin. *J. Adhes. Sci. Technol.* **2020**, *34*, 2389–2404. [CrossRef]
13. Giv, A.N.; Ayatollahi, M.R.; Ghaffari, S.H.; da Silva, L.F.M. Effect of Reinforcements at Different Scales on Mechanical Properties of Epoxy Adhesives and Adhesive Joints: A Review. *J. Adhes.* **2018**, *94*, 1082–1121. [CrossRef]
14. Saraç, İ.; Adin, H.; Temiz, Ş. Experimental Determination of the Static and Fatigue Strength of the Adhesive Joints Bonded by Epoxy Adhesive Including Different Particles. *Compos. Part B Eng.* **2018**, *155*, 92–103. [CrossRef]
15. Banea, M.D.; da Silva, L.F.M.; Carbas, R.J.C.; Campilho, R.D.S.G. Mechanical and Thermal Characterization of a Structural Polyurethane Adhesive Modified with Thermally Expandable Particles. *Int. J. Adhes. Adhes.* **2014**, *54*, 191–199. [CrossRef]

16. Caldona, E.B.; De Leon, A.C.C.; Pajarito, B.B.; Advincula, R.C. A Review on Rubber-Enhanced Polymeric Materials. *Polym. Rev.* **2017**, *57*, 311–338. [CrossRef]
17. Riew, C.K.; Siebert, A.R.; Smith, R.W.; Fernando, M.; Kinloch, A.J. Toughened Epoxy Resins: Preformed Particles as Tougheners for Adhesives and Matrices. In *Toughened Plastics II*; Advances in Chemistry; Riew, C.K., Kinloch, A.J., Eds.; American Chemical Society: Washington, DC, USA, 1996; Volume 252, pp. 33–44. ISBN 978-0-8412-3151-1.
18. Williams, R.J.J.; Rozenberg, B.A.; Pascault, J.-P. Reaction-Induced Phase Separation in Modified Thermosetting Polymers. In *Polymer Analysis Polymer Physics*; Advances in Polymer Science; Springer: Berlin/Heidelberg, Germany, 1997; pp. 95–156. ISBN 978-3-540-68374-2.
19. Tsang, W.L.; Taylor, A.C. Fracture and Toughening Mechanisms of Silica- and Core–Shell Rubber-Toughened Epoxy at Ambient and Low Temperature. *J. Mater. Sci* **2019**, *54*, 13938–13958. [CrossRef]
20. Wise, C.W.; Cook, W.D.; Goodwin, A.A. CTBN Rubber Phase Precipitation in Model Epoxy Resins. *Polymer* **2000**, *41*, 4625–4633. [CrossRef]
21. Bagheri, R.; Marouf, B.T.; Pearson, R.A. Rubber-Toughened Epoxies: A Critical Review. *Polym. Rev.* **2009**, *49*, 201–225. [CrossRef]
22. Kinloch, A.J. Toughening Epoxy Adhesives to Meet Today's Challenges. *MRS Bull.* **2003**, *28*, 445–448. [CrossRef]
23. Ghabezi, P.; Farahani, M. Effects of Nanoparticles on Nanocomposites Mode I and II Fracture: A Critical Review. In *Progress in Adhesion and Adhesives*; John Wiley & Sons, Ltd: Hoboken, NJ, USA, 2018; pp. 391–411. ISBN 978-1-119-52644-5.
24. Ghabezi, P.; Farahani, M. A Cohesive Model with a Multi-Stage Softening Behavior to Predict Fracture in Nano Composite Joints. *Eng. Fract. Mech.* **2019**, *219*, 106611. [CrossRef]
25. Yang, G.; OuYang, Q.; Ye, J.; Liu, L. Improved Tensile and Single-Lap-Shear Mechanical-Electrical Response of Epoxy Composites Reinforced with Gridded Nano-Carbons. *Compos. Part A Appl. Sci. Manuf.* **2022**, *152*, 106712. [CrossRef]
26. NajiMehr, H.; Shariati, M.; Zamani, P.; da Silva, L.F.M.; Ghahremani Moghadam, D. Investigating on the Influence of Multi-Walled Carbon Nanotube and Graphene Nanoplatelet Additives on Residual Strength of Bonded Joints Subjected to Partial Fatigue Loading. *J. Appl. Polym. Sci.* **2022**, *139*, 52069. [CrossRef]
27. Demir, K.; Gavgali, E.; Yetim, A.F.; Akpinar, S. The Effects of Nanostructure Additive on Fracture Strength in Adhesively Bonded Joints Subjected to Fully Reversed Four-Point Bending Fatigue Load. *Int. J. Adhes. Adhes.* **2021**, *110*, 102943. [CrossRef]
28. Zamani, P.; Jaamialahmadi, A.; da Silva, L.F.M. The Influence of GNP and Nano-Silica Additives on Fatigue Life and Crack Initiation Phase of Al-GFRP Bonded Lap Joints Subjected to Four-Point Bending. *Compos. Part B Eng.* **2021**, *207*, 108589. [CrossRef]
29. Takeda, T.; Narita, F. Fracture Behavior and Crack Sensing Capability of Bonded Carbon Fiber Composite Joints with Carbon Nanotube-Based Polymer Adhesive Layer under Mode I Loading. *Compos. Sci. Technol.* **2017**, *146*, 26–33. [CrossRef]
30. Khoramishad, H.; Khakzad, M. Toughening Epoxy Adhesives with Multi-Walled Carbon Nanotubes. *J. Adhes.* **2018**, *94*, 15–29. [CrossRef]
31. Akpinar, I.A.; Gürses, A.; Akpinar, S.; Gültekin, K.; Akbulut, H.; Ozel, A. Investigation of Mechanical and Thermal Properties of Nanostructure-Doped Bulk Nanocomposite Adhesives. *J. Adhes.* **2018**, *94*, 847–866. [CrossRef]
32. Jojibabu, P.; Zhang, Y.X.; Rider, A.N.; Wang, J.; Gangadhara Prusty, B. Synergetic Effects of Carbon Nanotubes and Triblock Copolymer on the Lap Shear Strength of Epoxy Adhesive Joints. *Compos. Part B Eng.* **2019**, *178*, 107457. [CrossRef]
33. Cha, J.; Kim, J.; Ryu, S.; Hong, S.H. Comparison to Mechanical Properties of Epoxy Nanocomposites Reinforced by Functionalized Carbon Nanotubes and Graphene Nanoplatelets. *Compos. Part B Eng.* **2019**, *162*, 283–288. [CrossRef]
34. Xu, L.R.; Li, L.; Lukehart, C.; Kuai, H. Mechanical Characterization of Nanofiber-Reinforced Composite Adhesives. *J. Nanosci. Nanotechnol.* **2007**, *7*, 2546–2548. [CrossRef]
35. Sam-Daliri, O.; Farahani, M.; Araei, A. Condition Monitoring of Crack Extension in the Reinforced Adhesive Joint by Carbon Nanotubes. *Weld. Technol. Rev.* **2020**, *91*, 7–15. [CrossRef]
36. Burch, K.; Doshi, S.; Chaudhari, A.; Thostenson, E.; Higginson, J. Estimating Ground Reaction Force with Novel Carbon Nanotube-Based Textile Insole Pressure Sensors. *Wearable Technol.* **2023**, *4*, e8. [CrossRef] [PubMed]
37. Sam-Daliri, O.; Farahani, M.; Faller, L.-M.; Zangl, H. Structural Health Monitoring of Defective Single Lap Adhesive Joints Using Graphene Nanoplatelets. *J. Manuf. Process.* **2020**, *55*, 119–130. [CrossRef]
38. Çakır, M.V.; Özbek, Ö. Mechanical Performance and Damage Analysis of GNP-Reinforced Adhesively Bonded Joints under Shear and Bending Loads. *J. Adhes.* **2023**, *99*, 869–892. [CrossRef]
39. Stetco, C.; Sam-Daliri, O.; Faller, L.-M.; Zangl, H. Piezocapacitive Sensing for Structural Health Monitoring in Adhesive Joints. In Proceedings of the 2019 IEEE International Instrumentation and Measurement Technology Conference (I2MTC), Auckland, New Zealand, 20–23 May 2019; pp. 1–5.
40. Radshad, H.; Khoramishad, H.; Nazari, R. The Synergistic Effect of Hybridizing and Aligning Graphene Oxide Nanoplatelets and Multi-Walled Carbon Nanotubes on Mode-I Fracture Behavior of Nanocomposite Adhesive Joints. *Proc. Inst. Mech. Eng. Part L J. Mater. Des. Appl.* **2022**, *236*, 1764–1776. [CrossRef]
41. Çakır, M.V. The Synergistic Effect of Hybrid Nano-Silica and GNP Additives on the Flexural Strength and Toughening Mechanisms of Adhesively Bonded Joints. *Int. J. Adhes. Adhes.* **2023**, *122*, 103333. [CrossRef]
42. Zamani, P.; FM da Silva, L.; Masoudi Nejad, R.; Ghahremani Moghaddam, D.; Soltannia, B. Experimental Study on Mixing Ratio Effect of Hybrid Graphene Nanoplatelet/Nano-Silica Reinforcement on the Static and Fatigue Life of Aluminum-to-GFRP Bonded Joints under Four-Point Bending. *Compos. Struct.* **2022**, *300*, 116108. [CrossRef]

43. Zamani, P.; Alaei, M.H.; da Silva, L.F.M.; Ghahremani-Moghadam, D. On the Static and Fatigue Life of Nano-Reinforced Al-GFRP Bonded Joints under Different Dispersion Treatments. *Fatigue Fract. Eng. Mater. Struct.* **2022**, *45*, 1088–1110. [CrossRef]
44. Özbek, Ö.; Çakır, M.V. MWCNT and Nano-Silica Hybrids Effect on Mechanical and Fracture Characterization of Single Lap Joints of GFRP Plates. *Int. J. Adhes. Adhes.* **2022**, *117*, 103159. [CrossRef]
45. Huang, Z.-M.; Zhang, Y.-Z.; Kotaki, M.; Ramakrishna, S. A Review on Polymer Nanofibers by Electrospinning and Their Applications in Nanocomposites. *Compos. Sci. Technol.* **2003**, *63*, 2223–2253. [CrossRef]
46. Palazzetti, R.; Zucchelli, A. Electrospun Nanofibers as Reinforcement for Composite Laminates Materials—A Review. *Compos. Struct.* **2017**, *182*, 711–727. [CrossRef]
47. Hamer, S.; Leibovich, H.; Intrater, R.; Zussman, E.; Siegmann, A.; Sherman, D. Mode I Interlaminar Fracture Toughness of Nylon 66 Nanofibrilmat Interleaved Carbon/Epoxy Laminates. *Polym. Compos.* **2011**, *32*, 1781–1789. [CrossRef]
48. Moroni, F.; Palazzetti, R.; Zucchelli, A.; Pirondi, A. A Numerical Investigation on the Interlaminar Strength of Nanomodified Composite Interfaces. *Compos. Part B Eng.* **2013**, *55*, 635–641. [CrossRef]
49. Beckermann, G.W.; Pickering, K.L. Mode I and Mode II Interlaminar Fracture Toughness of Composite Laminates Interleaved with Electrospun Polyamide Nanofibre Veils. *Compos. Part A Appl. Sci. Manuf.* **2015**, *72*, 11–21. [CrossRef]
50. Saghafi, H.; Palazzetti, R.; Zucchelli, A.; Minak, G. Influence of Electrospun Nanofibers on the Interlaminar Properties of Unidirectional Epoxy Resin/Glass Fiber Composite Laminates. *J. Reinf. Plast. Compos.* **2015**, *34*, 907–914. [CrossRef]
51. Daelemans, L.; van der Heijden, S.; De Baere, I.; Rahier, H.; Van Paepegem, W.; De Clerck, K. Using Aligned Nanofibres for Identifying the Toughening Micromechanisms in Nanofibre Interleaved Laminates. *Compos. Sci. Technol.* **2016**, *124*, 17–26. [CrossRef]
52. Daelemans, L.; van der Heijden, S.; De Baere, I.; Rahier, H.; Van Paepegem, W.; De Clerck, K. Improved Fatigue Delamination Behaviour of Composite Laminates with Electrospun Thermoplastic Nanofibrous Interleaves Using the Central Cut-Ply Method. *Compos. Part A Appl. Sci. Manuf.* **2017**, *94*, 10–20. [CrossRef]
53. Beckermann, G.W. Nanofiber Interleaving Veils for Improving the Performance of Composite Laminates. *Reinf. Plast.* **2017**, *61*, 289–293. [CrossRef]
54. Goodarz, M.; Bahrami, S.H.; Sadighi, M.; Saber-Samandari, S. Low-Velocity Impact Performance of Nanofiber-Interlayered Aramid/Epoxy Nanocomposites. *Compos. Part B Eng.* **2019**, *173*, 106975. [CrossRef]
55. Oh, H.J.; Kim, H.Y.; Kim, S.S. Effect of the Core/Shell-Structured Meta-Aramid/Epoxy Nanofiber on the Mechanical and Thermal Properties in Epoxy Adhesive Composites by Electrospinning. *J. Adhes.* **2014**, *90*, 787–801. [CrossRef]
56. On, S.Y.; Kim, M.S.; Kim, S.S. Effects of Post-Treatment of Meta-Aramid Nanofiber Mats on the Adhesion Strength of Epoxy Adhesive Joints. *Compos. Struct.* **2017**, *159*, 636–645. [CrossRef]
57. Razavi, S.M.J.; Neisiany, R.E.; Ayatollahi, M.R.; Ramakrishna, S.; Khorasani, S.N.; Berto, F. Fracture Assessment of Polyacrylonitrile Nanofiber-Reinforced Epoxy Adhesive. *Theor. Appl. Fract. Mech.* **2018**, *97*, 448–453. [CrossRef]
58. Ekrem, M.; Avcı, A. Effects of Polyvinyl Alcohol Nanofiber Mats on the Adhesion Strength and Fracture Toughness of Epoxy Adhesive Joints. *Compos. Part B Eng.* **2018**, *138*, 256–264. [CrossRef]
59. Musiari, F.; Pirondi, A.; Moroni, F.; Giuliese, G.; Belcari, J.; Zucchelli, A.; Brugo, T.M.; Minak, G.; Ragazzini, C. Feasibility Study of Adhesive Bonding Reinforcement by Electrospun Nanofibers. *Procedia Struct. Integr.* **2016**, *2*, 112–119. [CrossRef]
60. Musiari, F.; Pirondi, A.; Zucchelli, A.; Menozzi, D.; Belcari, J.; Brugo, T.M.; Zomparelli, L. Experimental Investigation on the Enhancement of Mode I Fracture Toughness of Adhesive Bonded Joints by Electrospun Nanofibers. *J. Adhes.* **2018**, *94*, 974–990. [CrossRef]
61. Brugo, T.; Musiari, F.; Pirondi, A.; Zucchelli, A.; Cocchi, D.; Menozzi, D. Development and Fracture Toughness Characterization of a Nylon Nanomat Epoxy Adhesive Reinforcement. *Proc. Inst. Mech. Eng.* **2018**, *233*, 465–474. [CrossRef]
62. Cocchi, D.; Musiari, F.; Brugo, T.M.; Pirondi, A.; Zucchelli, A.; Campanini, F.; Leoni, E.; Mazzocchetti, L. Characterization of Aluminum Alloy-Epoxy Bonded Joints with Nanofibers Obtained by Electrospinning. *J. Adhes.* **2020**, *96*, 384–401. [CrossRef]
63. *ASTM D3039/D3039M-08*; Standard Test Method for Tensile Properties of Polymer Matrix Composite Materials. ASTM International: West Conshohocken, PA, USA, 2014.
64. Han, X.; Jin, Y.; da Silva, L.F.M.; Costa, M.; Wu, C. On the Effect of Adhesive Thickness on Mode I Fracture Energy—An Experimental and Modelling Study Using a Trapezoidal Cohesive Zone Model. *J. Adhes.* **2020**, *96*, 490–514. [CrossRef]
65. Yan, C.C.; Ma, J.L.; Zhang, Y.X.; Wu, C.W.; Yang, P.; Wang, P.; Zhang, W.; Han, X. The Fracture Performance of Adhesively Bonded Orthodontic Brackets: An Experimental-FE Modelling Study. *J. Adhes.* **2020**, *98*, 180–206. [CrossRef]
66. Fernández, M.V.; de Moura, M.F.S.F.; da Silva, L.F.M.; Marques, A.T. Composite Bonded Joints under Mode I Fatigue Loading. *Int. J. Adhes. Adhes.* **2011**, *31*, 280–285. [CrossRef]
67. Gunnion, A.J.; Herszberg, I. Parametric Study of Scarf Joints in Composite Structures. *Compos. Struct.* **2006**, *75*, 364–376. [CrossRef]
68. Xu, Y.; He, Q.; Yang, W.; Sun, T.; Tang, Q. Study on Relationships between Curing Pressures and Mechanical Properties for Epoxy Adhesive Films. *Chem. Eng. Trans.* **2018**, *66*, 43–48. [CrossRef]
69. Xie, Z.; Wang, S.; Li, X. *Composite Tapered Scarf Joint Repair: Analytical Model and Experimental Validation*; Atlantis Press: Amsterdam, The Netherlands, 2016; pp. 720–726.
70. Krenk, S. Energy Release Rate of Symmetric Adhesive Joints. *Eng. Fract. Mech.* **1992**, *43*, 549–559. [CrossRef]

71. Minosi, S.; Cocchi, D.; Pirondi, A.; Zucchelli, A.; Campanini, F. Integration of Nylon Electrospun Nanofibers into Structural Epoxy Adhesive Joints. *IOP Conf. Ser. Mater. Sci. Eng.* **2021**, *1038*, 012048. [CrossRef]
72. Minosi, S.; Cocchi, D.; Maccaferri, E.; Pirondi, A.; Zucchelli, A.; Mazzocchetti, L.; Ambrosini, D.; Campanini, F. Exploitation of Rubbery Electrospun Nanofibrous Mat for Fracture Toughness Improvement of Structural Epoxy Adhesive Bonded Joints. *J. Adv. Join. Process.* **2021**, *3*, 100050. [CrossRef]

Disclaimer/Publisher's Note: The statements, opinions and data contained in all publications are solely those of the individual author(s) and contributor(s) and not of MDPI and/or the editor(s). MDPI and/or the editor(s) disclaim responsibility for any injury to people or property resulting from any ideas, methods, instructions or products referred to in the content.

Article

Determination of the Bonding Strength of Finger Joints Using a New Test Specimen Geometry

Hannes Stolze *, Michael Gurnik, Sebastian Kegel, Susanne Bollmus and Holger Militz

Wood Biology and Wood Products, Faculty of Forest Sciences and Forest Ecology, University of Goettingen, Buesgenweg 4, 37077 Goettingen, Germany
* Correspondence: hannes.stolze@uni-goettingen.de; Tel.: +49-551-39-33562

Abstract: In this study, a specimen geometry for testing finger joints was developed using finite element simulation and proofed by experimental testing. Six different wood species and three adhesives were used for finger-jointing specimens. With the test specimen geometry, the bonding strength of the finger joints was determined without the usual self-locking of the joint. Under load, the test specimen geometry introduces maximum stress at the beginning of the bond line (adhesive zone). However, the test specimen geometry does not generate a symmetric stress state. The main difficulty here is the flank angle of the finger joint geometry. The wood species and adhesives significantly influenced the performance of the finger joints.

Keywords: adhesive joint design; bonding strength; finger joints; finite element simulation; hardwoods; softwoods

---

Citation: Stolze, H.; Gurnik, M.; Kegel, S.; Bollmus, S.; Militz, H. Determination of the Bonding Strength of Finger Joints Using a New Test Specimen Geometry. *Processes* **2023**, *11*, 445. https://doi.org/10.3390/pr11020445

Academic Editor: Raul D.S.G. Campilho

Received: 25 November 2022
Revised: 20 January 2023
Accepted: 28 January 2023
Published: 2 February 2023

Copyright: © 2023 by the authors. Licensee MDPI, Basel, Switzerland. This article is an open access article distributed under the terms and conditions of the Creative Commons Attribution (CC BY) license (https:// creativecommons.org/licenses/by/ 4.0/).

1. Introduction

Finger joints are longitudinal bonded timber joints used in non-load-bearing and load-bearing applications. They play a key role in the load-bearing capacity of bonded engineered wood products (EWP) [1,2]. The formation of the bond line during the finger-jointing process differs from that of surface bonding [3–7]. Factors such as the structure of the bonding surface, pressing pressure, pressing and assembly time, and application quantity of the adhesives are different for finger-jointing and surface bonding. When testing adhesives for load-bearing applications according to EN 302-1 [8] by means of lap joints, the characteristics of finger joints according to EN 15497 [9], such as very short pressing times or bonding of end-grain wood, are not taken into account. Thus, the adhesives are only tested for surface bonding. Recently, it has become possible to produce more EWP based on hardwoods [10–16], which have a high strength potential. Compared with softwoods, the manufacture and testing of hardwood EWP's are rarely standardised. The strength of finger joints results from the joint's geometry-related self-locking (clamping effect) and the bonding strength [17,18]. Currently, there is no test standard to determine the bonding strength of a finger joint independent of the self-locking. This study aimed to develop a test specimen geometry for mechanical performance tests of finger joint bond lines without the usual self-locking of the joint. In previous tests, different geometries were assessed using the finite element method (FEM) [19–21] and the behaviour of the finger joint bond line was predicted. Experimental tests were carried out to validate the simulations. Different wood species were bonded with commercial adhesives, and the bonding strength was determined with the developed test specimen geometry.

Figure 1 shows a standard finger joint from EN 15497 [9] and a small test specimen with a finger joint geometry for EWP. Typical finger joints have a finger length between 7 mm and 50 mm and a flank angle between 3° and 8°. A general principle is that the strength of the finger joint increases when the flank angle is decreased as the bonding surface becomes larger [18,22–24]. The bonding strength, including the self-locking effect, is tested using the shown test specimen geometry (Figure 1). This study aimed to test the bonding strength of a single finger joint bond line without the usual self-locking of the joint. Therefore, a new test specimen geometry was developed which differs from a finger joint test specimen with self-locking and several bond lines. Furthermore, the clamping of small test specimens is difficult, as they often slip out of the clamping jaws when a tensile load is applied or break at the clamping in the case of wedge grips. This was considered in the development of the test specimen geometry of this study.

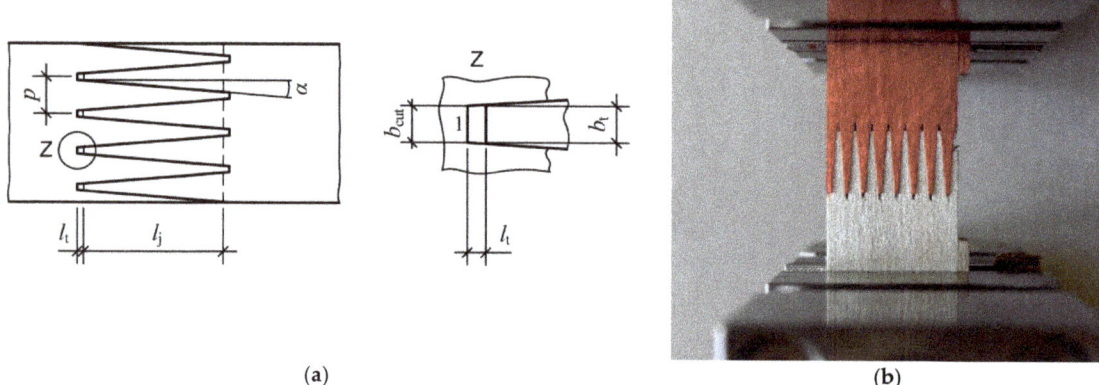

Figure 1. Standard finger joint from EN 15497 [9]: 1 finger base, l_j finger length, p finger pitch, α flank angle, l_t fingertip gap, b_{cut} width of cutter, b_t width of fingertip (**a**); small test specimen with standard finger joint, which is not suitable for the determination of bonding strength of a finger joint (**b**).

2. Materials and Methods

2.1. Wood and Adhesives

The wood species used in this study are shown in Table 1. The specimens were made from plain sawn boards with predominantly tangential grain. Before manufacture of the specimens, the boards were conditioned at 20 °C and 65% relative humidity until the equilibrium moisture content (EMC) was reached.

Table 1. Affiliation, origin, density, and EMC of tested wood species.

Wood Species	Affiliation	Origin	Density [g cm^{-3}]	EMC [%]
Beech *Fagus sylvatica*, L.	Hardwood	Germany	0.68 ± 0.03	11.8 ± 0.3
Birch *Betula pendula*, Roth.	Hardwood	Latvia	0.64 ± 0.05	11.5 ± 0.3
Poplar *Populus tremula*, L.	Hardwood	Latvia	0.49 ± 0.04	12.1 ± 0.3
Pine *Pinus sylvestris*, L.	Softwood	Germany	0.63 ± 0.05	13.5 ± 1.0
Larch *Larix decidua*, Mill.	Softwood	Germany	0.57 ± 0.04	13.8 ± 0.3
Spruce *Picea abies*, L.	Softwood	Germany	0.46 ± 0.02	12.4 ± 0.5

Commercially available melamine–urea–formaldehyde (MUF), phenol–resorcinol–formaldehyde (PRF), and 1-component polyurethane (PUR) adhesive systems were used (Table 2). They were processed according to the technical data sheets of the manufacturers.

Table 2. Properties and processing parameters of the adhesive systems.

Adhesives	Density [g cm^{-3}] R [1]	H	Viscosity [mPas] R	H	Mixing Ratio (R:H)	Application [g m^{-2}]
MUF	1.27	1.10	10,000–25,000	1700–3500	100:50	280, one-sided
PRF	1.16	1.18	5000–10,000	5000–8000	100:20	380, on both sides
PUR	1.16		24,000		1-comp., no primer	140, one-sided

[1] Resin (R) and hardener (H).

2.2. Finite Element Simulations and Shear–Tensile Tests

In an iterative process of finite element simulations and experimental testing, a shear–tensile test specimen for finger joints was developed. EN 302-1 [8] and a standard testing machine with 5×5 mm^2 clamping jaws were used as a basis for the design of the test specimen geometry. Essential test criteria were the location of the stresses and the location of the specimen failure. Stress concentrations and specimen failure were to be localized in the bond line. The following requirements were defined as important for the test specimen design:

- Complete transmission of the test load into the bond line during the test;
- Testing of a single bond line without self-locking;
- Centric force transmission and shear-tensile stress as only stress state;
- Consideration of the usual manufacturing process of finger joint bonding.

Ansys 2022 (Academic/Students) analysis software was used for a static-mechanical FE simulation of two test specimen geometries. The material properties were defined as follows:

- Linear–elastic behaviour;
- Orthotropic stiffness matrix for beech wood according to Schaffrath (2015) [25] (Table 3).

Table 3. Material parameters according to Schaffrath (2015) [25] for input to the finite element simulations.

Material Parameters	Direction	Beech Wood
Modulus of elasticity [N mm^{-2}]	E_X-longitudinal E_Y-tangential E_Z-radial	14,000 1160 2200
Transverse contraction coefficient Poisson	XY YZ XZ	0.043 0.71 0.073
Modulus of shear [N mm^{-2}]	XY YZ XZ	1080 460 1640

- In the contact area of the joints, the cohesive zone model (CZM) is based on the fracture energy of the PUR adhesive according to Serrano and Enquist (2005) [26] (Table 4).

Table 4. Contact properties based on strength and fracture energy of PUR adhesive according to Serrano and Enquist (2005) [26].

Mode I		Mode II	
Strength [N mm^{-2}]	Fracture energy [J m^{-2}]	Strength [N mm^{-2}]	Fracture energy [J m^{-2}]
6	550	12	1230

- Meshing: hexahedral elements in the shear region (Figure 2) and SOLID186 as main elements were used; hexahedral elements were preferred over tetrahedral elements because the hexahedral elements exhibited less stiff behaviour and showed more satisfactory convergence behaviour;
- Further conditions: fixed clamping at end face, area load in tensile direction (both test specimen geometries with 5 kN load at opposite end face). In the following, the relative stresses to the stress maximum are shown, so that they are independent of the applied load.

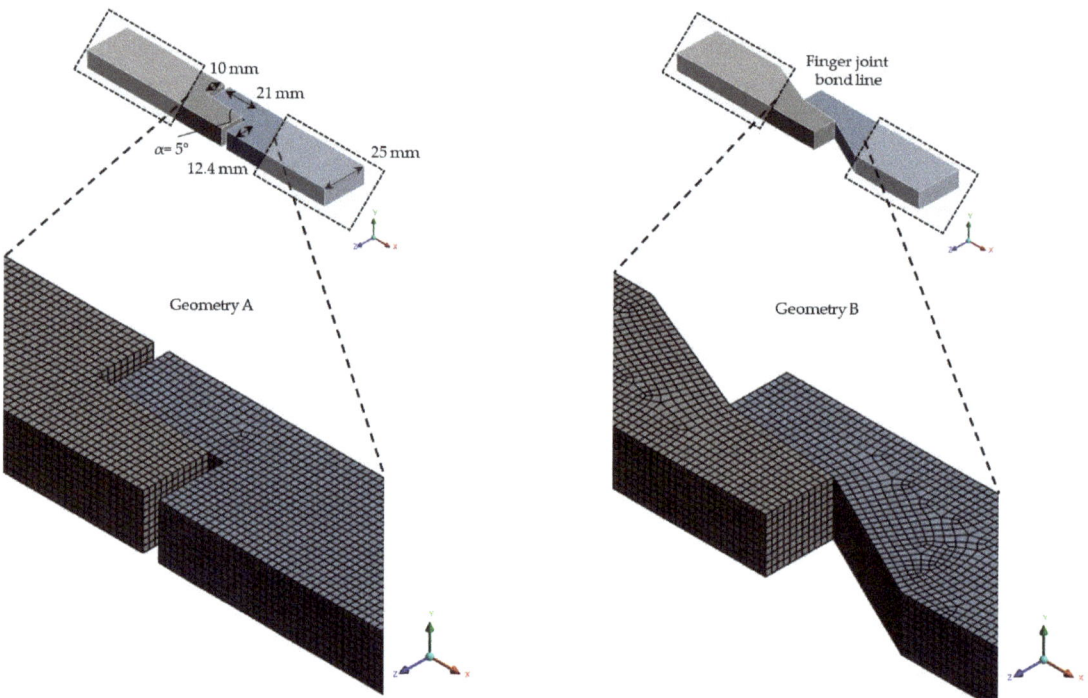

Figure 2. Mesh design of the test specimen geometries for the FE simulation. Geometry A with 104,083 nodes and 22,779 elements (**left**) and geometry B with 99,180 nodes and 21,681 elements (**right**). In the simulations the tangential surfaces were bonded with PUR and the flank angle of the finger joints was 5°. The notches of both test specimen geometries were cut asymmetrically because of the flank angle (shown for geometry A).

The resulting relative von Mises equivalent stresses of the simulated geometries were compared under the previously described load case.

The test specimen geometry B (Figure 2) was selected for the experiments. The specimens were produced according to EN 14080 [27] with a finger jointing line type of Ultra TT (Weinig Grecon GmbH & Co. KG, Alfeld/Leine, Germany). The following process parameters were used in this study. The pressing pressure was set for beech and adjusted proportionally to the lower density of the other wood species, so that the bonding surface was similar for all wood species (Table 1):

- Finger joint geometry: 21.0 mm finger length and 6.2 mm finger pitch;
- Cutting feed rate: 25 m min^{-1};
- Cutting direction: vertical profile, perpendicular to annual rings;

- Adhesive application: manual application, processing of PUR, MUF, PRF (Table 2), and bonding of radial surfaces;
- Pressing pressure: beech 12.5 N mm^{-2}, birch 11.8 N mm^{-2}, pine 11.6 N mm^{-2}, larch 10.5 N mm^{-2}, poplar 9.0 N mm^{-2}, spruce 8.5 N mm^{-2};
- Pressing time: 5 s.

The shear–tensile test specimens were made of lamellae with the dimensions 360 × 100 × 30 mm^{-3} (L × T × R) (Figure 3). The lamellae were cut in the middle and, with a few exceptions, reconnected by finger jointing as in their original state.

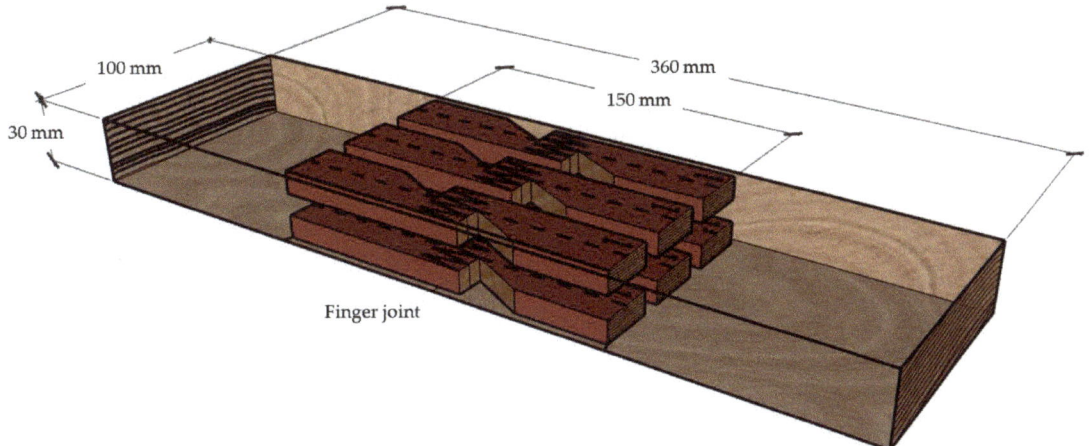

Figure 3. Manufacture of shear–tensile test specimens made from a lamella with a vertical finger joint profile.

After curing of the adhesives, the shear–tensile tests were carried out with an universal testing machine (Zwick Roell GmbH & Co. KG, Ulm, Germany) using a 5 kN load cell, a total clamping length of 100 mm, and a test speed of 0.5 mm min^{-1}. The bond line length (Figure 4, hypotenuse c) was measured representatively for each wood–adhesive combination microscopically using a digital microscope VHX-5000 (Keyence, Osaka, Japan). The bond line height corresponds to the specimen thickness and was measured on each test specimen using a digital calliper. Due to the flank angle of the finger joint, the load direction is neither parallel (requirement for shear stress) nor perpendicular (requirement for tensile stress) to the observed section area. A mixed mode loading of shear stress f_{vb} and tensile stress f_{va} was measured (Figure 4). Both stresses were calculated depending on the flank angle of the finger joint (in this study $\alpha = 5°$). The shear–tensile strength f_{vc} of the finger joint bond line was calculated according Equation (1) from EN 302-1 [8] as it was carried out in [3] for scarf joints:

$$\begin{aligned} f_{vc} &= \tfrac{F_{max}}{A} = \tfrac{F_{max}}{l \times h} = \sqrt{f_{vb}^2 + f_{va}^2} \\ f_{vb} &= \tfrac{F_{max}}{A} \times \cos(\alpha) \\ f_{va} &= \tfrac{F_{max}}{A} \times \sin(\alpha) \end{aligned} \quad (1)$$

f_{vc} = shear-tensile strength [N mm^{-2}]
f_{vb} = shear strength [N mm^{-2}]
f_{va} = tensile strength [N mm^{-2}]
F_{max} = applied breaking load [N]
A = finger-jointed area [mm^2]
l = length of bond line [mm]
h = height of bond line [mm]

On each tested specimen, the percentage of wood failure was estimated in 10% steps by visual inspection according to EN 302-1 [8].

Figure 4 shows a simplified drawing of the test specimen geometry of the experiments, its dimensions, asymmetric notches, and the setup of the shear–tensile test. The width of the fingertip and fingertip gap are not shown.

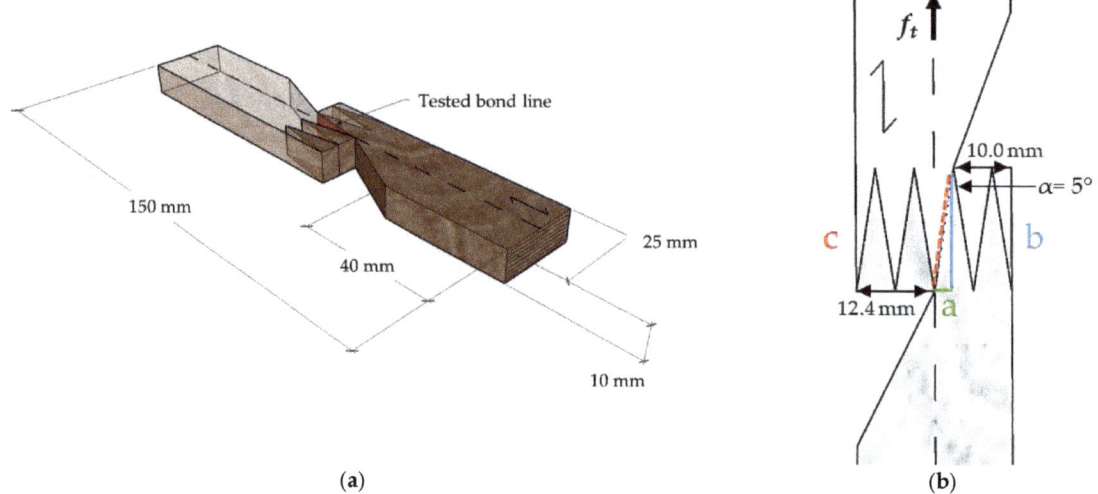

Figure 4. Test specimen geometry for the determination of the strength of a finger joint bond line (marked in red) (**a**); and shear–tensile test with asymmetric notches: cathetus a shows section area of tensile stress, cathetus b shows section area of shear stress, and hypotenuse c shows section area of shear–tensile stress (tested bond line) (**b**).

2.3. Data Processing

To evaluate the effects and interactions of the parameter settings, the following data were processed:

- Wood species and adhesive on the resulting parameters;
- Shear–tensile strength and wood failure percentage: two full factorial designs were set up (Table 5).

Table 5. Full factorial designs to evaluate effects and interactions of the parameter settings: n is number of specimens, f_{vc} is the shear–tensile strength, and WFP is wood failure percentage, each with 18 parameter settings.

No.	n f_{vc} and WFP	Wood Species	Adhesive
1	36	Beech	MUF
2	36	Beech	PRF
3	36	Beech	PUR
4	26	Birch	MUF
5	36	Birch	PRF
6	36	Birch	PUR
7	36	Poplar	MUF

Table 5. Cont.

No.	n f_{vc} and WFP	Wood Species	Adhesive
8	36	Poplar	PRF
9	36	Poplar	PUR
10	34	Pine	MUF
11	30	Pine	PRF
12	34	Pine	PUR
13	34	Larch	MUF
14	33	Larch	PRF
15	31	Larch	PUR
16	36	Spruce	MUF
17	36	Spruce	PRF
18	33	Spruce	PUR

In the following, statements on the wood species are to be interpreted in combination with the wood species-specific pressing pressure.

The interaction plots represent the mean values of all settings of one factor as a function of the setting of another factor [28]. The significance of the main effects and interactions was tested using an ANOVA [29]. The significance level was set to the value of 0.05.

3. Results and Discussion

3.1. Results of Finite Element Simulations and Shear–Tensile Tests

The von Mises equivalent stress for a section path through the adhesive joint of the specimen geometries A and B resulting from the finite element (FE) simulations is shown (Figure 5). These relative stresses shown in the graph refer to the stress maximum found in both calculations of the geometries, as mentioned above. The stress maximum of geometry A is in the "notch bottom area" next to the adhesive zone, and the initial failure of the specimen is expected to be in the wood and not in the bond line. The stress maximum of geometry B is located at the beginning of the adhesive zone and it is higher than that of geometry A. Higher stresses are expected at the entry point of the bond line, which proved to be more appropriate for testing the bonding strength. Otherwise, a pure wood fracture failure is very likely to occur in the notch base next to the adhesive zone. The initial failure in the bond line of the geometry B predicted by the FE simulations was confirmed in the experimental shear–tensile tests of this study. The stress distribution for both geometries is not symmetrical due to the asymmetric notch depths, and differences along the path are evident. Comparing the notches on both sides of the specimens, the deeper notched side shows higher stresses at the beginning of the adhesive joint than the other side.

Due to the flank angle of the finger joint and the exact specimen geometry, a mixed mode loading with more complex stress state than pure shear stress is present in the adhesive zone (for example, tensile stress needs to be considered). The shear stresses determined in the experiments of this study are expected to be lower compared with the pure shear stress. Nevertheless, geometry B is a suitable test specimen geometry for an evaluation and relative comparison of finger joint bondings. It fulfils all requirements mentioned at the beginning of this study except for the shear–tensile stress as the only stress state.

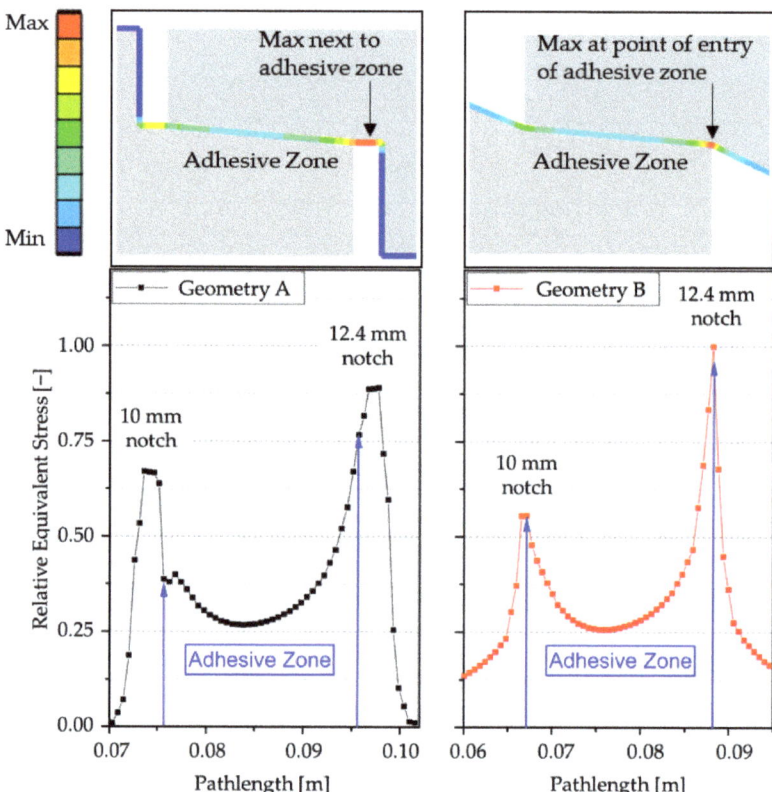

Figure 5. Relative von Mises equivalent stresses of the geometries in the adhesive zone and notches; geometry B was selected for the main experiments of this study.

The determined shear–tensile strength and the wood failure percentage of the finger joint bondings are shown (Figure 6). The highest shear–tensile strengths were achieved by beech and birch bonded with MUF adhesive. Some of these test specimens were able to fulfil the required strength values of EN 301 (for thin beech adhesive joints, 10 N mm^{-2}) [30]. However, the results of this study are not comparable to the test according to EN 301 [30]. With the test specimen geometry of this study, an overstressing at the tip of the finger joint is expected. This makes failure at lower loads more probable compared with the standard test by means of lap joints. Accordingly, a standard is necessary to be able to test and compare finger joint bondings with a defined test setup. Finger joints bonded with PUR achieved the lowest shear–tensile strengths for all wood species. The wood failure percentages show a large scattering overall. Compared with hardwoods, softwoods showed a higher wood failure percentage; spruce bonded with MUF showed almost complete wood failure. The high wood failure percentage is not equivalent to better bonding [31]. The inherent strength of the wood, which is density related, is an important factor for the evaluation of the bonding strength [32] and must be considered when comparing bondings. Hardwoods showed adhesion and adhesive failure as the main failure mode. This is reflected in the low wood failure percentages. To better utilize their high strength potential, the hardwood bondings need to be further improved. Using hardwoods, more significant differences were found between the adhesive systems than with the softwoods. In the case of the softwoods, it was mainly the strength of the wood that was tested.

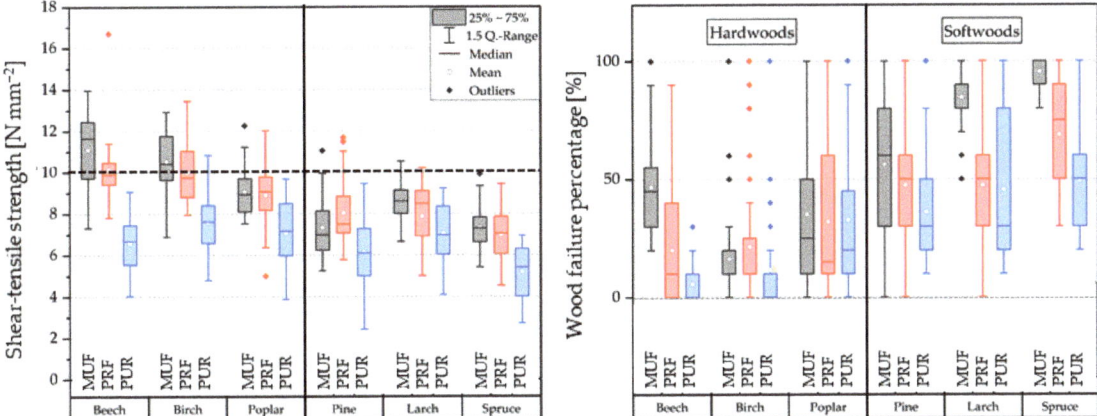

Figure 6. Boxplots of shear–tensile strength and wood failure percentage of the finger joint bondings: separated by hardwoods and softwoods and sorted by decreasing density of wood species; horizontal dotted black line marks standards requirement following EN 301 (10 N mm^{-2}) [30] for thin surface bonded joints of beech test specimens.

The specimen geometry for finger joint bondings used in this study was similar to the scarf joints used in [3,33]. As for scarf joints in [33], the tests in this study revealed relative differences between the bonding strengths of the adhesive systems and additionally between the wood species. These could not be shown with lap joint specimens in [33]. The differences in bonding strengths of the scarf joints were much more influenced by the adhesive systems than by the wood species [33]. As concluded for the scarf joints in [33], the test specimen geometry of this study is not suitable for the determination of absolute adhesive shear strength due to the mixed mode loading and possibly enhanced penetration of the adhesive into the end-grain wood. The effect of adhesive penetration on bonding strength needs to be verified in further studies.

The use of MUF tends to lead to the highest bonding strength and wood failure percentage, which may be explained by its high stiffness (less ductility). However, in this study much lower shear–tensile strengths were achieved with MUF-bonded finger joints compared with scarf joints in [3] (12 ± 1 N mm^{-2}) and similar to scarf joints in [33] (7.5 N mm^{-2}). The wood failure percentages were lower in [3] for several reasons, for example, wood properties, adhesive system, double-sided adhesive application, longer pressing time, and angle of the scarf joint could explain the higher shear–tensile strengths in [3]. This needs to be further investigated. Specimens bonded with PUR tended to show lower bonding strength and wood failure percentages, possibly due to the fact that it is more elastic (more ductile) [34]. As mentioned above, the penetration of the adhesive systems could be one more reason for the different bonding strengths of the finger joints. The MUF, in comparison with the PUR, is expected to penetrate deeper into the wood structure and can penetrate the cell wall [35]. Despite its low density (Table 1), poplar shows a high shear–tensile strength compared with the softwoods.

In [36], lap joints (surface bondings) were proofed using shear–tensile tests with the same or similar wood species, treatment, and adhesive systems. For most of the wood bondings, similar shear–tensile strengths were achieved as in this study. This initially indicates that finger joint bondings have the potential to achieve strengths like those of surface bondings. In this study, the finger-jointed poplar was able to achieve higher shear–tensile strengths. In [36], wood bonded with PUR was treated with a primer beforehand. All wood species could achieve significantly higher shear–tensile strengths compared with this study. There is currently no system on the market to apply primers on finger joints. An improvement in bonding performance is expected with the use of a primer [37]. The wood failure percentages in [36] were significantly higher. This can be explained by the subjective method to assess the wood failure percentage [38] or by the different bonding methods (surface and finger joint bonding). Furthermore, it is pointed out that the finger joints were tested in a standard climate state (20 °C and 65% rel. humidity). Pre-treatment according to EN 302-1 [8], e.g., water storage, can significantly influence the performance of the adhesives and wood [39].

The shear–tensile strength (f_{vc}), shear strength (f_{vb}), and tensile strength (f_{va}) of the tested finger joint bond lines are shown in Table 6. The splitting of f_{vc} into f_{vb} and f_{va} shows that under tensile load and with the used test specimen geometry, much greater shear stresses than tensile stresses were applied to the finger joint bond lines. A change in the flank angle of the finger joint geometry would lead to a change in the stress components.

Table 6. Average values of shear–tensile strength f_{vc}, shear strength f_{vb}, and tensile strength f_{va} of the tested finger joint bond lines. ± shows the standard deviation of the values.

Adhesive	MUF			PRF			PUR		
Wood Species	f_{vc} [N mm^{-2}]	f_{vb} [N mm^{-2}]	f_{va} [N mm^{-2}]	f_{vc} [N mm^{-2}]	f_{vb} [N mm^{-2}]	f_{va} [N mm^{-2}]	f_{vc} [N mm^{-2}]	f_{vb} [N mm^{-2}]	f_{va} [N mm^{-2}]
Beech	11.1 (±1.8)	11.1 (±1.8)	1.0 (±0.15)	10.1 (±1.4)	10.1 (±1.4)	0.9 (±0.12)	6.6 (±1.2)	6.6 (±1.2)	0.6 (±0.11)
Birch	10.5 (±1.5)	10.5 (±1.5)	0.9 (±0.13)	10.0 (±1.4)	9.9 (±1.4)	0.9 (±0.12)	7.6 (±1.3)	7.6 (±1.3)	0.7 (±0.11)
Poplar	9.1 (±1.1)	9.0 (±1.1)	0.8 (±0.10)	8.9 (±1.3)	8.9 (±1.3)	0.8 (±0.12)	7.1 (±1.7)	7.1 (±1.7)	0.6 (±0.15)
Pine	7.4 (±1.3)	7.3 (±1.3)	0.6 (±0.12)	8.0 (±1.6)	8.0 (±1.6)	0.7 (±0.14)	6.2 (±1.7)	6.1 (±1.7)	0.5 (±0.15)
Larch	8.6 (±1.0)	8.6 (±1.0)	0.8 (±0.08)	7.9 (±1.5)	7.8 (±1.5)	0.7 (±0.13)	7.1 (±1.4)	7.0 (±1.4)	0.6 (±0.12)
Spruce	11.1 (±1.8)	11.1 (±1.8)	1.0 (±0.15)	10.1 (±1.4)	10.1 (±1.4)	0.9 (±0.12)	6.6 (±1.2)	6.6 (±1.2)	0.6 (±0.11)

3.2. Two-Way Interactions and Analysis of Variance

The two-way interactions of the finger joint bondings for shear–tensile strength and wood failure percentage are shown (Figure 7). With a few exceptions, PUR bondings achieved the lowest mean shear–tensile strength and lowest mean wood failure percentage for all wood species. The key message of Figure 7 is that only a few interactions between the wood species and adhesives were found. This is indicated by the parallel course of the lines. A comparatively large drop in shear–tensile strength was observed for beech finger joints bonded with PUR, whereas birch finger joints achieved similar bonding strengths with the PUR as with the other adhesives. The density of beech and birch were similar which indicates that other reasons caused the differences. Further studies on the structure and chemistry of the bonding surfaces of the wood species should be carried out.

All main effects and interactions are significant at the predefined level of 0.05 (Table 7). As mentioned above, the factor wood species had a significant effect on the shear–tensile strength and the wood failure percentage. This is probably due to the different densities and inherent strength of the wood species. The adhesive had a lower effect on the shear–tensile strength and wood failure percentage than the wood species. The interaction of wood and adhesive were significant but not very pronounced as already shown in Figure 7.

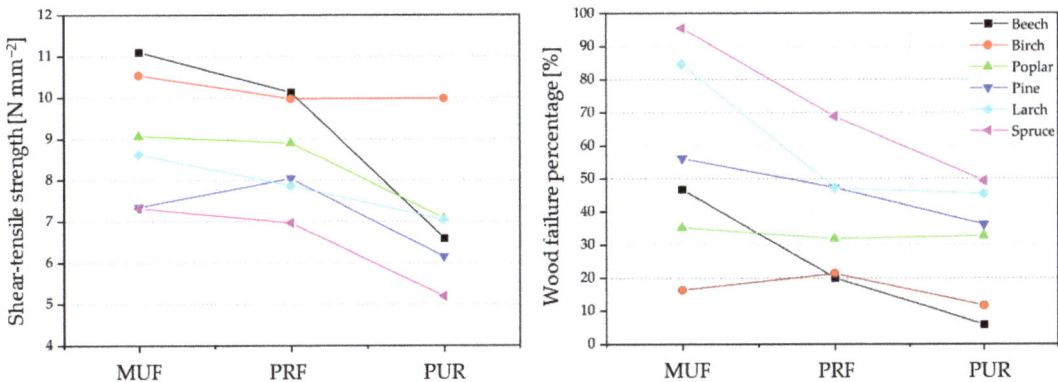

Figure 7. Two-way interactions of the finger joint bondings for mean of shear–tensile strength and mean of wood failure percentage.

Table 7. ANOVA results table (sig. level 0.05) based on statistical designs from Table 5.

Main Effect/ 2W Interaction	Shear–Tensile Strength		Wood failure Percentage	
	F-Value	p-Value	F-Value	p-Value
Wood species	95.79	2.50×10^{-74}	81.26	3.41×10^{-65}
Adhesive	117.08	9.93×10^{-44}	61.30	5.74×10^{-25}
Wood × adhesive	21.27	1.09×10^{-16}	7.05	1.56×10^{-10}

4. Conclusions

The present study proposes a specimen geometry for a finger-jointed wood bonding strength test. Furthermore, it presents the experiment results of tests employing the preferred test specimen geometry for a combination of six wood species specimens and three adhesives. The following conclusions can be drawn:

- A test specimen geometry for finger joints was identified using finite element simulations and proved by experimental testing. The test specimen geometry has a stress maximum at the beginning of the bond line (adhesive zone) and on the deeper notched side. Different finger joint bondings could be evaluated with the geometry and relative differences of the bondings were found. However, the geometry does not generate a symmetric stress state.
- A standard for testing finger joint bondings should be developed. The angle and length of the finger joint geometry affect the force transmission at the bond line and the resulting stress distribution. Different geometries should be tested, and geometry-dependent adjustment factors should be developed.
- Statements about the bonding strength are difficult since it is a combination of wood and adhesive failure. In this study, it was observed that the performance of the adhesives can be assessed more precisely when the wood species have higher strengths and can withstand loads closer to the limit of the adhesives.
- Further investigations, for example, roughness or wetting analyses, should be considered to be able to explain differences between the tested bondings.
- To improve finger-jointing and the high strength potential of hardwoods, adhesives and finger joint geometry should be further investigated.

Author Contributions: Conceptualization, H.S. and M.G.; methodology, H.S., M.G. and S.K.; formal analysis, M.G. and H.S.; investigation, S.K., H.S. and M.G.; resources, H.M.; writing—original draft preparation, H.S., S.K. and M.G.; writing—review and editing, all; visualization, H.S. and M.G.; supervision, H.M. and S.B.; funding acquisition, H.M. and S.B. All authors have read and agreed to the published version of the manuscript.

Funding: This research was funded by the German Federal Ministry for Economic Affairs and Energy (BMWi) through the Central Innovation Programme for small and medium-sized enterprises (SMEs) grant number FKZ 16KN042025.

Data Availability Statement: Not applicable.

Acknowledgments: The authors would like to express their sincere thanks to Weinig Grecon GmbH & Co. KG (Alfeld, Germany) for providing the finger-jointing line and expertise; Leitz GmbH & Co. KG (Oberkochen, Germany) for providing the cutting tools; the adhesive manufacturers for providing the adhesive systems; and ANSYS, Inc. (Canonsburg, U.S.) for providing the free student version of their software. We would also like to thank Philipp Schlotzhauer for initiating and supporting the project. We acknowledge support by the Open Access Publication Funds of Göttingen University.

Conflicts of Interest: The authors declare no conflict of interest.

References

1. Serrano, E. *Adhesive Joints in Timber Engineering: Modelling and Testing of Fracture Properties*; Lund University: Lund, Sweden, 2000.
2. Frangi, A.; Bertocchi, M.; Clauß, S.; Niemz, P. Mechanical behaviour of finger joints at elevated temperatures. *Wood Sci. Technol.* **2012**, *46*, 793–812. [CrossRef]
3. Tran, A.; Mayr, M.; Konnerth, J.; Gindl-Altmutter, W. Adhesive strength and micromechanics of wood bonded at low temperature. *Int. J. Adhes. Adhes.* **2020**, *103*, 102697. [CrossRef]
4. Stolze, H.; Gurnik, M.; Koddenberg, T.; Kröger, J.; Köhler, R.; Viöl, W.; Militz, H. Non-Destructive Evaluation of the Cutting Surface of Hardwood Finger Joints. *Sensors* **2022**, *22*, 3855. [CrossRef] [PubMed]
5. Gong, M.; Rao, S.; Li, L. Effect of Machining Parameters on Surface Roughness of Joints in Manufacturing Structural Finger-Jointed Lumber. *J. For. Eng.* **2017**, *2*, 10–18.
6. Follrich, J.; Vay, O.; Veigel, S.; Müller, U. Bond strength of end-grain joints and its dependence on surface roughness and adhesive spread. *J. Wood Sci.* **2010**, *56*, 429–434. [CrossRef]
7. Jokerst, R.W. *Finger-Jointed Wood Products*; Forest Products Laboratory: Madison, WI, USA, 1981; p. 26.
8. EN 302-1:2013-06; Adhesives for Load-Bearing Timber Structures—Test Methods—Part 1: Determination of Longitudinal Tensile Shear Strength. Beuth Verlag GmbH: Berlin, Germany, 2013; p. 15.
9. EN 15497:2014-07; Structural Finger Jointed Solid Timber—Performance Requirements and Minimum Production Requirements. Beuth Verlag GmbH: Berlin, Germany, 2014; p. 58.
10. Weimar, H.; Jochem, D. *Holzverwendung im Bauwesen—Eine Marktstudie im Rahmen der "Charta für Holz"*; Johann Heinrich von Thünen-Institut: Braunschweig, Germany, 2013.
11. Wehrmann, W.; Torno, S. *Laubholz für Tragende Konstruktionen-Zusammenstellung zum Stand von Forschung und Entwicklung*; Cluster—Initiative Forst und Holz in Bayern gGmbH: Freising, Germany, 2015; p. 18.
12. Linsenmann, P. *European Hardwoods for the Building Sector (EU Hardwoods)*; WoodWisdom-Net Research Programme; Holzforschung Austria: Wien, Austria, 2016; p. 57.
13. Informationsverein Holz e.V. *Konstruktive Bauprodukte Aus Europäischen Laubhölzern*; Spezial; Informationsverein Holz e.V.: Düsseldorf, Germany, 2017; ISBN 0446-2114.
14. Ehrhart, T. *European Beech—Glued Laminated Timber*; ETH Zurich: Zürich, Switzerland, 2019.
15. Knauf, M.; Frühwald, A. *Laubholz-Produktmärkte Aus Technisch-Wirtschaftlicher Und Marktstruktureller Sicht*; Fachagentur Nachwachsende Rohstoffe e.V.: Gülzow, Germany, 2020.
16. Obernostererer, D.; Jeitler, G.; Schickhofer, G. *Birke: Holzart Der Zukunft Im Modernen Holzbau*; University of Stuttgart: Stuttgart, Germany, 2022; p. 8.
17. Aicher, S.; Radovic, B. Untersuchungen zum Einfluß der Keilzinkengeometrie auf die Zugfestigkeit keilgezinkter Brettschichtholz-Lamellen. *Holz Roh Werkst.* **1999**, *57*, 1–11. [CrossRef]
18. Rao, S.; Gong, M.; Chui, Y.H.; Mohammad, M. Effect of Geometric Paramters of Finger Joint Profile on Ultimate Tensile Strenght of Single Finger-Joined Board. *Wood Fiber Sci.* **2012**, *44*, 8.
19. Timbolmas, C.; Rescalvo, F.J.; Portela, M.; Bravo, R. Analysis of poplar timber finger joints by means of Digital Image Correlation (DIC) and finite element simulation subjected to tension loading. *Eur. J. Wood Wood Prod.* **2022**, *80*, 555–567. [CrossRef]
20. Campilho, R.D.S.G. (Ed.) *Strength Prediction of Adhesively-Bonded Joints*; A Science Publishers Book; CRC Press: Boca Raton, FL, USA, 2017; ISBN 978-1-4987-2246-9.
21. Tran, V.-D.; Oudjene, M.; Méausoone, P.-J. FE analysis and geometrical optimization of timber beech finger-joint under bending test. *Int. J. Adhes. Adhes.* **2014**, *52*, 40–47. [CrossRef]

22. Selbo, M.L. Effect of Joint Geometry on Tensile Strength of Finger Joints. *For. Prod. J.* **1963**, *13*, 390–400.
23. Gehri, E. Verbindungstechniken mit hoher leistungsfähigkeit—Stand und entwicklung. *Holz Roh Werkst.* **1985**, *43*, 83–88. [CrossRef]
24. Radovic, B. Über Die Festigkeit von Keilzinkenverbindungen Mit Unterschiedlichen Verschwächungsgrad. *Bau. Mit Holz* **1993**, *3*, 196–201.
25. Schaffrath, J. Untersuchungen Zu Feuchtetransportvorgängen Und Feuchteinduzierten Verformungen Sowie Spannungen Bei Betrachtung Verschiedener Holzarten Und Unterschiedlicher Klimatischer Randbedingungen. Dissertation, Technische Universität München, München, Germany, 2015.
26. Serrano, E.; Enquist, B. Contact-free measurement and non-linear finite element analyses of strain distribution along wood adhesive bonds. *Holzforschung* **2005**, *59*, 641–646. [CrossRef]
27. EN 14080:2013-09; Timber Structures—Glued Laminated Timber and Glued Solid Timber—Requirements. Beuth Verlag GmbH: Berlin, Germany, 2013; p. 110.
28. Technical Statistics—Experimental Methodology (DoE). In *Quality Management in the Bosch-Group*; Robert Bosch GmbH: Gerlingen, Germany, 2020.
29. Montgomery, D.C. *Design and Analysis of Experiments*, 10th ed.; Wiley: Hoboken, NJ, USA, 2020; ISBN 978-1-119-49247-4.
30. EN 301:2018-01; Adhesives, Phenolic and Aminoplastic, for Load-Bearing Timber Structures—Classification and Performance Requirements. Beuth Verlag GmbH: Berlin, Germany, 2018.
31. Aicher, S.; Ahmad, Z.; Hirsch, M. Bondline shear strength and wood failure of European and tropical hardwood glulams. *Eur. J. Wood Wood Prod.* **2018**, *76*, 1205–1222. [CrossRef]
32. Iždinský, J.; Reinprecht, L.; Sedliačik, J.; Kúdela, J.; Kučerová, V. Bonding of Selected Hardwoods with PVAc Adhesive. *Appl. Sci.* **2020**, *11*, 67. [CrossRef]
33. Konnerth, J.; Gindl, W.; Harm, M.; Müller, U. Comparing dry bond strength of spruce and beech wood glued with different adhesives by means of scarf- and lap joint testing method. *Holz Roh Werkst.* **2006**, *64*, 269–271. [CrossRef]
34. Sebera, V.; Pečnik, J.G.; Azinović, B.; Milch, J.; Huč, S. Wood-adhesive bond loaded in mode II: Experimental and numerical analysis using elasto-plastic and fracture mechanics models. *Holzforschung* **2020**, *75*, 655–667. [CrossRef]
35. Herzele, S.; van Herwijnen, H.W.; Griesser, T.; Gindl-Altmutter, W.; Rößler, C.; Konnerth, J. Differences in adhesion between 1C-PUR and MUF wood adhesives to (ligno)cellulosic surfaces revealed by nanoindentation. *Int. J. Adhes. Adhes.* **2019**, *98*, 102507. [CrossRef]
36. Konnerth, J.; Kluge, M.; Schweizer, G.; Miljković, M.; Gindl-Altmutter, W. Survey of selected adhesive bonding properties of nine European softwood and hardwood species. *Eur. J. Wood Wood Prod.* **2016**, *74*, 809–819. [CrossRef]
37. Clerc, G.; Lehmann, M.; Gabriel, J.; Salzgeber, D.; Pichelin, F.; Strahm, T.; Niemz, P. Improvement of ash (*Fraxinus excelsior* L.) bonding quality with one-component polyurethane adhesive and hydrophilic primer for load-bearing application. *Int. J. Adhes. Adhes.* **2018**, *85*, 303–307. [CrossRef]
38. Künniger, T. A semi-automatic method to determine the wood failure percentage on shear test specimens. *Holz Roh Werkst.* **2008**, *66*, 229–232. [CrossRef]
39. Bockel, S.; Harling, S.; Grönquist, P.; Niemz, P.; Pichelin, F.; Weiland, G.; Konnerth, J. Characterization of wood-adhesive bonds in wet conditions by means of nanoindentation and tensile shear strength. *Eur. J. Wood Wood Prod.* **2020**, *78*, 449–459. [CrossRef]

Disclaimer/Publisher's Note: The statements, opinions and data contained in all publications are solely those of the individual author(s) and contributor(s) and not of MDPI and/or the editor(s). MDPI and/or the editor(s) disclaim responsibility for any injury to people or property resulting from any ideas, methods, instructions or products referred to in the content.

Review

Design and Experimental Analysis of an Adhesive Joint for a Hybrid Automotive Wheel

Jens-David Wacker [1,*], Tobias Kloska [2], Hannah Linne [3], Julia Decker [1], Andre Janes [2], Oliver Huxdorf [3] and Sven Bose [2]

[1] Fraunhofer Institute for Structural Durability and System Reliability LBF, 64283 Darmstadt, Germany
[2] OTTO FUCHS KG, 58540 Meinerzhagen, Germany
[3] INVENT GmbH, 38112 Braunschweig, Germany
* Correspondence: jens-david.wacker@lbf.fraunhofer.de; Tel.: +49-6151-705-8356

Abstract: When it comes to lightweight design of automotive wheels, hybrid designs consisting of a carbon composite wheel rim and a metallic, e.g., aluminum alloy, wheel disc offer significant potential. However, the conventionally used bolted joint between the two parts is complex and requires compromises in lightweight design due to the additional mechanical elements. Within this research, an adhesive joint for a hybrid wheel is developed in order to demonstrate its performance and lightweight potential. The main challenges are the reliable resistance against high structural loads during different load cases, as well as the residual stresses in the joint due to different thermal expansion rates of the composite and aluminum material. The developed joint combines an adhesive bond with a form-fitted geometry while still enabling an assembling process of the wheel disc in rotational direction. In addition, adaptations of the fiber layup in the rim area significantly reduce the thermal residual stresses in the joint by 47%. Subcomponent specimens, which represent the joint of an aluminum spoke with the composite rim, are manufactured and tested at different temperatures and load cases. The test results show sufficient strength of the adhesive joint as well as an improvement of the developed form-fitted joint compared to a basic adhesive bond. The adhesively joined wheel offers a lightweight potential of 6% compared to the bolted wheel.

Keywords: adhesive joint design; hybrid joint; lightweight wheel; composites; thermal expansion; experimental analysis; structural analysis

1. Introduction

Hybrid automotive wheels, consisting of a carbon composite (CFRP) wheel rim and an aluminum alloy wheel disc, have been state of the art for several years, offering a lightweight potential of 15 to 20% compared to monolithic aluminum wheels [1–4]. The large wheel rim represents the greatest portion of the wheel. Therefore, its composite design effectively reduces the rotational mass and improves the damping behavior of the wheel. In addition, the cylindrical geometry enables more efficient manufacturing processes such as braiding [5] and resin transfer molding [6,7], offering advantages regarding mass production compared to full composite wheels. The wheel disc with its complex spoke design, on the other hand, is best realized in metal manufacturing processes such as casting or forging [8], with high strength and fatigue values and precise processing of the hub intersection. The joint between the two parts is conventionally realized as a bolted joint [9,10], due to the high structural and thermal loads. However, the realization of bolted joints is complex. Milling processes of the composite part and integration of threaded holes in the aluminum disc are necessary, as well as sealing measures. In addition, the bolted joint usually needs to be combined with a form-fitted sleeve design as resistance against the high resulting shear loads. These additional mechanical elements lead to compromises in lightweight design.

Within this federal research project [11], an adhesive joint for a hybrid automotive wheel is developed in order to demonstrate its performance and lightweight potential. In general, adhesive bonds offer several advantages when it comes to joining metal and composite parts. The load introduction into the composite part can be realized without damaging fiber structures, dimensional deviations of the parts can be compensated in the adhesive thickness, and the overall mass of the joint can be reduced [12,13], (pp. 2–4). In case of a hybrid wheel, the joint design faces several challenges. As a safety component, the reliable resistance against high structural loads during different load cases such as straight driving, cornering, accelerating, and braking must be assured, as well as electric conductivity and resistance to ageing. In addition, the materials must withstand a large temperature range from low ambient temperatures to high braking temperatures. In case of composite wheels, measures to shield the wheel components from the braking heat are usually taken, such as coatings or layers for heat reflection, heat distribution or insulation [14,15]. However, the high temperature difference still leads to residual stresses in the joint, due to the different thermal expansion rates of the composite and aluminum material, and needs to be considered in the design process.

In the review of literature, several design parameters regarding adhesive bonding of dissimilar materials can be identified. Apart from the selection of adhesive and adherend material with suitable mechanical properties [12], (pp. 694–696), the geometrical design of the joint has a significant influence on the stresses in the adhesive and adherend, e.g., a single-lap design compared to a double-lap design [12], (pp. 713–714). In [16], a review on design techniques to improve the strength of adhesive joints is given. Examples are form-fitted configurations such as a wavy adherend design, transverse reinforcements such as pinning or stitching, or specific design of the adhesive edges such as adherend tapers or adhesive fillets. In addition, dual-adhesive concepts can be used [13], combining a high-temperature and a low-temperature adhesive. The hybrid wheel, however, demonstrates a more unique application compared to the often described overlap joints in the literature, due to its circular geometry, its specific deformation during thermal expansion, and loading situation.

When developing adhesive joints, computational and experimental methods are commonly used for structural validation. In the case of hybrid joints, different failure modes need to be considered. Apart from cohesive failure of the adhesive, several studies with composite adherends show delamination of the surface layer within the joint area, dependent, e.g., on the load introduction or thermal exposure of the joint [17–19].

Figure 1 shows the test program for the hybrid joint development for the research. Material properties for the simulation models are evaluated using coupon specimens [20,21]; the hybrid bond is validated on single lap joints [22]. Tests on subcomponent specimens first give experimental joint validation, and tests on wheel prototypes give validation on a component level.

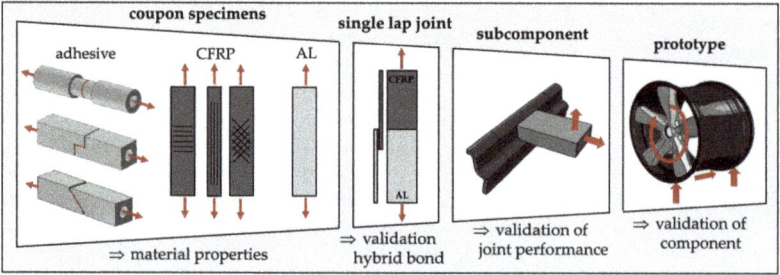

Figure 1. Test program for the experimental validation of the hybrid adhesive joint.

This paper elaborates on the preliminary design development of the hybrid adhesive joint and its first experimental validation on the subcomponent level. This includes the analysis of requirements, material selection, geometrical joint design, as well as the manufacturing and testing of subcomponent specimens. Other related topics, such as the characterization of the adhesive and adherend materials, structural simulation and strength analysis of the adhesive and adherends, as well as elaborations on the manufacturing concept of the wheel, may be presented in future publications.

2. Design of an Adhesive Joint for the Hybrid Wheel

2.1. Wheel Requirements

For the research, a 11.5 J × 20 EH2 ET 56 hybrid wheel with a five-spoke design and a max. wheel load of 575 kg was chosen as the reference wheel. In order to identify specific structural requirements for the joint, the different wheel load cases need to be considered. Within Table 1, the load cases such as straight driving, cornering, braking/accelerating and their respective maximum load values are listed. Maximum radial loads occur during straight driving, with maximum lateral load during cornering and maximum torsional moment during braking or accelerating, and equal values in opposite directions. Maximum temperature within the wheel rim is defined as 200 °C and within the joint as 150 °C, due to the greater distance to the brake. Lowest ambient temperature is defined as −40 °C.

Table 1. Load requirements for the hybrid wheel within the research project [11] according to OTTO FUCHS KG and Fraunhofer LBF.

No.	Load Case	Load	Value	Unit	Sketch
L1	straight driving (incl. rough road driving)	max. radial load	14.02	kN	
		max. lateral load	3.84	kN	
L2	cornering	max. radial load	10.59	kN	
		max. lateral load	12.71	kN	
L3	braking/accelerating	max. torsional moment	±1.91	kNm	
L4.1	thermal loading	max. temperature joint	150	°C	
L4.2		min. temperature joint	−40	°C	
L4.3		max. temperature wheel rim	200	°C	

For the objective of this research, the evaluations are limited to the selection of load cases listed in Table 1. The consideration of further combinations of load cases and temperatures can be addressed in a future detailed design stage.

2.2. Material Selection

Main requirements for the selection of materials for the hybrid wheel are high structural performance as well as thermal and corrosion resistance.

The adhesive selected for the project is a newly developed, one-component, heat-curing, epoxy-based structural adhesive by the associated project partner DuPont Specialty Products GmbH & Co KG. It has a high glass-transition temperature of 174 °C and a good capability of bonding dissimilar materials such as composites and metals. The adhesive will be further labeled as "BETAMATE™ HTG".

The aluminum alloy chosen for the wheel disc is EN AW-6082 T6 [23], a standard forging alloy by OTTO FUCHS KG with high strength and good corrosion resistance.

When it comes to selecting the composite material, the manufacturing process needs to be considered. As fabric, bidirectional woven carbon fabric WELA GG-245-1000T [24] is used, offering advantages regarding draping of complex geometries compared to non-crimp fabrics. Local reinforcements are realized with unidirectional carbon fiber WELA GV-303-0500UTFX [25]. The selected resin system is Araldite® LY 1560 [26], a toughened epoxy resin with a high glass-transition temperature of 205 °C, made for resin transfer molding (RTM) or infusion.

2.3. Analysis of Joint Requirements

The wheel loads described in Table 1 are introduced into the wheel at the tire–wheel intersection and are supported by the wheel hub. The load distribution within the wheel depends on the load case and the orientation of the spokes during the 360° rollover. In order to identify the critical loads occurring at the joint intersections, a finite element (FE) simulation using Ansys Workbench 2020 R1 software is carried out. The objective is the analysis of joint requirements by evaluating critical force and moment resultants within the joint.

2.3.1. Finite Element Model

The joint is modeled as a "basic joint", in which the outer surfaces of the aluminum spokes are joined with the wheel rim by a simple adhesive layer, as shown in Figure 2a. The composite rim with its specific fiber layup is modeled using Ansys Composite PrepPost (ACP) with shell elements. The aluminum disc, as well as the adhesive layer are modeled with solid elements. The material behavior is defined as linear elastic, using material data partly generated on coupon specimens within the project by Fraunhofer LBF. The most important material parameters are listed in Table 2.

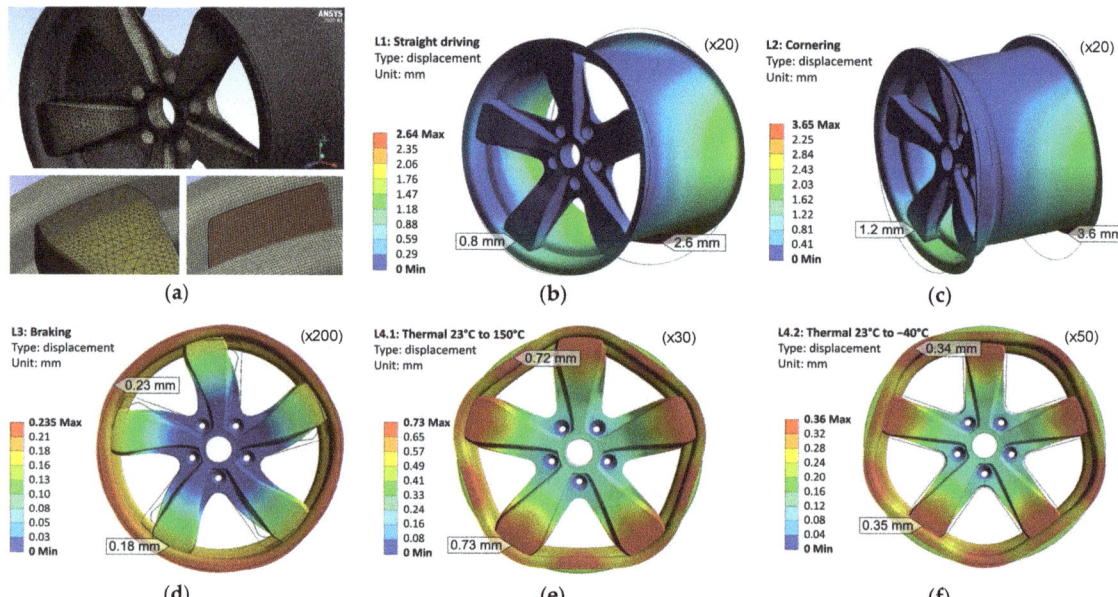

Figure 2. Simulation of the hybrid wheel in different load cases: (**a**) finite element model of wheel and joint with adhesive layer; (**b**) wheel displacement during L1: straight driving; (**c**) displacement during L2: cornering; (**d**) displacement during L3: braking; (**e**) displacement during L4.1: thermal load case 23 to 150 °C; (**f**) displacement during L4.2: thermal load case 23 to −40 °C.

Table 2. Selection of material parameters (for room temperature) used for linear elastic simulation of the hybrid wheel for the analysis of force and moment resultants within the joint.

Carbon/Epoxy Composite Orthotropic Ply WELA GG-245 [24]/ Araldite® LY 1560 [26]			Carbon/Epoxy Composite Unidirectional Ply WELA GV-303-0500 [25]/ Araldite® LY 1560 [26]			Aluminum Alloy, Isotropic EN AW-6082 T6 [23]			Adhesive, Isotropic BETAMATE™ HTG		
Property	Value	Unit	Property	Value	Unit	Property	Value	Unit	Property	Value	Unit
E_x	66.39 [1]	GPa	E_x	124.24 [1]	GPa	E	70.00 [3]	GPa	E	2.54 [1]	GPa
E_y	66.39 [1]	GPa	E_y	8.78 [1]	GPa	ν	0.33 [3]	-	ν	0.40 [4]	-
G_{xy}	15.76 [1]	GPa	G_{xy}	4.70 [2]	GPa	α	23.4 [3]	$10^{-6}/K$	α	40.0 [4]	$10^{-6}/K$
ν_{xy}	0.30 [1]	-	ν_{xy}	0.27 [2]	-						
α_x	2.2 [2]	$10^{-6}/K$	α_x	−0.5 [2]	$10^{-6}/K$						
α_y	2.2 [2]	$10^{-6}/K$	α_y	30.0 [2]	$10^{-6}/K$						

[1] data determined in coupon tests by Fraunhofer LBF. [2] data from similar material within Ansys Workbench 2020 R1 data base. [3] data from product data sheet [23]. [4] data according to DuPont Specialty Products GmbH & Co KG (Macquarie Park, Australia).

2.3.2. Wheel Deformation in Different Load Cases

For the interpretation of the structural behavior of the joint, examination of the simulated wheel deformation of the different load cases is helpful, as shown in Figure 2b–f.

The load cases "L1: straight driving" (Figure 2b) and "L2: cornering" (Figure 2c) lead to asymmetrical deformation of the wheel with a maximum deformation of 2.6 mm in L1 and 3.6 mm in L2 on the inboard side of the wheel rim. Deformations greater than 5 mm can result in critical tyer leakage. The outboard side of the wheel is less deformed, due to the stiffness of the aluminum wheel disc. However, especially in the case of "cornering", the high lateral load leads to maximum deformation of the spoke of 1.2 mm.

The load case "braking" (Figure 2d) is rotationally symmetric, due to the symmetrical introduction of the torsional moment. Only little deformation of max. 0.23 mm occurs in this load case.

For the thermal load cases, a stress neutral temperature at 23 °C is assumed, without consideration of possible residual stresses from the manufacturing process. The temperature rise in "L4.1: Thermal 23 to 150 °C" (Figure 2e) leads to an expansion of the wheel components. Due to the greater thermal expansion rate of the aluminum alloy compared to the composite material, the wheel disc compresses the wheel rim into a polygon-like shape with a maximum deformation of 0.73 mm. The temperature drop in "L4.2: Thermal 23 to −40 °C" (Figure 2e) leads to a greater contraction of the wheel disc, pulling the wheel rim interfaces toward the center axis. Here, the maximum deformation is 0.35 mm.

2.3.3. Force and Moment Resultants within the Joint

In order to evaluate the critical force and moment resultants within the joint, different spoke positions during the 360° rollover need to be considered. Therefore, each load case is simulated in different orientations of the wheel, allowing for a joint evaluation in 18° increments along the 360° rollover. The force and moment resultants are evaluated at the intersection between the outer surface of the aluminum spoke and the inner surface of the adhesive layer, as shown in Figure 3a. The resultants are orientated in a cylindrical coordinate system with a radial (R), lateral (L) and circumferential (φ) direction. A positive radial force resultant $+F_R$ can be interpreted as tensional loading of the adhesive layer, a negative radial force resultant $-F_R$ as compression loading.

An exemplary evaluation of the force and moment resultants over the 360° wheel rollover for the load case "L2: cornering" is shown in Figure 3b,c. The graphical course shows maxima and minima in different spoke positions. Extreme radial and lateral forces occur at 180° spoke position, with extreme circumferential forces as well as all extreme moment resultants at 126° and 234° spoke position.

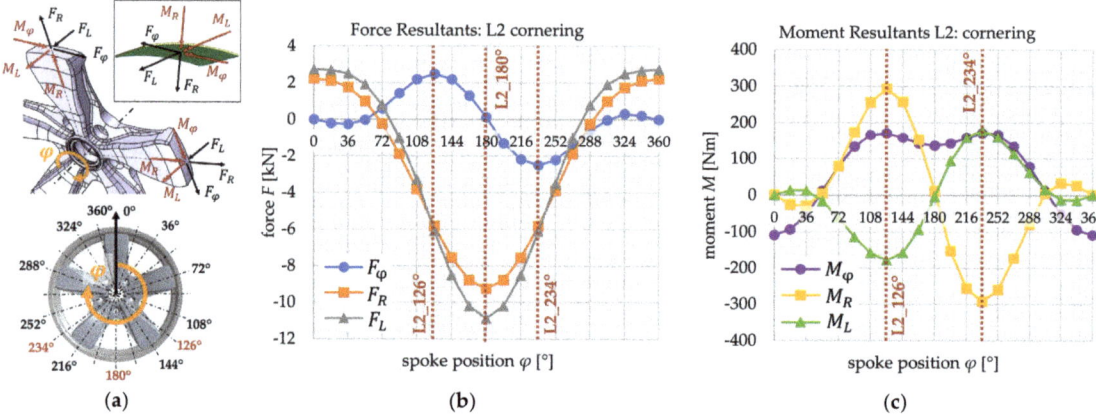

Figure 3. (a) Visualization of force and moment resultants at the intersection between the spoke and the adhesive layer during 360° rollover; (b) force resultants for the load case "L2: cornering"; (c) moment resultants for the load case "L2: cornering".

In Table 3, the selected critical force and moment resultants for each load case are listed, evaluated from the respective extrema in the 360° rollover. In the load case "L1: straight driving", maximum force resultants within the joint occur at 180° spoke position, with maximum moment resultants at 126° and 234°, similar to the load case "L2: cornering". However, the load values during cornering appear to be more extreme, with a high radial force resultant of −9.25 kN and a high lateral force resultant of -10.89 kN. The high radial moment resultants of 293 Nm can be explained by to the deformation reaction of the spoke, caused by the circumferential force resultant of 2.50 kN. Due to the open C-shaped cross-section of the spoke, the resulting bending deformation is coupled by a drilling deformation around the radial axis.

Table 3. Selection of critical force and moment resultants within the wheel joint for different load cases and spoke positions.

No.	Load Case	Spoke Position	Force Resultants			Moment Resultants		
			F_φ (kN)	F_R (kN)	F_L (kN)	M_φ (Nm)	M_R (Nm)	M_L (Nm)
L1	straight driving	126°	0.87	−3.67	−2.84	88	193	−205
		180°	0.08	−8.39	−6.64	43	8	−5
		234°	−0.87	−3.67	−2.84	88	−193	205
L2	cornering	126°	2.50	−5.83	−6.10	171	293	−178
		180°	0.13	−9.25	**−10.89**	136	11	−3
		234°	−2.50	−5.83	−6.10	171	−293	178
L3	braking	all pos.	−1.52	0	0	0	−78	9
L4.1	23 to 150 °C	all pos.	0	−28.61	0.02	−67	0	0
L4.2	23 to −40 °C	all pos.	0	**14.08**	−0.01	33	0	0

The same effect can be observed in the load case "L3: braking". The braking torque leads to a circumferential force resultant of −1.52 kN within the joint, which then results in a radial moment resultant of −78 Nm, due to the spokes' cross-sectional design. However, the braking load condition appears to be less critical for the joint, with significantly lower load values compared to the other load cases.

The thermal load cases lead to high residual radial force resultants, as described before. The temperature rise in "L4.1: Thermal 23 to 150 °C" leads to radial force resultants of

−28.61 kN, and the temperature drop in "L4.2: Thermal 23 to −40 °C" of +14.08 kN shows far more extreme values than in the other load cases.

When developing the adhesive joint for the hybrid wheel, all force and moment resultants and their interactions need to be considered as structural requirements. The resulting stress state within the adhesive layer depends on the final chosen geometrical design of the joint area. However, the evaluation of the principal stresses of the adhesive layer of this preliminary "basic design" give first conclusions about the joint loading:

- The high radial force resultant of +14.08 kN in L4.2 leads to critical tensional loading of the joint, due to the significantly lower tensional strength compared to the compressional strength of the adhesive.
- The high lateral force resultant of −10.89 kN in L2 leads to critical shear loading.
- The circumferential and lateral moment resultants M_φ and M_L can be considered more critical than the radial moment resultant M_R, because they lead to out-of-plane pealing stresses rather than in-plane shear stresses within the adhesive layer.
- The braking/accelerating load case can be considered as the least critical load case, resulting in rather low stress states.

2.4. Joint Design

For the development of the adhesive joint design, several concepts considering the review of literature are generated, analyzed via finite element simulation and evaluated according to their estimated structural performance, reliability, manufacturability and lightweight potential. The concepts include different approaches regarding design parameters such as geometrical design, material selection, bonding direction, as well as adaptations of the rim and spoke design. The final selected design features two main characteristics, which are elaborated in the following:

- the adaption of the fiber layup in the composite rim flange;
- the geometrical joint design with a form-fitted radial and lateral support.

2.4.1. Adaption of the Fiber Layup for the Composite Rim

The analysis of the joint requirements shows that critical radial force resultants occur in the thermal load cases due to the different thermal expansion rates of the aluminum wheel disc and the composite wheel rim. In the case of the composite rim, the thermal expansion rate as well as the rim stiffness result from the fiber layup and therefore offer the potential of more convenient design adaptations. In the original rim design (Figure 4a), unidirectional reinforcements are inserted in the rim flange areas with the objective of increasing the rim stiffness in the circumferential direction, as well as realizing thicker areas with a specific surface geometry. However, the reinforcements are primarily necessary for the in-board flange side. On the out-board side, the aluminum wheel disc increases the rim stiffness. This allows for the replacement of the unidirectional reinforcements with foam core segments (Figure 4b), while still achieving sufficient strength and stiffness of the rim flange. This sandwich design leads to a more flexible behavior of the rim flange in circumferential directing, as well as to a reduction of the difference in thermal expansion between the composite and aluminum components.

Figure 4c,d show the changes in deformation behavior between the original and the adapted design for the load case "L4.2: Thermal 23 to −40 °C". In both variations, the shrinkage of the aluminum disc due to the temperature drop is similar. The deviation of the composite rim to its undeformed shape, on the other hand, is less pronounced in the adapted variation, with only a 0.25 mm deviation between the spokes, compared to 0.35 mm in the original design.

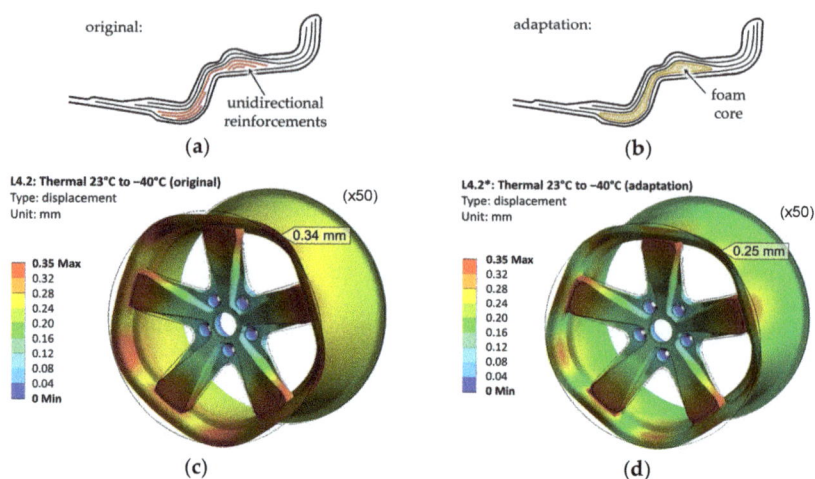

Figure 4. (**a**) Sketch (no detailed design) of the original fiber layup of the out-board rim flange with unidirectional reinforcements; (**b**) sketch of the adapted fiber layup with foam core; (**c**) deformation plot of original design under thermal load 23 to −40 °C; (**d**) deformation plot of adapted design.

Table 4 contains the force and moment resultants for the thermal load cases L4.1* and L4.2*, evaluated with the new design adaptation in the rim flange. The evaluation shows a significant reduction of the radial force resultants by 47%, with −15.18 kN instead of the former −28.61 kN in L4.1, and 7.47 kN instead of the former 14.08 kN in L4.2. Therefore, the design adaptation significantly improves the load requirement for the joint design.

Table 4. Force and moment resultants within the wheel joint for the thermal load cases, evaluated with the design adaptation in the rim flange.

No.	Load Case	Spoke Position	Force Resultants			Moment Resultants		
			F_φ (kN)	F_R (kN)	F_L (kN)	M_φ (Nm)	M_R (Nm)	M_L (Nm)
L4.1 *	23 °C to 150 °C	all pos.	0	−15.18	−0.10	53	0	0
L4.2 *	23 °C to −40 °C	all pos.	0	7.47	0.05	26	0	0

* Evaluated from model with design adaptation in the out-board rim flange.

Further investigation of the other load cases "L1*: straight driving", "L2*: cornering", and "L3*: braking" with the new design adaptation shows no significant change in force and moment resultants. This can be explained due to the rather force-controlled loading in these load cases, instead of the rather deformation-controlled loading in the thermal load cases.

2.4.2. Geometrical Joint Design

As a critical structural component, the hybrid wheel has high requirements regarding safety and reliability. In the case of adhesive joints, the combination of an adhesive bond with a form-fitted design can improve the joint performance and reduce critical tensional or pealing stresses, as well as enable a fail-safe mechanism in case of adhesive failure. However, the realization of a form-fitted adhesive design requires consideration of the bonding process, e.g., the application of the adhesive and the bonding direction.

Within an iterative design process, considering structural finite element analyses and manufacturing limits, a geometrical joint design is developed, in which a form-fitted radial and lateral support is implemented, as shown in Figure 5. The design is realized by adaptation of the foam core segments in the rim flange in the joint areas. The cross-sectional view shows the "claw-like" fit of the joint, supporting the adhesive bond regarding critical radial and lateral force resultants F_R and F_L, as well as radial and lateral moment resultants M_R and M_L. The overall design of the adhesively joined wheel offers a lightweight potential of 6% compared to the bolted hybrid wheel.

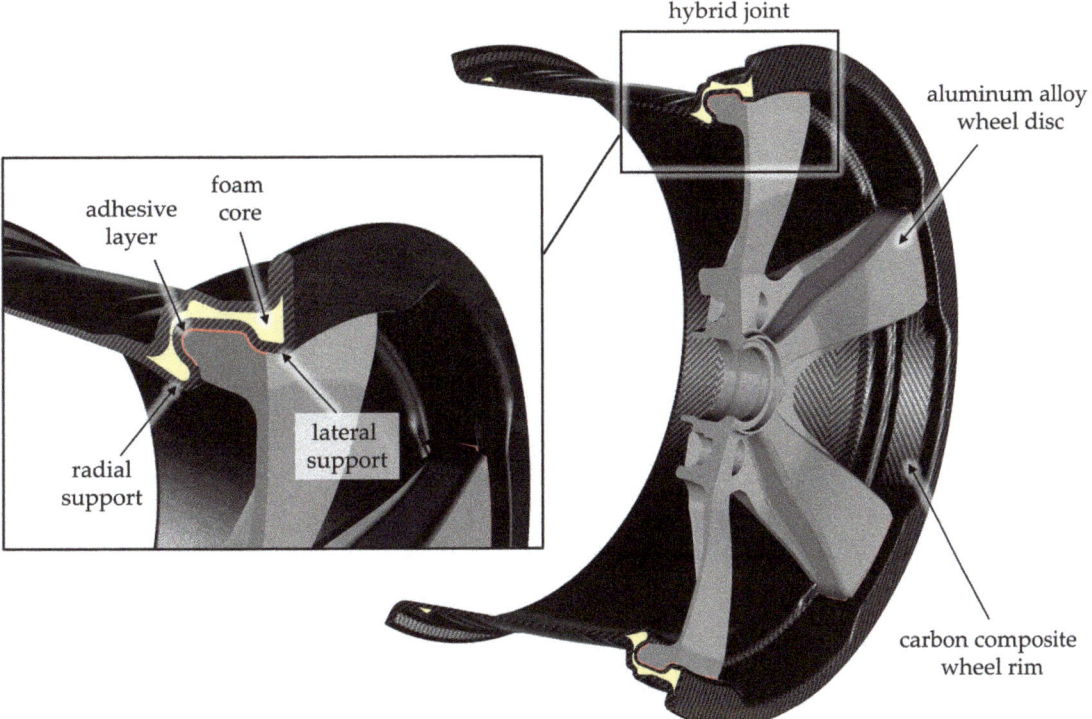

Figure 5. Form-fitted adhesive joint for the hybrid wheel, containing a radial and lateral support.

2.4.3. Manufacturing Concept

The least critical force resultants occur in the circumferential direction (Tables 3 and 4). Therefore, this degree of freedom is chosen as the bonding direction in which the adhesive bond without a form-fitted lock is considered sufficient. In the developed assembling process, the aluminum wheel disc is fixed on a rotation axis. The disc part is first positioned in between the joint areas of the composite rim. After application of the adhesive, the aluminum disc is rotationally moved into its final position, creating the form-fitted adhesive bond, as visualized in Figure 6a. In order to assure sufficient distribution of the adhesive over the bonding area and the realization of a defined adhesive thickness, the joint is designed in a wedge shape, as shown in Figure 6b. When locking the aluminum disc into position, an out-of-plane contact pressure is inserted, generating an evenly distributed adhesive layer with a constant thickness.

The wedge shape of the joint also enables the manufacturing of the composite rim via resin transfer molding (RTM). The geometry in the joint area can be realized in a multi-part tool, in which the respective tool segment can be demounted in a rotational direction.

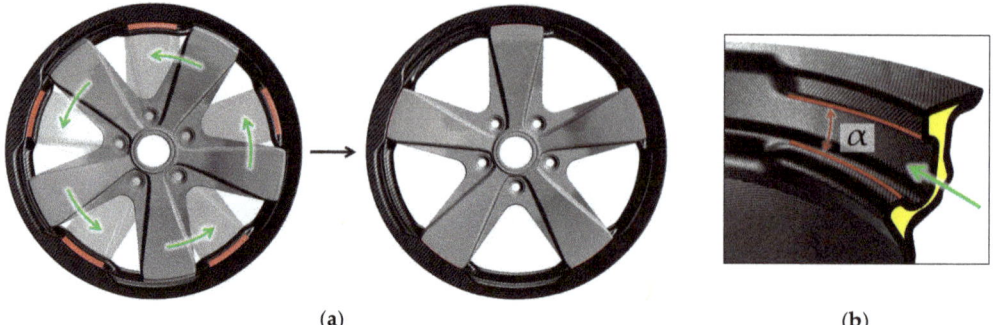

Figure 6. (a) Visualization of the bonding process in rotational direction (green arrow); (b) wedge shape (α) of the joint intersection.

3. Design and Manufacturing of Subcomponent Specimens
3.1. Design of Subcomponent Specimens
3.1.1. Geometrical Design of Subcomponent Specimen

Apart from the computational strength analysis of the joint, which may be presented in future publications, first, experimental validation of the joint performance is an important step in the preliminary design stage. Therefore, subcomponent specimens are designed, which represent the joint of an aluminum spoke with the composite rim, as shown in Figure 7a. The joint geometry is projected from the circular layout to a linear layout, reducing the manufacturing efforts of the subcomponent. However, the cross-sectional design with the foam core, as well as the bonding area and the wedge shape, stay similar to the joint design in the wheel. In order to demonstrate the benefit of the form-fitted adhesive design compared to a basic adhesive design, two variations of subcomponent specimen are realized, as shown in Figure 7b,c.

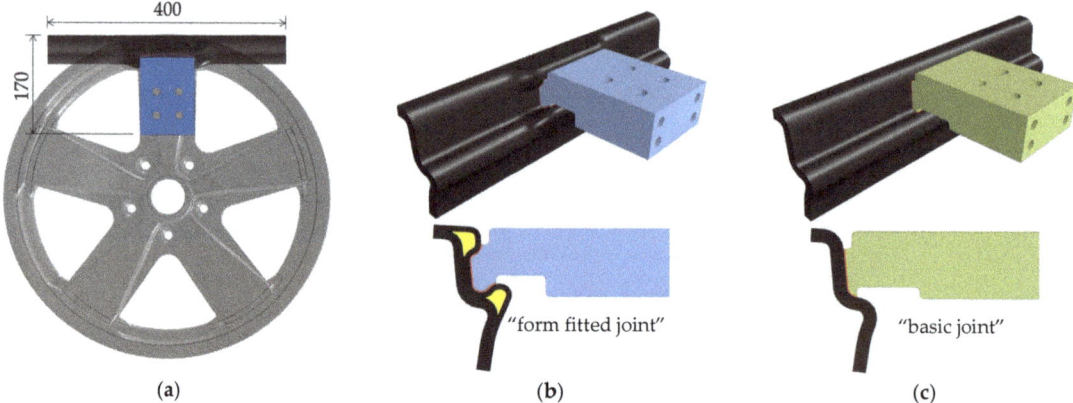

Figure 7. (a) Visualization of the subcomponent specimen as representation of the joint between the aluminum spoke and composite rim; (b) subcomponent specimen as a variation "form-fitted joint"; (c) subcomponent specimen as a variation "basic joint".

The interfaces of the specimens are designed in a way that they can be loaded in a test bench in a radial and a lateral orientation, as further described in Section 4.1. Both ends of the composite part can be fixed in a clamping device. The aluminum part can be joined to the test stand via bolted joints.

3.1.2. Comparative Evaluation of the Subcomponent Design

The subcomponent specimens represent a simplified design of the wheel joint, enabling a first experimental validation with comparatively little manufacturing and testing efforts. The finite element simulation of the basic joint in the wheel model and the subcomponent model enable a comparative evaluation.

The evaluated force and moment resultants are shown in Table 5. The radial loading of the subcomponent up to 7.47 kN shows similar resultants compared to the critical thermal load case of the wheel model. The lateral loading of the subcomponent up to −10.89 kN compared to the critical cornering load case of the wheel shows greater deviations of the radial force resultant and the circumferential moment resultant. However, it can be argued that the additional compressional radial loading of −9.25 kN in the wheel model has a supporting effect regarding lateral strength of the joint, due to the greater surface pressure.

Table 5. Force and moment resultants within the wheel joint for the thermal load cases, evaluated with the design adaptation in the rim flange.

Model	Load Case	Force Resultants			Moment Resultants		
		F_φ (kN)	F_R (kN)	F_L (kN)	M_φ (Nm)	M_R (Nm)	M_L (Nm)
subcomponent	radial loading: 7.47 kN	0	7.47	0.09	5	0	0
wheel	thermal: 23 °C to −40 °C	0	7.47	0.05	26	0	0
subcomponent	lateral loading: 10.89 kN	0.20	−0.86	−10.89	63	0	0
wheel	cornering: 180° position	0.13	−9.25	−10.89	136	11	−3

The comparative evaluation of the joint deformation for the exemplary radial load case is shown in Figure 8a,b. The composite part of the subcomponent specimen is designed with a specific resulting stiffness; thus, the deformation in the joint area in the model (0.39 mm) is roughly similar to the deformation in the wheel model (0.35 mm).

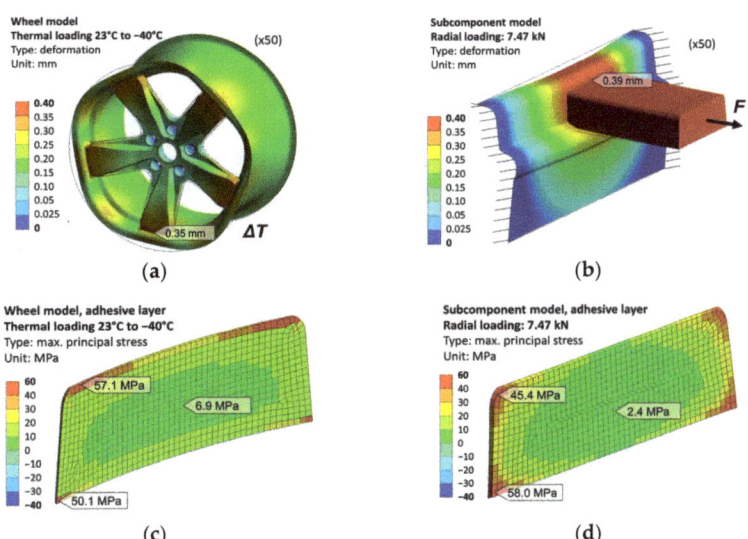

Figure 8. Comparative finite element analyses; (**a**) deformation of wheel; (**b**) deformation of subcomponent; (**c**) adhesive stresses in wheel; (**d**) adhesive stresses in subcomponent.

The comparative evaluation of the maximum principal stress in the adhesive layer for the exemplary radial load case is shown in Figure 8c,d. In both models, the adhesive area and thickness are similar. However, the simplified linear layout, as well as the difference in stiffness of the aluminum adherend, lead to a slightly different stress state in the subcomponent model, with more pronounced stresses up to 58.0 MPa at the bottom corners and less pronounced stresses of 2.4 MPa in the center. Still, the deviations are considered acceptable for the objective of first experimental validation with subcomponents. More realistic validations can be generated by tests on wheel prototypes after a detailed design stage.

3.2. Manufacturing of Subcomponent Specimens

The manufacturing of subcomponent specimens is performed in several different steps. The composite adherend is realized via a vacuum infusion process. Therefore, the dry layup, including the woven fabrics and the foam core segments, is placed into a mold and sealed with a vacuum bag, as shown in Figure 9a. The resin is then infused into the mold cavity by a vacuum pump. For the curing process of the resin, the specimens are placed in an oven at 120 °C for 40 min and then at 190 °C for 2 h. In Figure 9b, a cross-sectional cut of the "form-fitted" composite adherend with its foam core is shown.

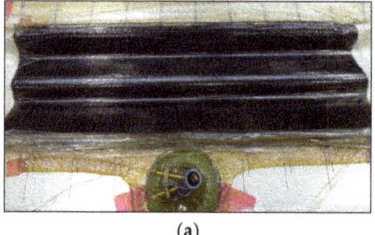
(a)

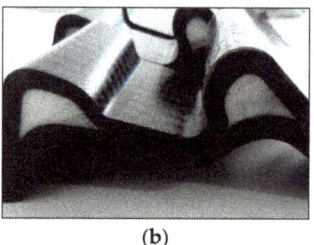
(b)

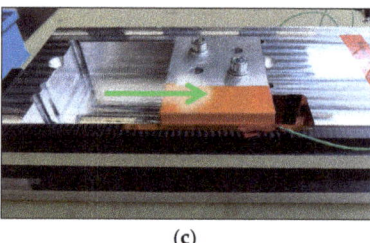
(c)

Figure 9. Manufacturing of subcomponent specimens: (**a**) vacuum infusion process of composite adherend; (**b**) cross-sectional view of form-fitted composite adherend; (**c**) bonding process (green arrow) within a mounting tool.

The aluminum adherend is made from a forged aluminum alloy within a milling and drilling process. For the adhesive bonding process, a mounting tool is realized, as shown in Figure 9c. Here, the adherends are joined in a similar bonding direction as planned for the wheel joint, creating contact pressure via the wedge-shaped form of the interfaces. For the curing process of the adhesive, the assembly is exposed to 180 °C for 30 min. Figure 10 shows the finished subcomponent specimens in variations such as "basic joint" and "form-fitted joint".

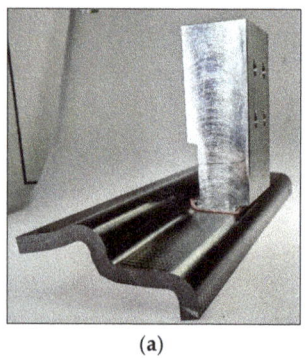
(a)

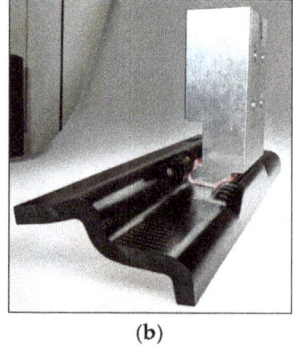
(b)

Figure 10. Subcomponent specimens as (**a**) "basic joint"; (**b**) "form-fitted joint".

4. Experimental Analysis of Subcomponent Specimens

The objective of the tests on subcomponent specimens is a first experimental evaluation of the performance of the joint design. Therefore, a test bench is realized, a test program is defined, and the test results are discussed. The main aspects of the investigations are the following:

- Can the joint withstand the required maximum loading?
- What are the failure modes of the joint?
- How does the "form-fitted joint" perform compared to the "basic joint"?
- What is the influence of temperature on the joint performance?

4.1. Test Bench

For the experimental evaluation of the joint, a test bench is realized, which allows for the testing of the joint in different load cases and at different temperatures. Figure 11a shows the CAD design of the test bench with its main components. The specimen can be mounted in a clamping device, connecting the composite side of the specimen to a fixed support. The loads are introduced on the aluminum side of the specimen by a hydraulic cylinder with a maximum limit of 25 kN. The movement of the cylinder is guided by a linear carriage. The specimen is placed within a climate chamber, so that loading at different temperatures can be realized by a hot-air unit. The resulting forces can be measured by a 3D load cell, placed outside the climate chamber. The displacement as well as the temperature are measured close to the adhesive joint.

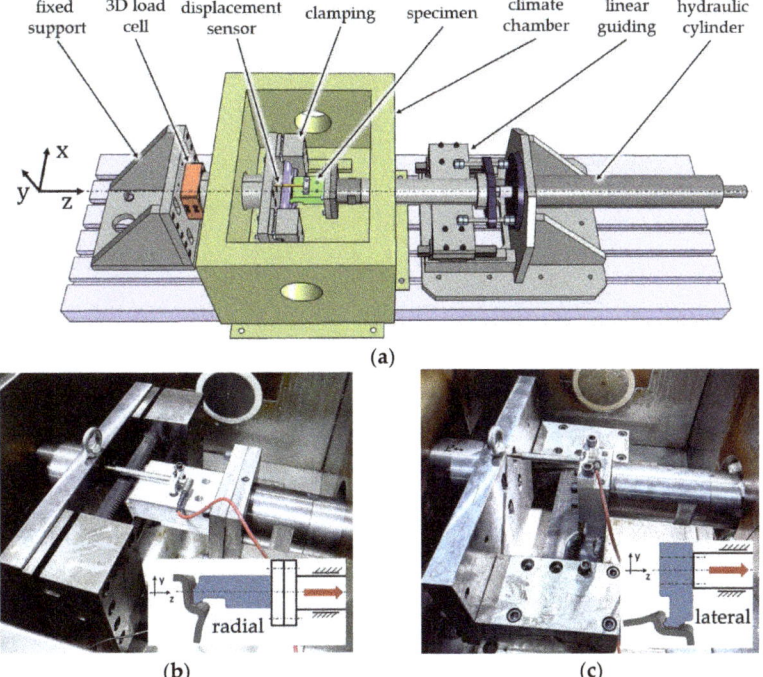

Figure 11. (a) CAD design of the test bench for the experimental evaluation of subcomponent specimens; (b) radial orientation of the specimen within the test bench; (c) lateral orientation.

The specimens can be mounted in two different orientations, as shown in Figure 11b,c. In the first orientation, the load can be introduced in the radial direction of the joint; in the second orientation, the load is introduced in the lateral direction.

4.2. Test Program

The selected test program is shown in Table 6. Due to a limited number of specimens, the test parameters are limited to two variations of specimens ("basic" and "form-fitted"), two load cases (radial and lateral), and two temperatures (23 and 150 °C). Each parameter set contains a sample size of two to three specimens. The load is applied as quasi-static loading, with a constant displacement of 1 mm/min.

Table 6. Test program for subcomponent specimens with different load cases and temperatures.

Specimens	Type of Test	Load Case	Number of Specimens	
			23 °C	150 °C
"basic specimen"	quasi-static test	radial	3	2
		lateral	3	2
"form-fitted specimen"	quasi-static test	radial	3	3
		lateral	3	3
	residual fatigue test	radial	1	
		lateral	1	

On two specimens, a residual fatigue test is performed under tension/tension loading with a stress ratio of $R = 0.1$ and a frequency of 4 Hz.

4.3. Test Results and Discussion

4.3.1. Quasi-Static Tests

The test results of the quasi-static tests on the subcomponent specimens are shown in Figures 12 and 13 as force–displacement curves. For each test, a first-crack initiation, located at the edge of the adhesive layer, can be identified in the trend of the curve and confirmed by the visual observation of a video recording of the test. After a phase of crack propagation, total failure of the specimens occurs as a rupture. However, due to the limit of the hydraulic cylinder at 25 kN, some specimens are not tested until total failure.

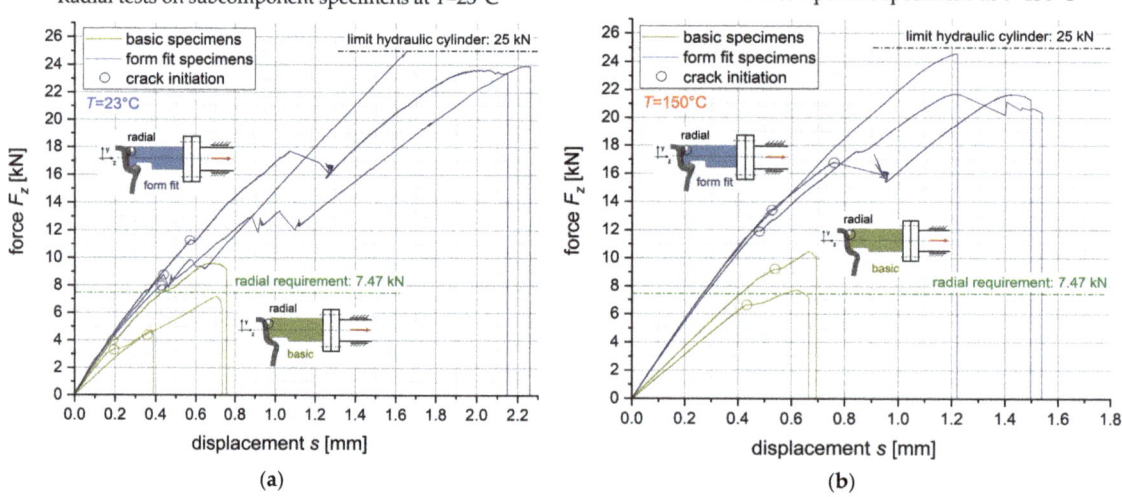

Figure 12. Force–displacement curves of radial tests on subcomponent specimens: (a) 23 °C; (b) 150 °C.

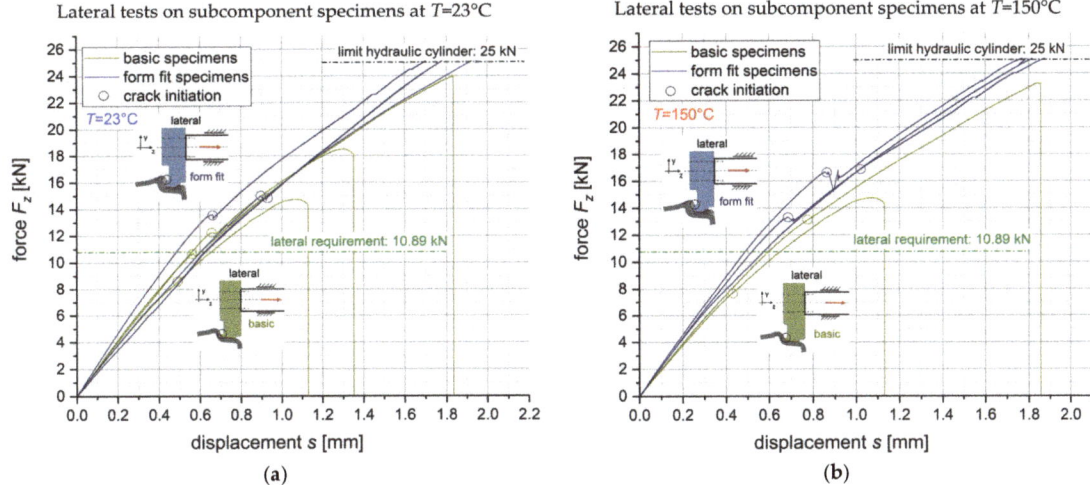

Figure 13. Force–displacement curves of lateral tests on "basic" and "form-fitted" subcomponent specimens at (**a**) 23°C; (**b**) 150°C.

The test results of the radial load case are shown in Figure 12a for 23 °C and Figure 12b for 150 °C. For both temperatures, the results show significant improvement of the maximum bearable force of the "form-fitted joint" compared to the "basic joint". The crack initiation of the "form-fitted joint" occurs after the required radial load of 7.47 kN (Table 4), while the "basic joint" does not meet the requirement.

Table 7 shows the adherends after total failure with their characteristic failure patterns. At 23 °C, the failure predominantly occurs in the surface layer of the composite adherend, due to the high out-of-plane stresses in the radial load case. At 150 °C, the failure pattern shows more pronounced cohesive failure within the adhesive layer.

Table 7. Characteristic failure pattern in radial tests for the "basic joint" and the "form-fitted joint".

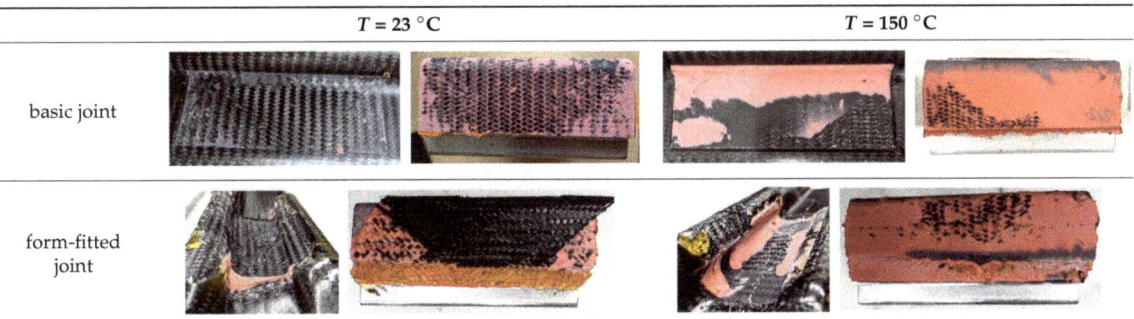

The test results of the lateral load case are shown in Figure 13a for 23 °C and Figure 13b for 150 °C. Similar to the radial tests, the "form-fitted joint" shows improved performance regarding strength at crack initiation as well as total failure, compared to the "basic joint". For both temperatures, the required maximum lateral load of 10.89 kN (Table 3) is met by the "form-fitted joint". Table 8 shows the facture pattern of the "basic joint" from the lateral tests with predominant cohesive failure within the adhesive layer. For the "form-fitted joint", rupture does not occur within the range of 25 kN of the hydraulic cylinder. However, the visual crack observation indicates a failure mode within the composite surface layer.

Table 8. Characteristic failure pattern in the lateral test for the "basic joint".

	T = 23 °C		T = 150 °C	
basic joint			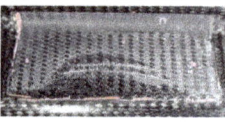	

For both load cases, radial and lateral, the variations in temperatures of 23 and 150 °C do not show significant influences in absolute strength of the joint. Still, variations in failure modes occur, with predominant failure of the composite surface layer at 23 °C and predominant cohesive failure within the adhesive layer at 150 °C. This indicates different effects of temperature influence for the composite adherend and the adhesive, which have been observed in other studies in the review of literature [19,20].

The distinctive phase of crack propagation before total failure of the "form-fitted joint" indicates a high margin of safety after crack initiation. In addition, further design optimization of the adhesive edge, which has been identified in the literature [18], might result in further improvement of the joint strength. However, the low sample size and the scattering of the test results need to be considered.

4.3.2. Residual Fatigue Tests

For each load case, one of the "form-fitted specimens", which did not rupture after 25 kN quasi-static loading, is tested in a residual fatigue test at a very high maximum load of 24 kN at 23 °C. The specimen under cyclic radial loading fails after 5639 cycles, and the specimen under cyclic lateral loading fails after 691,763 cycles. This preliminary fatigue evaluation indicates good fatigue strength of the joint.

5. Conclusions

Within this research, an adhesive joint for a hybrid automotive wheel is developed, joining the aluminum wheel disc with the composite wheel rim. The development includes the analysis of joint requirements, the generation of a joint design and the experimental evaluation of the joint performance in tests on subcomponent specimens.

The structural joint requirements are obtained via a finite element simulation of the hybrid wheel in different load cases (Table 1). The force and moment resultants at the interface between the aluminum spoke and composite rim are evaluated, identifying several critical load combinations (Table 3).

Within a design phase considering different design parameters from the literature, an adhesive joint design is developed (Figure 5), which offers a lightweight potential of 6% compared to a conventionally bolted wheel design, and which contains two main characteristics:

- The adaption of the fiber layup in the composite rim flange, which reduces the radial force resultants during the thermal load cases significantly;
- The geometrical joint design with a form-fitted radial and lateral support

For the experimental evaluation of the joint design, subcomponent specimens (Figure 10) are manufactured. The specimens represent the joint of the aluminum spoke with the composite rim and are realized in two variations, as a "basic joint" and as a "form-fitted joint". The experimental evaluation offers the following conclusions:

- The newly developed "form-fitted joint" meets the required critical radial and lateral load and shows significant strength increasement compared to the "basic joint".
- After a first-crack initiation, the joint shows a distinctive crack propagation phase before final rupture, offering advantages regarding safety design.

- The variation in temperature influences the failure mode of the joint, with a predominant failure of the composite surface layer at 23 °C and a more pronounced cohesive failure within the adhesive layer at 150 °C.
- Residual fatigue tests on the subcomponent specimens indicate good fatigue strength.

The experimental evaluation shows promising results regarding the structural performance of the joint design. However, further investigations within a detailed design phase and experimental phase need to be carried out:

- optimization via detailed structural analyses of adhesive and adherend failure;
- optimization of the joint design regarding crack initiation at the edge of the adhesive;
- further evaluation of critical load cases in multiaxial loading at different temperatures with a larger sample size of specimens;
- fatigue tests on wheel prototypes.

Author Contributions: Conceptualization, J.-D.W., T.K., H.L., J.D. and S.B.; methodology, J.-D.W., T.K., J.D. and S.B.; validation, J.-D.W. and J.D.; formal analysis, J.-D.W.; investigation, J.-D.W. and T.K.; data curation, J.-D.W.; writing—original draft preparation, J.-D.W.; writing—review and editing, T.K., H.L., J.D., A.J., O.H. and S.B.; visualization, J.-D.W.; supervision, J.D. and S.B.; project administration, T.K. and S.B.; funding acquisition, J.D. and S.B. All authors have read and agreed to the published version of the manuscript.

Funding: This research was funded by the German Federal Ministry of Education and Research (BMBF) within the framework "Hybrid Materials–New Possibilities, New Market Potentials (HyMat)" and was managed by the Project Management Agency Jülich (PTJ). The authors are responsible for the content of this publication.

Acknowledgments: The authors thank Alexander Droste, Research Investigator of DuPont, for the donation of adhesive material and his technical advice as associated partner of the project GOHybrid [11].

Conflicts of Interest: The authors declare no conflict of interest.

References

1. Hybrid Composite Wheel Reduces Fuel Consumption. Available online: https://www.reinforcedplastics.com/content/products/hybrid-composite-wheel-reduces-fuel-consumption/ (accessed on 4 January 2023).
2. Every Gram Counts-M Carbon Compound Wheels for the BMW M4 GTS. Available online: https://www.bmw-m.com/en/topics/magazine-article-pool/every-gram-counts.html (accessed on 4 January 2023).
3. Wheels from the Highest Standard-Mubea Performance Wheels. Available online: https://www.mubea.com/en/mubea-performance-wheels (accessed on 4 January 2023).
4. CFK-Räder Gehen 2016 in Serie. Available online: https://www.kfz-betrieb.vogel.de/cfk-raeder-gehen-2016-in-serie-a-505562/ (accessed on 4 January 2023).
5. Porsche and the Braided Carbon Fiber Wheel. Available online: https://www.compositesworld.com/articles/porsche-and-the-braided-carbon-fiber-wheel (accessed on 4 January 2023).
6. Rondina, F.; Taddia, S.; Mazzocchetti, L.; Donati, L.; Minak, G.; Rosenberg, P.; Bedeschi, A.; Dolcini, E. Development of full carbon wheels for sport cars with high-volume technology. *Compos. Struct.* **2018**, *192*, 368–378. [CrossRef]
7. Wacker, J.-D.; Laveuve, D.; Contell Asins, C.; Büter, A. Design of a composite nose wheel for commercial aircraft. *IOP Conf. Ser. Mater. Sci. Eng.* **2021**, *1024*, 012018. [CrossRef]
8. The Fuchsfelge-Forged, Not Cast. Available online: https://www.fuchsfelge.com/en/the-fuchsfelge.html (accessed on 4 January 2023).
9. Thyssenkrupp Carbon Components GmbH. Vehicle Wheel Comprising a Wheel Rim and a Wheel Disc. Patent WO2016/037611A1, 17 March 2016.
10. Mubea Carbo Tech GmbH. Wheel for a Vehicle. Patent WO2016/066769A1, 6 May 2016.
11. GOHybrid-Gestaltung und Optimierung von Hybridverbindungen unter Besonderer Berücksichtigung der Unterschiedlichen Wärmedehnungen der Werkstoffpartner. Available online: https://www.werkstoffplattform-hymat.de/Group/GOHybrid/Pages (accessed on 4 January 2023).
12. Da Silva, L.F.M.; Öchsner, A.; Adams, R.D. *Handbook of Adhesion*, 2nd ed.; Springer: Berlin/Heidelberg, Germany, 2011; pp. 2–4.
13. Marques, E.A.S.; da Silva, L.F.M.; Banea, M.D.; Carbas, R.J.C. Adhesive Joints for Low- and High-Temperature Use: An Overview. *J. Adhes.* **2015**, *7*, 556–585. [CrossRef]
14. Mubea Carbo Tech GmbH. Heat Shield Structure for a Wheel. Patent WO2016/097159A1, 17 March 2016.

15. Carbon Revolution PTY Ltd. Method of Producing Thermally Protected Composite. Patent WO2016/168899A1, 27 October 2016.
16. Shang, X.; Marques, E.A.S.; Machado, J.J.M.; Carbas, R.J.C.; Jiang, D.; da Silva, L.F.M. Review on techniques to improve the strength of adhesive joints with composite adherends. *Compos. Part B Eng.* **2019**, *177*, 107363. [CrossRef]
17. Wacker, J.-D.; Tittmann, K.; Koch, I.; Laveuve, D.; Gude, M. Fatigue life analysis of carbon fiber reinforced polymer (CFRP) components in hybrid adhesive joints. *Mater. Sci. Eng. Technol.* **2021**, *52*, 1230–1247. [CrossRef]
18. Qin, G.; Na, J.; Tan, W.; Mu, W.; Ji, J. Failure prediction of adhesively bonded CFRP-Aluminum alloy joints using cohesive zone model with consideration of temperature effect. *J. Adhes.* **2019**, *95*, 723–746. [CrossRef]
19. *ASTM D5573−99*; Standard Practice for Classifying Failure Modes in Fiber-Reinforced-Plastic (FRP) Joints. ASTM International: West Conshohocken, PA, USA, 2019.
20. *ISO 11003-2:2019-06*; Adhesives-Determination of Shear Behaviour of Structural Adhesives-Part 2: Tensile Test Method Using Thick Adherends. International Organization for Standardization: London, UK, 2019.
21. *ISO 527-5:2009*; Plastics-Determination of Tensile Properties-Part 5: Test Conditions for Unidirectional Fibre-Reinforced Plastic Composites. DIN Deutsches Institut für Normung e.V.: Berlin, Germany, 2009.
22. *ISO 4587:2003-03*; Adhesives-Determination of Tensile Lap-Shear Strength of Rigid-to-Rigid Bonded Assemblies. DIN Deutsches Institut für Normung e.V.: Berlin, Germany, 2003.
23. OTTO FUCHS KG. EN AW-6082 nach DIN EN 573 FUCHS AS15/AS11. Product Data Sheet, Rev. 1. Available online: https://www.otto-fuchs.com/fileadmin/user_upload/Infocenter/Werkstoffinformationen/Al-Datenblaetter/AS10-15.pdf (accessed on 4 January 2023).
24. WELA Handelsgesellschaft mbH. Kohlefasergewebe WELA GG-245-1000T (AKSA A-38). Product Data Sheet. 2020. Available online: https://wela-hamburg.de/wp-content/uploads/2019/01/datenblatt.pdf (accessed on 4 January 2023).
25. WELA Handelsgesellschaft mbH. UD-Kohlefasergewebe WELA GV-303-0500UTFX. Product Data Sheet. 2013. Available online: https://wela-hamburg.de/faserverstaerkungen/ (accessed on 4 January 2023).
26. Huntsman Advanced Materials. Araldite®LY 1560/Aradur®917-1/Accelerator DY 079. Product Data Sheet. 2016. Available online: https://www.huntsman-transportation.com/EN/products/all-products/composite-resin-systems.html/ (accessed on 4 January 2023).

Disclaimer/Publisher's Note: The statements, opinions and data contained in all publications are solely those of the individual author(s) and contributor(s) and not of MDPI and/or the editor(s). MDPI and/or the editor(s) disclaim responsibility for any injury to people or property resulting from any ideas, methods, instructions or products referred to in the content.

Article

Reuse of Carbon Fibers and a Mechanically Recycled CFRP as Rod-like Fillers for New Composites: Optimization and Process Development

José Antonio Butenegro [1,*], Mohsen Bahrami [1], Miguel Ángel Martínez [1] and Juana Abenojar [1,2]

[1] Materials Science and Engineering and Chemical Engineering Department, IAAB, University Carlos III Madrid, 28911 Leganés, Spain
[2] Mechanical Engineering Department, Universidad Pontificia Comillas, Alberto Aguilera 25, 28015 Madrid, Spain
* Correspondence: jbuteneg@ing.uc3m.es; Tel.: +34-655390804

Abstract: The rising amount of carbon fiber reinforced polymer (CFRP) composite waste requires new processes for reintroducing waste into the production cycle. In the present research, the objective is the design and study of a reuse process for carbon fibers and CFRP by mechanical recycling consisting of length and width reduction, obtaining rods and reintegrating them as fillers into a polymeric matrix. Preliminary studies are carried out with continuous and discontinuous unidirectional fibers of various lengths. The processing conditions are then optimized, including the length of the reinforcement, the need for a plasma surface treatment and/or for resin post-curing. The resin is thermally characterized by differential scanning calorimetry (DSC), while the composites are mechanically characterized by tensile strength tests, completed by a factorial design. In addition, the composites tested are observed by scanning electron microscopy (SEM) to study the fracture mechanics. Optimal processing conditions have been found to reduce the reinforcement length to 40 mm while maintaining the mechanical properties of continuous reinforcement. Furthermore, the post-curing of the epoxy resin used as matrix is required, but a low-pressure plasma treatment (LPPT) is not recommended on the reinforcement.

Keywords: polymer composites; carbon fiber reinforced polymers; recycling processes; properties optimization

1. Introduction

Carbon fiber reinforced polymer (CFRP) composite materials are made from carbon fibers embedded in a polymer matrix. These materials are renowned for their high strength-to-weight ratio, which makes them suitable for a wide range of applications, such as the construction of aircraft and other transportation vehicles, sporting goods, wind turbine blades, and other applications that require advanced composite materials [1–3]. CFRPs have high tensile strength and excellent fatigue resistance, allowing them to withstand high levels of stress and repeated stress without breaking. In recent years, there has been increasing interest in using CFRPs in the construction of buildings and other structures [4–6], as well as in medical devices and energy storage technologies, due to their potential for weight savings and improved performance compared to traditional materials. These materials are also utilized in the aerospace, automotive, and sporting goods industries due to their lightweight properties.

The limited service life of these CFRPs is one of today's environmental issues. The service life of CFRPs is about 50 years, which is the key reason for the recycling concept [7–9]. When the CFRPs reach the end of their service life, the carbon fibers are still able to retain their properties. Nevertheless, extensive use of CFRP leads to crucial waste disposal problems. The disposal of CFRPs has become a growing concern due to the increasing volume of material produced each year. The average cost per kilogram of CFRP produced using virgin carbon fibers is approximately between USD 30 and USD 60, with the majority of material being used in the aerospace and automotive sectors [10]. In these applications, turning CFRP waste into reusable materials and closing the loop in the CFRP life-cycle is the key challenge to increasing resource efficiency and continuing the use of materials [11].

There are three main routes for recycling CFRPs: mechanical, thermal, and chemical [12]. Mechanical recycling consists of shredding and grinding of CFRP components into smaller pieces, which can then be used as feedstock for the production of new composite materials [13,14]. Thermal recycling usually involves the use of pyrolysis, a process that breaks down the polymer matrix of CFRPs into smaller molecules through the application of heat [15–18]. Chemical recycling is related to the use of solvents to dissolve the polymer matrix, allowing for the separation and purification of the carbon fibers [19,20]. Despite the potential for recycling CFRPs, the current rate of recycling is relatively low, with only around 5% of CFRP material being recycled each year. The average cost per kilogram of recycled CFRP is much lower than that of virgin carbon fiber reinforced polymer, making it a cost-effective alternative to the production of new material [10]. However, the lack of infrastructure and specialized equipment for recycling CFRPs remains a significant barrier to increasing the rate of recycling. Furthermore, the development of new manufacturing techniques and the increasing demand for one-dimensional CFRPs are likely to drive down production costs in the future [21,22].

A significant concern associated with the recycling of carbon fibers and carbon fiber reinforced polymers (CFRPs) is the potential loss of added value resulting from the recycling process. The mechanical properties and performance of the final recycled material can be influenced by the length of the fibers present, with short fibers, typically less than 1 mm in length, and long fibers, typically longer than 1 mm in length, exhibiting different characteristics [23]. Short fibers are more compatible with traditional polymer matrix composite manufacturing techniques, but possess relatively low mechanical performance in comparison to long fibers [24,25]. Conversely, long fibers exhibit higher strength and stiffness, and are less prone to pullout or debonding from the matrix [26]. Longer fiber lengths allow for a near one-dimensional orientation, which improves the mechanical behavior of the composites. This implies that long fibers tend to preserve more added value and require less energy for mechanical reduction.

Fiber-matrix adhesion in CFRPs is another critical factor that directly affects the mechanical behavior of these materials. Plasma treatments are a commonly used method for surface modification of carbon fibers to improve fiber-matrix adhesion [27–29]. The use of plasma treatments represents a quick, environmentally friendly, and non-toxic dry process that modifies surfaces without altering bulk properties [30]. Particularly, low-pressure plasma (LPP) treatment is a cost-effective method for modifying material surfaces at the microscopic level without the need for labor-intensive processes or chemicals [31]. This technique allows for the controlled and reproducible modification of the surface of various materials to improve their bonding capabilities or to impart new surface properties. Additionally, LPP can be applied over a wider area to completely cover surfaces. As such, LPP offers a fast, clean, green, and efficient treatment option for carbon fibers in the production of CFRPs [32].

Many research studies are unable to attain definitive results due to a deficiency in design. This can be avoided through a properly implemented experiment, as it is a design error rather than a systematic issue [33]. An analysis of influence may be utilized to identify implicit problems in order to ascertain the suitability of a decision and to gain a comprehensive understanding of the obtained conclusions [34]. When planning the testing of discontinuous fiber composites, it is essential to determine whether to follow a standard or to consider alternative approaches.

Factorial experimental design, in general, involves the analysis of various factors that can influence an experiment. Factorial design is a powerful tool for understanding the complex interactions between processing or manufacturing parameters [35]. Therefore, factorial design allows for the identification of the optimal combination of levels given to the factors through the testing of relevant hypotheses related to the various factors and the estimation of their effect on the test results. It utilizes a linear statistical model for predicting responses for each factor through the addition of a common parameter for all combinations of factor levels. However, factorial design is also subject to several limitations. These include the limited number of factors that can be studied simultaneously, the potential for interactions between factors to impact the response, and the number of levels that can be used for each factor. Additionally, factorial design relies on the assumption of normality in the response, which may not always be valid [36]. Moreover, factorial design can become complex when dealing with large amounts of data and multiple factors and levels, requiring specialized software or statistical techniques for analysis.

The DOE (Design Of Experiments) method is a fractional factorial experiment in which only a carefully chosen subset of the treatment combinations necessary for a full factorial experiment to be conducted is chosen [37]. This method is more effective than other design of experiment methods, such as Taguchi, which is less time-consuming but ignores interactions between factors [38].

In the present research, a comprehensive characterization of the epoxy resin has been carried out through the thermal and chemical study of its curing. After manufacturing the composites, the DOE method has been applied to study the effect of various processing parameters, such as reinforcement length (L), post-curing (C), and plasma treatment (P), on the properties of the resulting composite materials. By comparing the results at two levels of each factor (high and low), the optimal processing conditions for the production of CFRPs have been identified. Figure 1 depicts factors in a three-dimensional space, with each factor being represented on a different axis. The results are enhanced with SEM micrographs showing the reinforcement-matrix interface in the case of carbon fiber and epoxy resin.

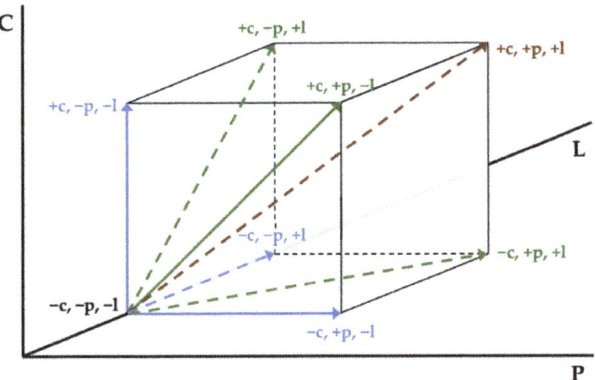

Figure 1. Variation of a single factor (blue), of two factors (green), and of three factors (brown) between the low value (−) and the high value (+). P, C, and L are represented in the axis X, Y, and Z, respectively. P corresponds to the application of a plasma treatment; C means post-curing was performed in the composites; L refers to the reinforcement length.

Therefore, the objective of this research is to design a process for the incorporation of recycled carbon fibers and composite materials into new composite materials. Mechanical recycling is selected as the recycling method due to its lower energy requirements, reduced generation of waste, and ability to maintain the integrity of the fibers in comparison to thermal and chemical recycling methods. A factorial design is employed to determine the most significant parameters impacting the properties of the manufactured composite material. The results obtained from the factorial design are then applied to a new composite material consisting of epoxy resin reinforced with a commercially available CFRP mechanically recycled in the shape of rods. This mechanical recycling process differs from current methods, which predominantly consist of grinding and are consequently more energy-intensive, by preserving the influence of fiber length to retain added value.

2. Materials and Methods

2.1. Materials

A rigid epoxy system consisting of an epoxy resin (SR 8500, Sicomin Epoxy Systems, Châteauneuf-les-Martigues, France) and a hardener (SD 8601, Sicomin Epoxy Systems, Châteauneuf-les-Martigues, France) was selected as matrix. The resin and hardener are mixed following a 100/35 weight ratio at room temperature conditions (23 °C/50% RH) The epoxy system has a clear liquid aspect and SD 8601 hardener is reported to have an ultra-slow reactivity. The manufacturer recommends performing a post-curing of 8 h at 80 °C after one day to achieve optimal properties. Table 1 presents a comparison between the epoxy system without post-curing and the epoxy system after undergoing the previously mentioned post-curing process. The last two rows in Table 1 present the data provided by the manufacturer and the data calculated experimentally, respectively. Hereafter, the experimental data from the last row will be considered for comparison purposes.

Table 1. Mechanical properties of the epoxy system SR 8500/SD 8601, as reported in the PDS for the first two rows [39] and experimental data obtained by the authors for the last row.

Curing Schedule of SR 8500/SD 8601	Tensile			Flexural			Charpy Impact Strength	Glass Transition
	Tensile Strength (MPa)	Elastic Modulus (Gpa)	Strain at Failure (%)	Flexural Strength (Mpa)	Flexural Modulus (Gpa)	Strain at Failure (%)	Resilience (J/m^2)	Glass Transition Temperature (°C)
14 days 23 °C	42	3.4	1.2	69	3.5	1.8	9	51
24 h 23 °C + 8 h 80 °C (manufacturer)	69	2.8	4.8	112	3.0	10.7	65	87
24 h 23 °C + 8 h 80 °C (experimental)	51 ± 5	1.6 ± 0.1	3.4 ± 1.0	154 ± 5	9.9 ± 0.8	2.4 ± 0.2		88 ± 1

Carbon fiber fabric (GG 600 T, MEL Composites, Barcelona, Spain) and pultruded carbon fiber plates (Carbodur S 512, Sika S.A.U. España, Alcobendas-Madrid, Spain) were selected as reinforcement. Carbon fibers were cut in the shape of bundles, manually separating the fibers inside the bundles aiming to keep their one-dimensional nature. Carbon fibers' length was reduced to the desired values mechanically by means of scissors. In the case of CFRP as reinforcement (see Table 2), the plates were mechanically cut to obtain 1–1.5 mm width rods, keeping the original thickness of 1.2 mm. Therefore, both reinforcements are examples of mechanically recycled fibers and CFRPs, respectively.

Table 2. Technical information of Sika Carbodur S 512, as reported in the PDS [40].

Material	Density (g/cm^3)	Fiber Volume Fraction (%)	Tensile Strength (MPa)	Elastic Modulus (GPa)	Transverse Modulus (GPa)	Longitudinal Poisson's Ratio	Strain at Failure (%)	Glass Transition Temperature (°C)
Carbodur S 512	1.60	>68	2900	165	9	0.28	1.80	>100

2.2. Manufacturing of Specimens

The specimens were manufactured in silicone molds in all cases, with a constant length/width ratio higher than 10 for all sets. In addition, a carbon fiber content of 13 ± 1% in weight was set. In the case of plasma treatment being required, it was carried out in a vacuum chamber with air atmosphere on the carbon fibers or on the mechanically recycled composite rods before the final composites were manufactured. The composites were demolded at least 7 days after the epoxy components were mixed and the specimens were fabricated. If post-curing was necessary, 24 h after mixing the components, the post-curing was carried out in an oven at 80 °C for 8 h in an air atmosphere. In this case, the demolding of composites was carried out 16 h after removal from the oven, when both the mold and the samples were at room temperature.

2.3. Surface Modification

Before manufacturing the composites and in some of the cases, carbon fibers were treated with LPP in a plasma cleaner chamber (Harrick, Ithaca, NY, USA) in an air atmosphere to produce plasma at a pressure of 300 mtorr. After achieving a stable vacuum in the chamber, carbon fibers were treated for 2 min at a power of 30 W.

To ensure the preservation of the surface modification caused by the LPP treatment, the composites were manufactured immediately following the removal of the carbon fibers from the vacuum chamber.

2.4. Morphology Study

Carbon fibers' surface, as well as the fracture surface of the composite specimens after tensile testing, were analyzed using a scanning electron microscope (SEM) (Philips XL-33 FEI EUROPE SEM, Eindhoven, The Netherlands) in order to study the effect of the LPP treatment. The specimens were coated with gold in a Polaron high-resolution sputter coater to serve as a conductor for the electrons and provide sufficient contrast in the SEM micrographs.

2.5. Mechanical Characterization

To evaluate the mechanical behavior of the epoxy and the composites manufactured, tensile tests were carried out on a universal testing machine (Microtest, Madrid, Spain). A load cell of 20 kN was used for test data acquisition. The test speed used was 1 mm/min since the objective is a quasi-static test. To avoid slippage during testing, P180 sandpaper was used between the specimens and the grips. At least five specimens were tested per set.

Stress is calculated, according to ISO 527-5, as the ratio between the force and the area of the specimen, the latter being assumed constant during the test. Strain is defined, according to the same standard, as the ratio between the increase in length experienced by the specimen and its initial value, the latter taken as the distance between grips. Finally, the elastic modulus is defined, similarly with respect to the aforementioned standard, as the ratio between the stress and strain increments in the strain range between 0.0005 and 0.0025.

2.6. Thermal and Chemical Characterization

The materials were characterized by differential scanning calorimetry (DSC) and Fourier-transform infrared spectroscopy (FTIR).

A DSC 822 (Mettler Toledo GmbH, Greifensee, Switzerland) was used to determine the curing kinetics of the epoxy. Samples with weights of 8 ± 1 mg were placed in 40 µL aluminum crucibles and under nitrogen atmosphere for testing.

To determine the curing kinetics of the epoxy, scans consisting of non-isothermal heating from -20 to $200\ °C$ at different heating rates of 5, 10, and $20\ °C/min$ were performed. By means of STARe software (Mettler Toledo GmbH, Greifensee, Switzerland), model-free kinetics (MFK) was used to calculate the degree of conversion at different temperatures based on the thermograms obtained at different heating rates. The analysis of these values enables the calculation of the activation energy as a function of the degree of conversion [41,42]. A comprehensive understanding of the curing process of the resin is of importance, as optimal curing conditions will yield the most favorable mechanical performance.

Additionally, the glass transition temperature (T_g) was studied by performing cycles at $20\ °C/min$ from -20 to $200\ °C$. T_g is a second-order transition, without a phase change, but rather a change in the volume of the sample due to the mobility of the chains.

A Bruker Tensor 27 spectrometer (Bruker Optik GmbH, Ettlingen, Germany) was used to obtain the infrared spectra of the samples. The attenuated total reflectance (ATR) technique was used to analyze the chemical modifications produced at about 5–10 µm depth of the sample. A diamond prism was used and the angle of incidence of the infrared radiation was 45°. Forty scans with a resolution of 4 cm^{-1} were obtained and averaged between 600 and 4000 cm^{-1}.

3. Results and Discussion

3.1. Matrix Characterization: Curing of the Epoxy Resin

Thermal characterization of the epoxy matrix was performed using DSC to study the curing kinetics and determine the glass transition temperature. In addition, a chemical analysis of the matrix was conducted using FTIR-ATR to examine the chemical changes that occurred during the curing process and assess other aspects related to the curing of the matrix [43].

3.1.1. Differential Scanning Calorimetry (DSC)

The curing process of the epoxy resin is studied using thermograms at different rates. The corresponding Sicomin 8500 data are shown in Figure 2. The curves shift to different temperatures (T_p) depending on the heating rate (β). The integrals of the curves have similar areas, representing the heat released during the curing process (ΔH).

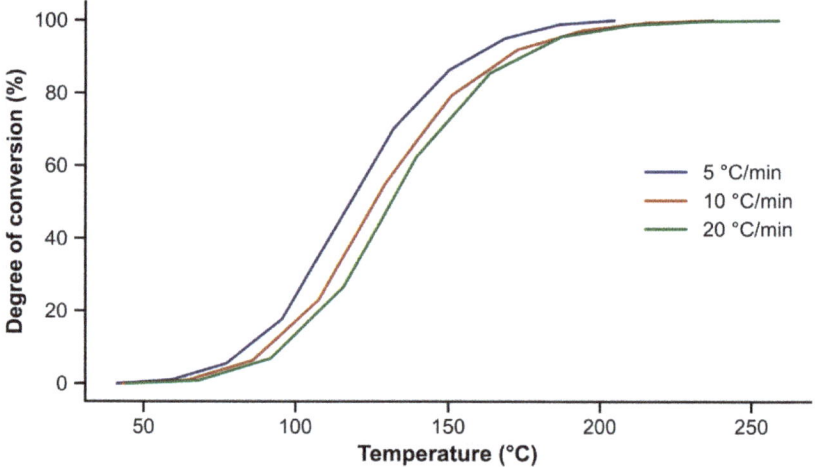

Figure 2. Temperature peaks as a function of heating rate for the epoxy resin.

Figure 2 illustrates the displacement of temperature peaks to higher temperatures as the heating rate increases. In addition, Table 3 presents the heat released by Sicomin 8500 during the curing process. Using the STARe software, the degree of conversion is calculated from the initial thermograms at different rates These curves are shown for Sicomin 8500 in Figure 3. It is necessary for the application of MFK that the curves do not intersect, and since this requirement is satisfied, the activation energy can be calculated using the STARe software as a function of the degree of conversion based on these curves.

Table 3. Curing peak temperatures and enthalpy of the epoxy resin.

β (°C/min)	5	10	20	$\Delta H \pm 1$ (J/g)
		$T_p \pm 1$ (°C)		
Sicomin 8500	113	127	143	444

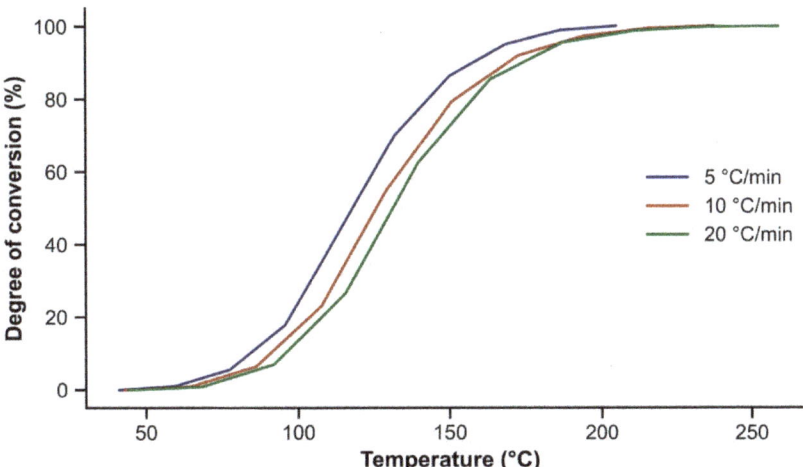

Figure 3. Degree of conversion as a function of temperature at different heating rates for the epoxy resin.

Figure 4 exhibits the activation energy for the epoxy resin as a function of the degree of conversion [44]. The activation energy of common industrial epoxy is known to be influenced by a range of factors. These include, but are not limited to, the chemical composition of the epoxy, the curing conditions (temperature, time, and humidity), the presence of impurities such as moisture or contaminants, the aging of the epoxy, even when properly stored, and variations in measurement conditions, including temperature and humidity. Three well-defined phases are observed. In the first zone, corresponding to the n-order reaction, a higher level of energy is required to initiate the reaction; in an intermediate zone, during the autocatalytic process, the energy needed to maintain the reaction is practically constant; finally, a greater energy input is required to complete the crosslinking reaction [41,45,46]. The necessity for a significantly higher energy input to complete the curing reaction confirms the necessity for post-curing the resin or, alternatively, a longer curing time at room temperature.

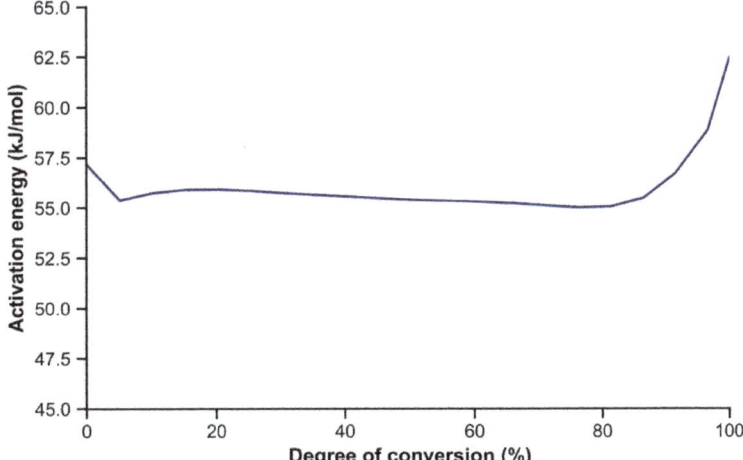

Figure 4. Activation energy as a function of the degree of conversion for the epoxy resin.

Based on the calculation of the activation energy, the isothermal curing at constant temperature for the epoxy resin can be simulated. Table 4 shows this simulation for Sicomin 8500. It should be noted that it is not possible to simulate the 100% curing environment as the error incurred would be very high. Similarly, the beginning of the reaction is taken as 0.1–0.2% cured. Table 4 reports that the curing time at room temperature is 10,140 min (7 days and 1 h), an excessively long time for the curing of an epoxy resin in an industrial process. At the proposed post-curing temperature (80 °C), 99% curing is achieved in 260 min (4 h and 20 min).

Table 4. Isothermal process simulation at different temperatures for the epoxy resin.

Temperature (°C)	25	50	75	80	90	100
Degree of Conversion (%)			Time (min)			
10	158.7	27.9	6.3	4.8	2.8	1.7
20	305.3	53.3	12.0	9.1	5.4	3.3
30	450.6	79.1	17.8	13.6	8.1	4.9
40	609.3	107.6	24.4	18.6	11.0	6.7
50	805.9	143.1	32.6	24.8	14.8	9.0
60	1089.6	193.9	44.2	33.7	20.1	12.3
70	1534.7	274.4	62.8	48.0	28.6	17.5
80	2336.6	419.5	96.4	73.6	44.0	27.0
90	4847.6	838.1	186.4	141.6	83.6	50.7
99	10,140.0	1644.3	346.3	260.4	150.8	89.9

Once the resin has been prepared, the T_g is calculated by DSC. In this case, two scans were made on the same sample to erase the thermal history and determine if the resin has finished curing. Sicomin 8500 presents an overlapped T_g with a relaxation enthalpy, followed by a curing peak in the first scan after curing at room temperature, performed 24 h after mixing the components. The presence of a curing peak indicates the epoxy resin is not fully cured. Once the thermal history has been erased and the adhesive has finished curing with post-curing, in the second scan it has a T_g of 88 ± 1 °C, as shown in Figure 5. A higher T_g indicates greater crosslinking and, consequently, higher stiffness, confirming what was observed in the mechanical properties in terms of higher stress [47].

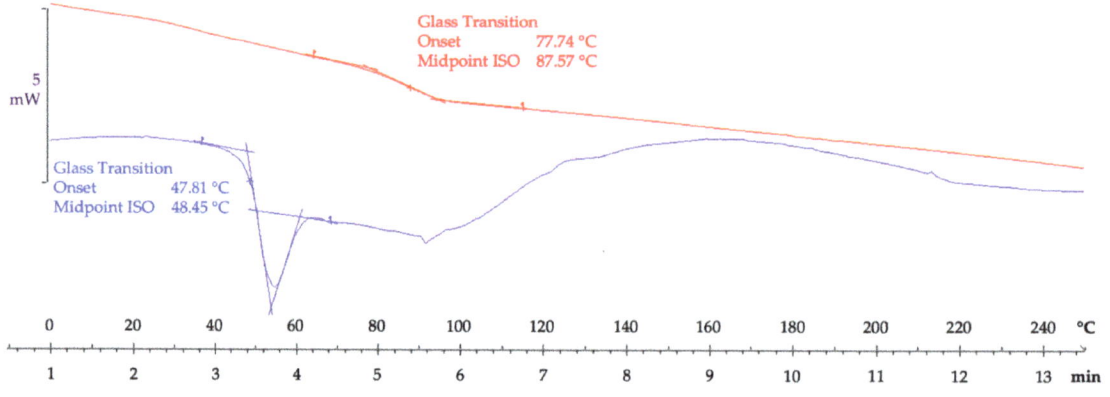

Figure 5. Glass transition temperature of the epoxy resin.

3.1.2. Fourier-Transform Infrared Spectroscopy (FTIR)

Figure 6 shows the spectra of components A and B of Sicomin 8500 resin before mixing, once mixed (uncured), cured, and after post-curing. It is clearly seen that the peak at 914 cm^{-1} disappears after post-curing. However, if post-curing is not carried out, after 24 h of curing at room temperature, the peak continues to appear. This peak corresponds to the oxirane group, which opens up with the reaction with the hardener or component B, giving rise to OH groups that autocatalyze the reaction and produce the crosslinking of the resin. The hardener (component B) is a polymer of lower complexity, where a greater presence of CH₂ and CH₃ aliphatic groups can be observed. The rest of the peaks can be observed in Table 5 and correspond mainly to C-C or CH bonds, with the presence of aromatic rings being important in this case, which increase the rigidity of the epoxy due to the lack of mobility in double bonds. These observations, along with those obtained by means of DSC, confirm that Sicomin 8500 is an epoxy with a reduced level of filler additives.

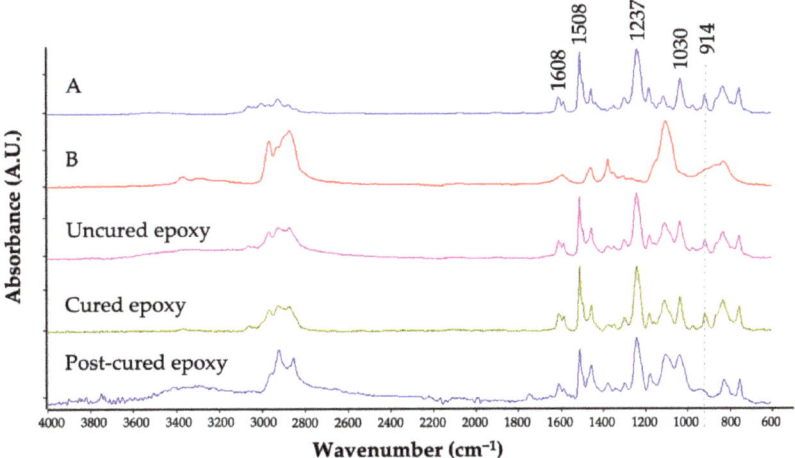

Figure 6. Infrared spectra of components A and B, and epoxy resin before curing, after curing, and after post-curing.

Table 5. Correspondences of the most characteristic infrared bands of the spectrum, adapt [48]. Reproduced with permission from authors.

Wavenumber (cm⁻¹)	Correspondence
755	–CH₂ γ, for C–(CH₂)n–C, n < 4
831	ArC–H δ oop
914	C–O–C st s, oxirane
971	CH₂ ω
1030	ArC–O–C–al st s
1110	C–OH st
1180	ArC–H d ip
1237	ArC–O–C–al st as
1297	C–O–C st as, oxirane
1348	–CH₃ δ st
1453	CH₂ ip
1508–1608	ArC–C
2813–3105	–CH₃, –CH₂, –CH st
3400–3600	–OH, –NH st
3700–3850	–OH

st: tension, ar: aromatic, δ: bending, s: symmetric, as: asymmetric, ω: flapping, t: torsion, al: aliphatic, ip: in-plane bending, oop: out-of-plane bending, γ: skeleton vibrations, n: number of CH₂ groups.

3.2. Carbon Fiber as a Reinforcement

Composite material specimens made of carbon fiber have been mechanically tested through tensile tests. Table 6 collects the design factors and the test plan chosen. Figure 1 represents the factors in a three-dimensional space, each factor being represented on a different axis. Eight sets of specimens have been designed to cover the cases shown in the table. On the one hand, it is possible to apply a LPP treatment to the carbon fibers before they are impregnated with epoxy (factor P). On the other hand, once the composite has been manufactured, it can be subjected to a post-curing process to achieve faster curing of the resin while improving certain mechanical properties (factor C). Finally, the length of the carbon fiber can be either 18 or 40 mm (factor L). It should be noted that certain factors, such as the porosity in the final composite, the width of the carbon fiber beams used, or the deviation of the fibers from the hypothesis of one-dimensionality (both in the horizontal plane and in the vertical plane), among others, have not been considered.

Table 6. Design factors and test plan. P corresponds to the application of a plasma treatment; C means post-curing was performed in the composites; L refers to the reinforcement length, in mm.

Configuration	Design Factors			Test Plan		
	P	C	L	P	C	L
1	−	−	−	No	No	18
p	+	−	−	Yes	No	18
c	−	+	−	No	Yes	18
l	−	−	+	No	No	40
cl	−	+	+	No	Yes	40
pl	+	−	+	Yes	No	40
pc	+	+	−	Yes	Yes	18
pcl	+	+	+	Yes	Yes	40

Table 7 exhibits the mechanical results obtained after tensile tests. The average of five samples is shown along with the standard deviation as a measure of variability. Furthermore, the coefficient of variation (CoV) is represented as the ratio between the standard deviation and the average, expressed in percentage. Due to the inherent inhomogeneity of the specimens, in addition to effects not considered, such as deviation from the hypothesis of one-dimensionality in the horizontal and vertical planes, the CoV shows high values in almost all the tested sets and for all the measured responses.

Table 7. Mechanical results obtained from testing. CoV represents the coefficient of variation.

Configuration	Stress (MPa) [CoV]	Strain at Maximum Stress (%) [CoV]	Elastic Modulus (GPa) [CoV]
1	31 ± 9 [27.9]	2.2 ± 0.7 [23.9]	1.4 ± 0.4 [32.0]
p	99 ± 13 [13.1]	4.1 ± 0.9 [22.6]	2.4 ± 0.5 [22.4]
c	89 ± 21 [23.2]	2.3 ± 0.2 [7.2]	3.9 ± 0.9 [24.4]
l	100 ± 13 [12.8]	5.6 ± 0.8 [15.1]	1.6 ± 0.4 [26.1]
cl	237 ± 47 [19.9]	5.5 ± 1.5 [26.6]	4.3 ± 0.9 [21.4]
pl	74 ± 12 [16.7]	2.4 ± 0.7 [29.4]	3.1 ± 0.5 [16.9]
pc	99 ± 24 [24.4]	3.1 ± 0.6 [19.0]	3.1 ± 0.5 [16.0]
pcl	148 ± 44 [29.7]	7.5 ± 1.5 [19.6]	2.0 ± 0.6 [32.5]

Table 8 presents a matrix representation of the individual effects of each factor and their interactions. By analyzing this matrix, in conjunction with the experimental data provided in Table 7, it is possible to extract several key data points: the average response, three main factor effects, three interaction effects between pairs of factors, and one interaction effect involving three factors. Utilizing Yates' algorithm allows for the calculation of influence values, represented in Table 8. To determine the Yates order for a fractional factorial design, it is necessary to understand the confounding structure of the design. The experimental data obtained for each combination of parameters are used to calculate the effect of each factor and interaction on the response, which is achieved by adding or subtracting the averages of the responses. Therefore, Table 8 lists the effects and interactions of the factorial experiment, with the influence factors serving as dimensionless indicators that can be compared based on their relationship to the order of magnitude of the response.

Table 8. Factors and their interactions after applying Yates' algorithm.

	P	C	L	C × L	P × L	P × C	P × C × L
Stress (MPa)	−4.6	33.6	30.1	19.1	−24.1	−15.1	−0.6
Strain at maximum stress (%)	0.2	0.5	1.2	0.7	−0.5	0.5	0.8
Elastic Modulus (GPa)	−0.1	0.6	0.0	−0.2	−0.1	−0.7	−0.3

In the sample space, it can be seen how the factor that presents the most positive impact in terms of stress is the application of a post-curing (C), followed by using long fibers (L), and the combination of those two (C × L); meanwhile, a LPP treatment (P) and its combination with any other factor leads to a reduction in this response.

The length of the carbon fiber (L) exhibits the most significant positive change in terms of strain, followed by a combination of LPP, post-curing, and long fibers (P × C × L), and a combination of post-curing and long fibers (C × L). In this case, the combination of LPP and long fiber leads to a decrease in strain at maximum failure.

Finally, the stiffness, represented by the elastic modulus, is favored by the application of post-curing, followed by the incorporation of long fibers (with a positive value close to zero). As in the case of stress, the use of LPP and its combination with the remaining factors leads to a decrease in the elastic modulus. Similarly, the combination of post-curing and long fibers results in a reduction in stiffness.

From the analysis of the influence values shown in Table 8, the key factors are the application of post-curing to the resin, an increase in fiber length, and a combination of both. Regarding post-curing, it is the factor that most improves tensile strength and provides greater stiffness to the material. This need for post-curing coincides with the information provided by the manufacturer in the technical data sheet referred to in Table 1. Additionally, this coincides with the results obtained from the chemical and thermal characterization of the epoxy resin, which indicated that post-curing improved the mechanical properties of the resin, increasing its T_g, and demonstrated that post-curing ensured the opening of the oxirane rings that confirm the completion of the epoxy crosslinking process. In terms of fiber length, increasing the length of the carbon fiber beams from 18 to 40 mm has an effect almost as positive as post-curing on tensile strength while having the most positive effect on strain at maximum stress. However, fiber length does not modify significantly (it should be noted that the obtained influence values are relative) the stiffness of the composites, as occurs with the application of post-curing. Finally, the combination of post-curing and fiber length also yields very positive results in terms of stress and strain at maximum stress, although stiffness is compromised. In general, the energy absorption capacity before breaking, known as toughness, of the composites that either cure with post-curing, incorporate long fibers, or combine post-curing and long fibers, is higher than that of the rest of the sets. Additionally, it seems clear that the LPP treatment leads to a general decrease in the mechanical behavior of the composites.

To gain a deeper understanding of the impact of LPP treatment on carbon fibers, Figure 7 exhibits samples that have been studied using SEM. The examination was focused on the samples that yielded the most promising results, as determined by a factorial design study. These samples included the combination of post-cured and long carbon fibers (C × L), as well as the combination of LPP treatment, post-cured and long carbon fibers (P × C × L). This analysis aimed to verify the potential positive effect of LPP treatment on the carbon fibers, and to evaluate the adhesion properties between the reinforcement and matrix.

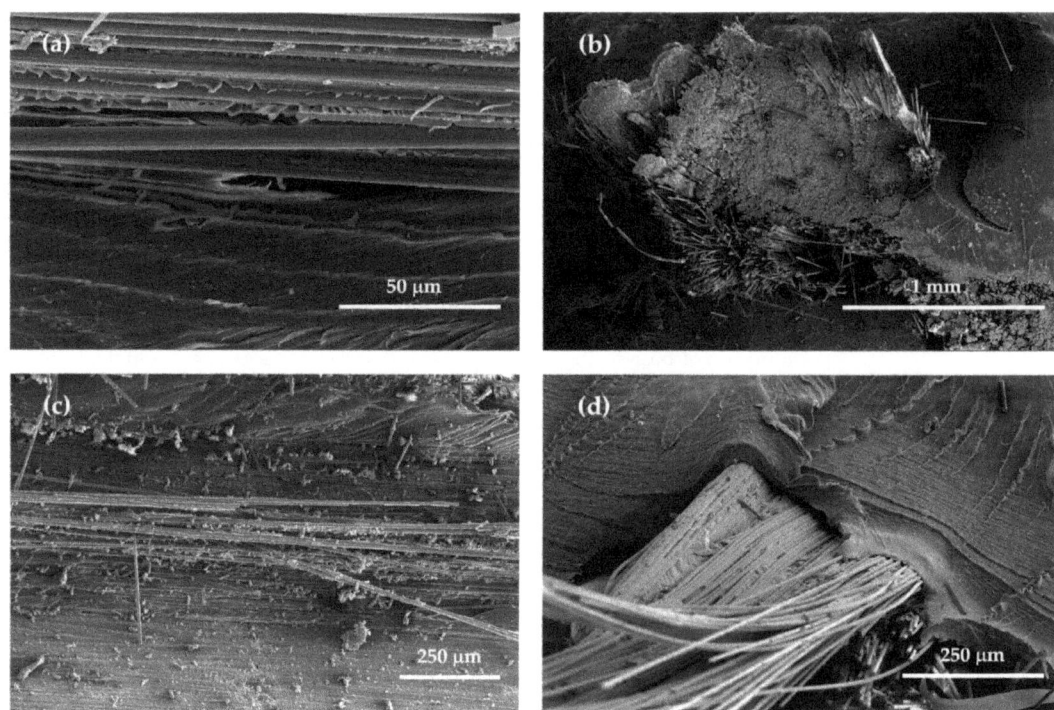

Figure 7. SEM micrographs at different magnification levels: (**a,b**) composites with untreated carbon fibers; (**c,d**) composites with LPP-treated carbon fibers.

Figure 7 illustrates composite specimens in which the LPP treatment was not applied to the carbon fibers (Figure 7a,b) and those in which it was (Figure 7c,d). No discernible difference was observed between the specimens, suggesting that the plasma treatment did not yield an improvement in the reinforcement-matrix adhesion. Consequently, it can be inferred that the plasma treatment, in contrast to its effects on natural fibers, did not enhance the mechanical performance of carbon fiber composites in this instance [49]. To address this issue, a more effective approach might be to enhance the surface contact of the carbon fibers by reducing their width or thickness, or by modifying the resin so that capillary action does not impede the wetting of the carbon fibers.

3.3. Use Case: A Mechanically Recycled CFRP as a Reinforcement

Following the characterization of the resin and the development of the process for the production of composite materials reinforced with mechanically cut carbon fiber, it is proposed to extend this process using rod-shaped mechanically recycled CFRP (Carbodur S512) as reinforcement in the same epoxy matrix. The composite materials reinforced with mechanically recycled CFRP were fabricated in the same manner as in the previous case, following the same process. In accordance with the reported results, the reinforcement of these new composite materials was long (40 mm), did not receive LPP treatment, and was subjected to post-curing in an oven at 80 °C for 8 h 24 h after manufacture. The composite materials were then tested through tensile tests. Table 9 shows the results obtained from tensile tests for the epoxy resin, carbon fiber reinforced composites and CFRP reinforced composites. The results were subjected to a Grubbs test (95% confidence interval) to detect and eliminate outliers. Furthermore, the results, compared in pairs, were subjected to an analysis of variance (ANOVA), which showed that the sets are not similar.

Table 9. Comparison between epoxy resin, composites manufactured with carbon fiber as reinforcement and composites manufactured with a mechanically recycled CFRP as reinforcement.

Reinforcement	Stress (MPa) [CoV]	Strain at Maximum Stress (%) [CoV]	Elastic Modulus (GPa) [CoV]
None	51 ± 5 [10.3]	3.4 ± 1.0 [28.2]	1.6 ± 0.1 [6.9]
Carbon fiber	237 ± 47 [19.9]	5.5 ± 1.5 [26.6]	4.3 ± 0.9 [21.4]
CFRP	88 ± 12 [13.6]	4.0 ± 0.6 [14.7]	1.6 ± 0.1 [6.8]

Table 9 compares the mechanical behavior of the epoxy resin, composites made with carbon fiber reinforcement, and those using mechanically recycled CFRP as reinforcement. It was observed that the use of mechanically recycled CFRP, which includes a cured epoxy resin, resulted in rods that were more constrained in the composite than carbon fibers. Furthermore, the contact surface between the epoxy matrix and the mechanically recycled CFRP rods was found to be lower than in the case of carbon fiber reinforcement, leading to decreased reinforcement-matrix adhesion and lower levels of sustained stress, strain, elastic modulus, and therefore, absorbed energy.

Figure 8 shows the stress–strain curves of the materials compared in Table 9, including the epoxy resin, carbon fiber reinforced composites, and CFRP reinforced composites.

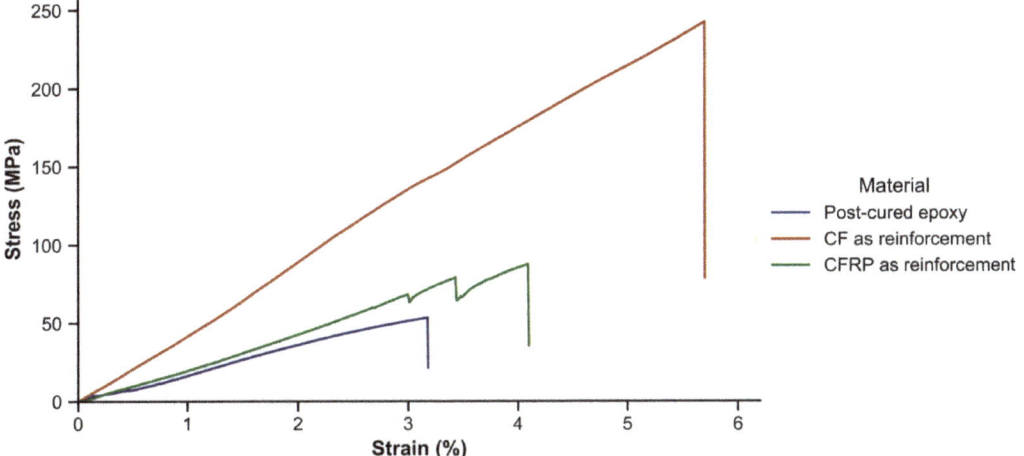

Figure 8. Stress–strain curves for the epoxy resin, carbon fiber reinforced composites and CFRP reinforced composites.

On the one hand, Figure 8 exhibits that the area under the curve of the carbon fiber reinforced composites is much larger than that of the epoxy resin. The maximum stress is 365% higher, the strain at maximum stress is 63% higher, and the elastic modulus is 172% higher.

On the other hand, CFRP-reinforced composites display a larger area under the curve than epoxy resin and compared to it show 75% higher stress, 18% higher strain at maximum stress and comparable elastic modulus. In this case, even with a reinforcement mass fraction of 13%, which means a fiber volume fraction of only 8%, there is a substantial improvement in mechanical properties. This confirms the feasibility of using small amounts of reinforcement in the epoxy resin for its use in non-structural applications.

Comparing now the composites among each other, it is observed that those reinforced with carbon fibers are able to absorb a greater amount of energy. Thus, it is observed that these composites, compared to those reinforced with CFRP, show a 170% higher stress, a 37% higher strain at maximum stress, and a 163% higher elastic modulus.

Potential solutions to this inferior mechanical behavior include further separating the CFRP rods to increase the contact surface with the epoxy or modifying the epoxy to improve the wetting of the reinforcement.

The results obtained demonstrate the validity of using epoxy as a matrix and recycled composite material as reinforcement in a process for reintroducing composite materials that have reached their end-of-life back into the production cycle. Otherwise, these CFRPs would likely have been destined for incineration or landfilling. In the present case, the epoxy–epoxy bond exhibits favorable adhesion properties and does not require further treatments for improvement.

4. Conclusions

The process of manufacturing composites using discontinuous carbon fibers and mechanically recycled commercial CFRP in the form of rods was developed in this study.

The thermal characterization of the epoxy matrix was carried out to investigate the curing kinetics and the T_g. The chemical characterization of the epoxy matrix was also conducted to examine the chemical changes in both components, as well as after mixing, after curing for 24 h, and after post-curing. The findings from this analysis confirmed the observations made using DSC and the recommendations made by the manufacturer concerning the necessity of post-curing. The proposed post-curing was 8 h at 80 °C to achieve 99% curing of the resin and to increase the T_g (88 ± 1 °C), which resulted in greater crosslinking and higher resin stiffness.

A fractional factorial design (DOE method) was utilized to examine the effect of three chosen factors on the mechanical behavior of composites using discontinuous carbon fibers as reinforcement: LPP treatment, post-curing, and fiber length. By varying the values of each factor between high and low, the eight possible combinations were tested and the effect of each factor individually, as well as pair interactions and all factors together, were analyzed. Analysis of the experimental data using Yates' algorithm revealed several general conclusions. Firstly, LPP treatment had a minimal impact on mechanical properties and tended to be unfavorable. Essentially, any combination of another factor with LPP treatment resulted in a decline in mechanical properties, except in the case of strain, which was enhanced when LPP treatment was accompanied by post-curing (up to 3.1%) or post-curing and long fiber (up to 7.5%). Secondly, the factors that most significantly improved mechanical behavior were post-curing in terms of stress and stiffness (89 MPa and 3.9 GPa, respectively), and fiber length in terms of strain at maximum stress (up to 5.6%). The optimal combination of factors, particularly considering stress, was found to be post-curing and fiber length (C x L). Therefore, this combination was selected as the most suitable for the manufacture of the composites, which led to achieving a maximum stress, stiffness, and strain at maximum stress of 237 MPa, 4.3 GPa, and 5.5%, respectively.

Finally, composites were manufactured using the same epoxy matrix but a commercially recycled CFRP mechanically in the form of rods. The epoxy–epoxy bond between the matrix and reinforcement leads to effective adhesion, making additional treatments unnecessary. While using mechanically recycled CFRP as a reinforcement is beneficial from a cost and weight perspective, the mechanical properties of the composite produced with carbon fibers are still superior to those of the CFRP, as demonstrated in this study with a 170% higher stress, a 37% higher strain at maximum stress, and a 163% higher stiffness. Potential solutions for future work may include further separating the CFRP rods to increase the contact surface with the epoxy or enhancing the wetting properties of the epoxy and reinforcement. While the results showed potential for improvement, they also illustrated the feasibility of the recycling process for CFRP and the manufacturing of composites using recycled CFRP.

Author Contributions: Conceptualization and methodology, J.A.B., M.B., M.Á.M. and J.A.; data curation, J.A.B. and M.Á.M.; writing—original draft preparation, J.A.B.; writing—review and editing, M.B., M.Á.M. and J.A.; supervision, M.Á.M. and J.A.; funding acquisition, J.A.B., M.B., M.Á.M. and J.A.; All authors have read and agreed to the published version of the manuscript.

Funding: This research received no external funding.

Data Availability Statement: Not applicable.

Acknowledgments: The authors gratefully acknowledge those colleagues who kindly contributed to the development of this project.

Conflicts of Interest: The authors declare no conflict of interest. The funders had no role in the design of the study; in the collection, analyses, or interpretation of data; in the writing of the manuscript; or in the decision to publish the results.

References

1. Lavayen-Farfan, D.; Butenegro-Garcia, J.A.; Boada, M.J.L.; Martinez-Casanova, M.A.; Rodriguez-Hernandez, J.A. Theoretical and Experimental Study of the Bending Collapse of Partially Reinforced CFRP–Steel Square Tubes. *Thin-Walled Struct.* **2022**, *177*, 109457. [CrossRef]
2. Galvez, P.; Abenojar, J.; Martinez, M.A. Effect of Moisture and Temperature on the Thermal and Mechanical Properties of a Ductile Epoxy Adhesive for Use in Steel Structures Reinforced with CFRP. *Compos. Part B Eng.* **2019**, *176*, 107194. [CrossRef]
3. Rubino, F.; Nisticò, A.; Tucci, F.; Carlone, P. Marine Application of Fiber Reinforced Composites: A Review. *J. Mar. Sci. Eng.* **2020**, *8*, 26. [CrossRef]
4. Xiong, Z.; Wei, W.; Liu, F.; Cui, C.; Li, L.; Zou, R.; Zeng, Y. Bond Behaviour of Recycled Aggregate Concrete with Basalt Fibre-Reinforced Polymer Bars. *Compos. Struct.* **2021**, *256*, 113078. [CrossRef]
5. Xiong, Z.; Wei, W.; He, S.; Liu, F.; Luo, H.; Li, L. Dynamic Bond Behaviour of Fibre-Wrapped Basalt Fibre-Reinforced Polymer Bars Embedded in Sea Sand and Recycled Aggregate Concrete under High-Strain Rate Pull-out Tests. *Constr. Build. Mater.* **2021**, *276*, 122195. [CrossRef]
6. Akbar, A.; Liew, K.M. Assessing Recycling Potential of Carbon Fiber Reinforced Plastic Waste in Production of Eco-Efficient Cement-Based Materials. *J. Clean. Prod.* **2020**, *274*, 123001. [CrossRef]
7. Zhu, J.-H.; Chen, P.; Su, M.; Pei, C.; Xing, F. Recycling of Carbon Fibre Reinforced Plastics by Electrically Driven Heterogeneous Catalytic Degradation of Epoxy Resin. *Green Chem.* **2019**, *21*, 1635–1647. [CrossRef]
8. Xian, G.; Guo, R.; Li, C.; Wang, Y. Mechanical Performance Evolution and Life Prediction of Prestressed CFRP Plate Exposed to Hygrothermal and Freeze-Thaw Environments. *Compos. Struct.* **2022**, *293*, 115719. [CrossRef]
9. Wu, J.; Li, C.; Hailatihan, B.; Mi, L.; Baheti, Y.; Yan, Y. Effect of the Addition of Thermoplastic Resin and Composite on Mechanical and Thermal Properties of Epoxy Resin. *Polymers* **2022**, *14*, 1087. [CrossRef]
10. Meng, F.; McKechnie, J.; Pickering, S.J. An Assessment of Financial Viability of Recycled Carbon Fibre in Automotive Applications. *Compos. Part A Appl. Sci. Manuf.* **2018**, *109*, 207–220. [CrossRef]
11. Pimenta, S.; Pinho, S.T. Recycling Carbon Fibre Reinforced Polymers for Structural Applications: Technology Review and Market Outlook. *Waste Manag.* **2011**, *31*, 378–392. [CrossRef] [PubMed]
12. Butenegro, J.A.; Bahrami, M.; Abenojar, J.; Martínez, M.Á. Recent Progress in Carbon Fiber Reinforced Polymers Recycling: A Review of Recycling Methods and Reuse of Carbon Fibers. *Materials* **2021**, *14*, 6401. [CrossRef] [PubMed]
13. Karuppannan Gopalraj, S.; Kärki, T. A Review on the Recycling of Waste Carbon Fibre/Glass Fibre-Reinforced Composites: Fibre Recovery, Properties and Life-Cycle Analysis. *SN Appl. Sci.* **2020**, *2*, 433. [CrossRef]
14. Butenegro, J.A.; Bahrami, M.; Swolfs, Y.; Ivens, J.; Martínez, M.Á.; Abenojar, J. Novel Thermoplastic Composites Strengthened with Carbon Fiber-Reinforced Epoxy Composite Waste Rods: Development and Characterization. *Polymers* **2022**, *14*, 3951. [CrossRef] [PubMed]
15. Abdallah, R.; Juaidi, A.; Savaş, M.A.; Çamur, H.; Albatayneh, A.; Abdala, S.; Manzano-Agugliaro, F. A Critical Review on Recycling Composite Waste Using Pyrolysis for Sustainable Development. *Energies* **2021**, *14*, 5748. [CrossRef]
16. Abdou, T.R.; Botelho Junior, A.B.; Espinosa, D.C.R.; Tenório, J.A.S. Recycling of Polymeric Composites from Industrial Waste by Pyrolysis: Deep Evaluation for Carbon Fibers Reuse. *Waste Manag.* **2021**, *120*, 1–9. [CrossRef]
17. Hao, S.; He, L.; Liu, J.; Liu, Y.; Rudd, C.; Liu, X. Recovery of Carbon Fibre from Waste Prepreg via Microwave Pyrolysis. *Polymers* **2021**, *13*, 1231. [CrossRef]
18. Nistratov, A.V.; Klimenko, N.N.; Pustynnikov, I.V.; Vu, L.K. Thermal Regeneration and Reuse of Carbon and Glass Fibers from Waste Composites. *Emerg. Sci. J.* **2022**, *6*, 967–984. [CrossRef]

19. Lee, M.; Kim, D.H.; Park, J.-J.; You, N.-H.; Goh, M. Fast Chemical Recycling of Carbon Fiber Reinforced Plastic at Ambient Pressure Using an Aqueous Solvent Accelerated by a Surfactant. *Waste Manag.* **2020**, *118*, 190–196. [CrossRef]
20. Jiang, J.; Deng, G.; Chen, X.; Gao, X.; Guo, Q.; Xu, C.; Zhou, L. On the Successful Chemical Recycling of Carbon Fiber/Epoxy Resin Composites under the Mild Condition. *Compos. Sci. Technol.* **2017**, *151*, 243–251. [CrossRef]
21. Kupski, J.; Teixeira de Freitas, S. Design of Adhesively Bonded Lap Joints with Laminated CFRP Adherends: Review, Challenges and New Opportunities for Aerospace Structures. *Compos. Struct.* **2021**, *268*, 113923. [CrossRef]
22. Rajak, D.K.; Wagh, P.H.; Linul, E. Manufacturing Technologies of Carbon/Glass Fiber-Reinforced Polymer Composites and Their Properties: A Review. *Polymers* **2021**, *13*, 3721. [CrossRef] [PubMed]
23. Roux, M.; Eguémann, N.; Dransfeld, C.; Thiébaud, F.; Perreux, D. Thermoplastic Carbon Fibre-Reinforced Polymer Recycling with Electrodynamical Fragmentation: From Cradle to Cradle. *J. Thermoplast. Compos. Mater.* **2017**, *30*, 381–403. [CrossRef]
24. Thomas, C.; Borges, P.H.R.; Panzera, T.H.; Cimentada, A.; Lombillo, I. Epoxy Composites Containing CFRP Powder Wastes. *Compos. Part B Eng.* **2014**, *59*, 260–268. [CrossRef]
25. Giorgini, L.; Benelli, T.; Brancolini, G.; Mazzocchetti, L. Recycling of Carbon Fiber Reinforced Composite Waste to Close Their Life Cycle in a Cradle-to-Cradle Approach. *Curr. Opin. Green Sustain. Chem.* **2020**, *26*, 100368. [CrossRef]
26. Wan, Y.; Takahashi, J. Tensile and Compressive Properties of Chopped Carbon Fiber Tapes Reinforced Thermoplastics with Different Fiber Lengths and Molding Pressures. *Compos. Part A Appl. Sci. Manuf.* **2016**, *87*, 271–281. [CrossRef]
27. Liston, E.M. Plasma Treatment for Improved Bonding: A Review. *J. Adhes.* **1989**, *30*, 199–218. [CrossRef]
28. Bahrami, M.; Abenojar, J.; Martínez, M.A. Comparative Characterization of Hot-Pressed Polyamide 11 and 12: Mechanical, Thermal and Durability Properties. *Polymers* **2021**, *13*, 3553. [CrossRef]
29. Arpagaus, C.; Oberbossel, G.; Rudolf von Rohr, P. Plasma Treatment of Polymer Powders—From Laboratory Research to Industrial Application. *Plasma Process Polym.* **2018**, *15*, 1800133. [CrossRef]
30. Bahrami, M.; Enciso, B.; Gaifami, C.M.; Abenojar, J.; Martinez, M.A. Characterization of Hybrid Biocomposite Poly-Butyl-Succinate/Carbon Fibers/Flax Fibers. *Compos. Part B Eng.* **2021**, *221*, 109033. [CrossRef]
31. Mandolfino, C.; Lertora, E.; Gambaro, C.; Pizzorni, M. Functionalization of Neutral Polypropylene by Using Low Pressure Plasma Treatment: Effects on Surface Characteristics and Adhesion Properties. *Polymers* **2019**, *11*, 202. [CrossRef] [PubMed]
32. Enciso, B.; Abenojar, J.; Martínez, M.A. Influence of Plasma Treatment on the Adhesion between a Polymeric Matrix and Natural Fibres. *Cellulose* **2017**, *24*, 1791–1801. [CrossRef]
33. Baird, S.; Bohren, J.A.; McIntosh, C. Optimal Design of Experiments in the Presence of Interference. *Rev. Econ. Stat.* **2018**, *100*, 844–860. [CrossRef]
34. Dosta, S.; Dale, S.; Antonyuk, S.; Wassgren, C.; Heinrich, S.; Litster, J.D. Numerical and Experimental Analysis of Influence of Granule Microstructure on Its Compression Breakage. *Powder Technol.* **2016**, *299*, 87–97. [CrossRef]
35. Yoozbashizadeh, M.; Chartosias, M.; Victorino, C.; Decker, D. Investigation on the Effect of Process Parameters in Atmospheric Pressure Plasma Treatment on Carbon Fiber Reinforced Polymer Surfaces for Bonding. *Mater. Manuf. Process.* **2019**, *34*, 660–669. [CrossRef]
36. Saleem, M.M.; Somá, A. Design of Experiments Based Factorial Design and Response Surface Methodology for MEMS Optimization. *Microsyst Technol* **2015**, *21*, 263–276. [CrossRef]
37. Martínez, M.A.; López de Armentia, S.; Abenojar, J. Influence of Sample Dimensions on Single Lap Joints: Effect of Interactions between Parameters. *J. Adhes.* **2021**, *97*, 1358–1369. [CrossRef]
38. Gao, L.; Adesina, A.; Das, S. Properties of Eco-Friendly Basalt Fibre Reinforced Concrete Designed by Taguchi Method. *Constr. Build. Mater.* **2021**, *302*, 124161. [CrossRef]
39. Sicomin. Sicomin SR 8500/SD 860x—Technical Data Sheet. 2014. Available online: http://sicomin.com/datasheets/product-pdf44.pdf (accessed on 6 December 2022).
40. Sika. Sika Carbodur S—Product Data Sheet. 2018. Available online: https://usa.sika.com/dms/getdocument.get/c8cdaca2-8860-4fff-9179-0fa2c1290128/sika_carbodur_s.pdf (accessed on 6 December 2022).
41. Abenojar, J.; Tutor, J.; Ballesteros, Y.; del Real, J.C.; Martínez, M.A. Erosion-Wear, Mechanical and Thermal Properties of Silica Filled Epoxy Nanocomposites. *Compos. Part B Eng.* **2017**, *120*, 42–53. [CrossRef]
42. Drzeżdżon, J.; Jacewicz, D.; Sielicka, A.; Chmurzyński, L. Characterization of Polymers Based on Differential Scanning Calorimetry Based Techniques. *TrAC Trends Anal. Chem.* **2019**, *110*, 51–56. [CrossRef]
43. Dong, A.; Zhao, Y.; Zhao, X.; Yu, Q. Cure Cycle Optimization of Rapidly Cured Out-Of-Autoclave Composites. *Materials* **2018**, *11*, 421. [CrossRef] [PubMed]
44. Vyazovkin, S.; Wight, C.A. Model-Free and Model-Fitting Approaches to Kinetic Analysis of Isothermal and Nonisothermal Data. *Thermochim. Acta* **1999**, *340–341*, 53–68. [CrossRef]
45. Abenojar, J.; Martínez, M.A.; Pantoja, M.; Velasco, F.; Del Real, J.C. Epoxy Composite Reinforced with Nano and Micro SiC Particles: Curing Kinetics and Mechanical Properties. *J. Adhes.* **2012**, *88*, 418–434. [CrossRef]
46. Abenojar, J.; Martínez, M.A.; Velasco, F.; Pascual-Sánchez, V.; Martín-Martínez, J.M. Effect of Boron Carbide Filler on the Curing and Mechanical Properties of an Epoxy Resin. *J. Adhes.* **2009**, *85*, 216–238. [CrossRef]

47. Carbas, R.J.C.; da Silva, L.F.M.; Marques, E.A.S.; Lopes, A.M. Effect of Post-Cure on the Glass Transition Temperature and Mechanical Properties of Epoxy Adhesives. *J. Adhes. Sci. Technol.* **2013**, *27*, 2542–2557. [CrossRef]
48. Socrates, G. *Infrared and Raman Characteristic Group Frequencies: Tables and Charts*, 3rd ed.; John Wiley & Sons, Ltd.: Hoboken, NJ, USA, 2014.
49. Enciso, B.; Abenojar, J.; Paz, E.; Martínez, M.A. Influence of Low Pressure Plasma Treatment on the Durability of Thermoplastic Composites LDPE-Flax/Coconut under Thermal and Humidity Conditions. *Fibers Polym.* **2018**, *19*, 1327–1334. [CrossRef]

Disclaimer/Publisher's Note: The statements, opinions and data contained in all publications are solely those of the individual author(s) and contributor(s) and not of MDPI and/or the editor(s). MDPI and/or the editor(s) disclaim responsibility for any injury to people or property resulting from any ideas, methods, instructions or products referred to in the content.

www.ingramcontent.com/pod-product-compliance
Lightning Source LLC
LaVergne TN
LVHW070429100526
838202LV00014B/1559

MDPI
St. Alban-Anlage 66
4052 Basel
Switzerland
www.mdpi.com

Processes Editorial Office
E-mail: processes@mdpi.com
www.mdpi.com/journal/processes

Disclaimer/Publisher's Note: The statements, opinions and data contained in all publications are solely those of the individual author(s) and contributor(s) and not of MDPI and/or the editor(s). MDPI and/or the editor(s) disclaim responsibility for any injury to people or property resulting from any ideas, methods, instructions or products referred to in the content.